8

APPUNTI
LECTURE NOTES

Umberto Zannier
Scuola Normale Superiore
Piazza dei Cavalieri, 7
56126 Pisa, Italy

Lecture Notes on Diophantine Analysis

Umberto Zannier

Lecture Notes on Diophantine Analysis

with an appendix
by Francesco Amoroso

EDIZIONI
DELLA
NORMALE

© 2014 Scuola Normale Superiore Pisa
Versione rivista e aggiornata

Prima edizione: 2009

ISBN 978-88-7642-341-3
ISBN 978-88-7642-517-2 (eBook)

Contents

Preface

The present lecture notes originate from an introductory course delivered at the Scuola Normale in Pisa during the academic year 2006/2007. Basically, they were addressed especially to a public of graduate students. Dr. Marco Illengo took some notes of the course, providing much help and good motivation for writing this book.

Also in view of the origin of these notes from a course addressed to a relatively wide audience, we have tried to be self-contained and detailed; generally speaking, the prerequisites do not go beyond basic mathematical material. A knowledge of the basic theory of algebraic numbers is useful, especially from Chapter 3 onwards. In some of the Supplements and in Chapter 4 a little knowledge of algebraic geometry is required.

The proofs of the theorems usually follow well-known lines, but occasionally some variations are introduced. A few theorems are stated but not proved here, which is indicated with a star. All of the chapters contain several exercises; they are not listed at the end, but appear along the text. Sometimes they consist of known (or new) facts whose complete proofs have not been included for brevity. In part for this reason, we have almost always provided hints for the solutions: this is not because we distrust the reader's skill, but rather has the purpose of keeping the whole almost self-contained and of giving the choice of whether spending any time in seeking a solution or read it directly.

We have borrowed freely from other sources, like Mordell's *Diophantine Equations* and especially the recent book by Bombieri and Gubler *Heights in Diophantine Geometry*, an invaluable source.

The chapters are concluded with 'Notes', of bibliographical and historical content. To limit the (already large) number of references, sometimes we have not given exact data for the original papers, but rather referred to some book containing fully detailed references.

Preface to the revised version

This volume represents a revised version of the former, rather than a "new edition"; indeed, there are no new chapters or even sections, and we have merely corrected inaccuracies of various nature and added occasionally some remarks, exercises and references.

Notations and conventions

The letters $\mathbb{N}, \mathbb{Z}, \mathbb{Q}, \mathbb{R}, \mathbb{C}, \mathbb{Q}_p$ will have the usual meaning. By 'positive' number we mean a real number > 0. By $[x]$ we denote the 'integral part' of the number $x \in \mathbb{R}$ *i.e.* the largest integer m with $m \leq x$.

If k is a field, we shall denote with $\bar{k}$ an algebraic closure.

For a (commutative) ring R, we shall denote by R^* the group of invertible elements in R.

By the *rank* of an abelian group Γ, written multiplicatively, we shall mean the maximum number of elements $\gamma_1, \ldots, \gamma_r \in \Gamma$ such that no relation exists of the shape $\gamma_1^{a_1} \cdots \gamma_r^{a_r} = 1$ with integer exponents a_i not all zero. For instance, a torsion group has rank 0.

By *algebraic variety* we mean a subset of an affine or projective space, defined by a set of algebraic equations; we do not insist that the variety is irreducible. By saying that a variety $\mathcal{X}$ is defined over a field k we mean that it may be defined by a set of equations with coefficients in k, and we sometimes write $\mathcal{X}/k$; we usually identify the variety with the set of its points over an algebraic closure of k. Usually we shall consider varieties defined over $\overline{\mathbb{Q}}$ (or occasionally over $\mathbb{C}$ or some finite field).

For an algebraic variety $\mathcal{X}$ defined over k, we denote by $k(X)$ the field of rational functions on $\mathcal{X}$ with coefficients in k. By $\mathcal{X}(k)$ we mean the set of points of $\mathcal{X}$ with coordinates in k. By *algebraic point* we usually mean a point in $\mathcal{X}(\overline{\mathbb{Q}})$.

Usually, capitals X_i will denote variables over a ground field k. If $f \in k[X_1, \ldots, X_n]$ and if σ is an automorphism of k, f^σ will denote the polynomial obtained by applying σ to the coefficients of f.

If $\mathcal{X}$ is a variety defined by equations $f_i = 0$ over a field k, by $\mathcal{X}^\sigma$ we mean the variety defined by the equations $f^\sigma = 0$. It is easy to see (using the Nullstellensatz) that this does not depend on the set of defining equations.

The symbol "$f \ll g$", for complex functions f, g of a variable x, will mean as usual that $|f(x)| \leq c|g(x)|$ for an unspecified number c, independent of x but which may depend on other data, thought as fixed as x varies. (This is occasionally indicated explicitly by writing *e.g.* "$f \ll_{S,\epsilon} g$".)

We have often used the word 'algorithm' to mean a procedure which leads to the solution of the relevant problem in finite time. We stress that all the algorithms presented in this book allow to estimate the running time for actual computation in terms of standard functions.

ACKNOWLEDGEMENTS. I wish to thank Marco Illengo for taking the notes of the course, for checking many details and for working out a number of examples; also, I owe him substantial help for the TeX preparation. In the same direction I thank Francesco Veneziano.
I have borrowed freely from several sources, but in particular the book [17] by Enrico Bombieri and Walter Gubler has been an invaluable reference on many occasions; I wish to thank the authors for this. I further thank David Masser, who among other things provided me with some of his unpublished notes in the context of Thue's theorem, and Pietro Corvaja for helpful remarks.

I thank Francesco Amoroso for writing the Appendix to the book, which is an updated reference for the subtlest recent results in the quantification of the lower bounds for heights treated in Chapter 4. This also gives a sketch of the delicate methods.

Finally, I thank the 'Centro Edizioni' of the Scuola Normale, especially Luisa Ferrini, for her precious help in the actual publication of the volume.

ACKNOWLEDGEMENTS for this revised edition. I express my thanks to Dr. Clemens Fuchs of the ETH, Zürich, for substantial comments on the previous edition and for pointing out a large number of typos which had survived despite repeated checking. In the same direction I thank Prof. Andrzej Schinzel. I further thank Dr. Luisa Ferrini, who took care and notice of the distribution of the first edition, also providing suitable information to the interested people.

Introduction

The denomination 'Diophantine' comes from the ancient mathematician Diophantus of Alexandria (about 250 a.D.), who wrote a treatise on mathematical problems in which solutions in integer or rational numbers were required.[1] So, an equation is *diophantine* when we seek solutions in integers or rationals. Naturally, the word 'seek' is not mathematically well-defined: by this we are thinking of problems like:

(i) Establish whether a given equation has or not solutions, or
(ii) Find all solutions, or
(iii) Describe in some simple way the distribution of solutions, if there are infinitely many.

Depending on the case, we shall meet all of these viewpoints, but, needless to say, only in rather special cases; actually, it has been proved by Matjiasevic in 1970 that already Question (i) does not admit a complete algorithmic solution (negative answer to X Hilbert Problem). Also, the general level of these lectures has to be considered introductory, elementary and self-contained.

Throughout, we shall often adopt a geometrical language, viewing the solutions of an equation (or system of equations) as points in a cartesian space, defining a *variety*. The points with integer (or rational) coordinates will be called integer (or rational) points. In this view, and because a rational number is defined by a *ratio* rather than a *pair* of integers, the distinction between integer and rational solutions roughly corresponds to consider varieties in affine and projective spaces.

We shall focus mainly on *affine* diophantine problems, and moreover on the special case of curves, in practice seeking integer solutions of equations in two variables. This case historically suggested some major

---

[1] This consisted of several books, of which only a part survived to our times.

ideas for more general problems. We shall present the important connections with Diophantine Approximation, and prove Thue's celebrated results. However we shall not provide details for subsequent deeper and more refined investigations, in order to better stress the main principles. In later chapters we shall also treat some (more modern) diophantine problems in a generalized sense, that is with algebraic points, not restricted to $\mathbb{Q}$ or a number field, and subject to new restrictions (for instance, having small height). The corresponding results, although apparently far from the 'classical' ones, will be shown however to have relevant applications also to ordinary diophantine equations.

More precisely, here is a description of the contents of the whole volume:

In Chapter 1 we shall recall the easy cases of equations in one variable and the linear case of two variables; then we shall give the general theory of quadratic equations in integers, which boils down to Pell Equation. We shall point out the link of these problems with Diophantine Approximation.

Chapter 2 will be mainly concerned with Thue's theorems for diophantine equations $f(X, Y) = c$, with homogeneous f. We shall present the relation of this with the rational approximations to algebraic numbers and we shall prove in full detail Thue's diophantine approximation theorem. As a corollary his finiteness theorem for equations will follow. We shall recall without proof some sharpenings that occurred in Diophantine Approximation after Thue, notably Roth Theorem; we have preferred to give the proof only of Thue's theorem, because it contains many of the basic principles, whereas the proof of more refined results tend to become rather complicated in detail, with the risk of obscuring some ideas to the beginner. Preliminary to the proofs, we shall explain the main steps of the underlying strategy. As in Chapter 1, we shall stick to 'classical' integer solutions, namely over $\mathbb{Z}$.

In Chapter 3 we shall start to formulate diophantine problems over number fields. We shall introduce the fundamental concept of (Weil) *height* and develop in detail some of its main properties. Then we shall formulate without proof the General Roth Theorem in diophantine approximation, and present some applications of it, to the Thue-Mahler Equation and to the S-unit Equation. Finally, also for later use, we shall study the height on finitely generated multiplicative groups of algebraic numbers, interpreting it as a norm on an euclidean space.

In Chapter 4 we shall discuss some kind of diophantine problems in which the variables are free to run over $\overline{\mathbb{Q}}$, but are subject to other arithmetical restrictions; an instance is provided by solutions of equations in roots of unity of arbitrarily high order; in geometrical language, this corresponds to the search for torsion points on subvarieties of $\mathbb{G}_m^n$. We shall explore the more general problem of algebraic points with 'small' height, which represents the toric case of a conjecture by F. Bogomolov. This was solved by S. Zhang around 1995, with rather sophisticated methods from Arakelov Theory. We shall present an elementary proof of Zhang's theorem, and develop some consequences. We shall also sketch an independent elementary approach by Y. Bilu, relying on his equidistribution theorem for Galois conjugates of algebraic numbers with small height.

In Chapter 5 we shall go back to the S-unit equation and S-unit Theorem, already discussed in Chapter 3. This theorem is important for several reasons; in particular, it implies a finiteness theorem for Thue-Mahler Equations, as shown in Chapter 3. In that Chapter the S-unit Theorem is deduced from the General Roth Theorem, not proved in these notes. To fill this gap, a complete independent proof of the S-unit Theorem will be now given. This proof (by Beukers and Schlickewei) is quantitative and yields a very sharp estimate for the number of solutions. As a corollary, we shall obtain a quantitative estimate for the number of solutions of Thue-Mahler Equations. Some fundamental ingredients for the uniformity of these estimates come from Chapter 4.

The first four chapters are concluded with some 'Supplements', consisting of various material related to the topics of the chapter.

The Supplements to the first chapter concern two applications of Dirichlet Lemma in Diophantine Approximation, a solution of the Pell Equation $x^2 - py^2 = -1$ (p a prime $\equiv 1$ (mod 4)) expressed in terms of p-th roots of unity, a Pell Equation in polynomials, a proof of the irrationality of e^n and of π, an algorithm for rational points on conics and Fermat's theorem on the non-existence of right angled triangles with rational sides and square area. All of these topics are treated in a self-contained way.

The Supplements to Chapter 2 assume a little more knowledge from the geometry of algebraic curves. Siegel's finiteness theorem for integral points is stated without proof. However certain interesting cases (like the 'double Pell Equation' and a case of S-unit equation) are dealt with (over $\mathbb{Z}$) by means only of Thue Theorem. The remaining Supplements consist of an algorithm to test the existence of infinitely many integral points on a rational curve, Runge's theorem on integral points and a very brief discussion of a version of Thue Equation for polynomials.

The Supplements to Chapter 3 contain a full treatment of the S-unit equation over function fields, proving the so-called Mason-Stothers abc-theorem; an alternative approach with respect to known proofs also appears, which leads to similar conclusions for curves more general than

$X + Y = 1$. Then, as applications of the elementary theory of heights, we have included an algorithm to compute the multiplicative dependence relations among given algebraic numbers, and a specialization theorem for multiplicative dependence of rational functions on curves and their values; this last issue has been the object of much recent work by several authors.

Finally, the Supplements to Chapter 4 contain the basic theory of closed subgroups of $\mathbb{R}^n$ (with an application to Kronecker's theorem in simultaneous diophantine approximation) and the Skolem-Mahler-Lech Theorem on the zeros of recurrences, together with a significant rephrasing and generalization to the context of algebraic groups. (An 'Open Question' is also stated for sequences arising from Taylor expansions of algebraic functions.)

The book is concluded with an Appendix by F. Amoroso, dealing with the problems of Chapter 4, from a quantitative viewpoint. This direction has been the object of much recent research. The Appendix mentions several recent results, and a few sketches of some methods of proof.

Chapter 1
Some classical diophantine examples

In this elementary chapter we shall deal with some classical diophantine equations, to be solved in ordinary integers of $\mathbb{Z}$. After a brief study of the case of a single variable and of the linear case, we shall go to quadratic equations in two variables, which represent conics in $\mathbb{A}^2$. The fundamental theory here comes from the *Pell Equation* $X^2 - dY^2 = 1$, where d is a fixed positive integer, not a square. This study also links diophantine equations with diophantine approximation, a theory which provides most important tools, that we shall meet throughout. After Pell Equation we shall give a complete effective treatment of the integral points for general conics, *i.e.* quadratic equations in two variables to be solved in $\mathbb{Z}^2$.

In the Supplements we shall discuss some topics related with diophantine approximation, a cyclotomic solution of a Pell Equation and a polynomial version of it. We shall also recall very briefly an effective treatment of rational points on conics in $\mathbb{P}_2$ and a theorem of Fermat.

1.1. The case of a single variable

We start with the simplest cases: one equation in one variable, given by $f(X) = 0$, where $f(X) = a_0 X^d + a_1 X^{d-1} + \ldots + a_d$ is a nonzero complex polynomial, for which we are interested in the rational or integral solutions. Note that there are at most d solutions in complex numbers, which can be approximated in terms of the a_i by known methods; however *a priori* we do not know how to 'calculate' the rational ones, if there are any.

For this question, let us assume that the coefficients $a_0, \ldots, a_d$ are given integers with $a_0 a_d \neq 0$. If $x = p/q$ is a rational solution in lowest terms (*i.e.* p, q are coprime integers) it follows from the equation that p divides $q^d a_d$ and q divides $p^d a_0$. Since p, q are coprime this implies that p divides a_d and q divides a_0. So this gives only finitely many possible fractions p/q among which there are all the possible rational roots of $f(X)$. We can explicitly check whether a given fraction is or is not a root, so all such roots can be written down if we actually know the coefficients, *i.e.* we can determine all the possible rational solutions in a finite number of steps.

All of this plainly takes care also of arbitrary systems of polynomial equations in one variable.

Remark 1.1. Similar considerations produce an algorithm for factoring over $\mathbb{Q}$ a polynomial $f(X) \in \mathbb{Z}[X]$. The essence is in Gauss Lemma, by which we may look only at the factors in $\mathbb{Z}[X]$; so, let $g(X) = b_0 X^r + \ldots + b_r$ be such a factor of degree r, where b_0 clearly must divide a_0. Recall now that any complex root ξ of f satisfies $|\xi| \leq \sum_{i=0}^{d}(|a_i|/|a_0|) =: A$ (see exercise below). Since every root of g is a root of f and since $\pm b_i/b_0$ is the i-th elementary symmetric function of the roots of g, we have $|b_i| \leq |b_0|2^r A^r \leq |a_0|2^r A^r$. This shows that the coefficients of g may be explicitly bounded only in terms of f and since they are integers this gives rise to only finitely many possibilities, which can be checked.

Exercise 1.1.   Find the irreducible factors over $\mathbb{Q}$ of the polynomial $6X^5 - 7X^4 + 2X^3 + 3X^2 + 13X - 10$.

Exercise 1.2. Let $h(X) = h_0 X^m + h_1 X^{m-1} + \ldots + h_m \in \mathbb{C}[X], h_0 \neq 0$, and let $\xi \in \mathbb{C}$ be a root of h. Prove that $|\xi| \leq \sum_{i=0}^{m}(|h_i|/|h_0|)$.

Exercise 1.3. Obtain another algorithm for factoring $f \in \mathbb{Z}[X]$ as follows: for $n \geq \deg f$, choose $a_1 < \ldots < a_n \in \mathbb{Z}$ so that $f(a_i) \neq 0$ for all i; now, for each choice of divisors $d_1|f(a_1), \ldots, d_n|f(a_n)$ there is at most one polynomial $g \in \mathbb{C}[X]$ of degree $< \deg f$ such that $g(a_j) = d_j$ for $j = 1, \ldots, n$. Prove that the factors of f in $\mathbb{Z}[X]$ must lie among such finitely many polynomials, and that they can be explicitly computed.

Exercise 1.4.

Let p be a prime number, $f \in \mathbb{Z}[x]$ be a monic polynomial irreducible modulo p, and $h \in \mathbb{Z}[x]$ be a polynomial not divisible by $f(x)$ modulo p. Prove that for all integers $m > \deg f/\deg h$ the polynomial $f(x)^m + ph(x)$ is irreducible over $\mathbb{Q}$. (Hint: by Gauss lemma we may assume by contradiction that $f(x)^m + ph(x) = A(x)B(x)$ with A, B monic nonconstant polynomials with integer coefficients. We find that the reductions modulo p of both A, B are powers of f, hence equations $A(x) = f(x)^a + pu(x), B(x) = f(x)^b + pv(x)$, $u, v \in \mathbb{Z}[x]$, where $a, b > 0, a + b = m$. Hence $u(x)f(x)^b + v(x)f(x)^a + pu(x)v(x) = h(x)$. However this implies that $f(x)$ divides $h(x)$ modulo p, against the assumptions.

We remark that setting $f(x) = x$ we obtain as a special case the celebrated Eisenstein irreducibility criterion.)

1.2. The linear case in two variables

We now go to the case of two variables. Suppose we have a system of equations $f_i(X, Y) = 0, i = 1, \ldots, r$, where $f_i \in \mathbb{Z}[X, Y]$, that we want to solve in integers. We may write $f_i(X, Y) = g(X, Y)h_i(X, Y)$ where $g, h_i \in \mathbb{Z}[X, Y]$ and the h_i have no common factor in $\mathbb{Q}[X, Y]$. The system $h_i = 0, i = 1, \ldots, r$ may be dealt with on eliminating separately X and Y (using resultants, or equivalently Euclid's algorithm with respect to X and Y). We obtain two nontrivial equations $R(X) = S(Y) = 0$, which reduces the problem to the case of a single variable. So it suffices to study the equations $g(X, Y) = 0$, which represent plane curves.

The simplest instance occurs with lines in a plane, represented by the equations of degree 1:

$$aX + bY = c, \qquad a, b, c \in \mathbb{Z}, \tag{1.1}$$

for which we seek integer solutions. Note that the rational solutions give no problem: if, say, $b \neq 0$, we can just choose any rational number x and define $y := (c - ax)/b$. For integers, things are just a little subtler.

The special choice $c = 0$ leads to the equation $ax = (-b)y$, which may be viewed as expressing two decompositions of a same integer into factors. The study of the linear equations thus leads in particular to the whole theory of factorization of integers (due in essence to Euclid); this fact alone suggests that our subject deserves most serious attention.

Now, note that in (1.1) we can divide out by a gcd of the coefficients a, b, c and so we can assume $(a, b, c) = 1$. After this normalization, if a and b are not coprime, their gcd cannot divide c, whence the equation has no integer solutions. Then let us assume that a and b are coprime, so, e.g. by Euclid's algorithm, there exist integers m and n such that $am + bn = 1$; in this case a solution of (1.1) is $(x_0, y_0) = (cm, cn)$. Now let (x_1, y_1) be another integer solution; by subtracting we find

$$a(x_1 - x_0) = b(y_0 - y_1).$$

Since $(a, b) = 1$, we see that b must divide $x_1 - x_0$, so $x_1 = x_0 + kb$ where k is an integer. If $b \neq 0$ (which we can assume) we find $y_1 = y_0 - ka$, and thus we obtain a linear *parametrization* of all the integer solutions, which always make up an infinite set:

$$(x_1, y_1) = (x_0, y_0) + k(b, -a) = (cm + kb, cn - ka), \qquad k \in \mathbb{Z}.$$

In particular, all of this shows that the fundamental equation $ax + by = 1$ allows to describe completely the general case.

Remark 1.2. (Euclid's algorithm) We just recall this celebrated algorithm for solving $ax + by = \gcd(a, b)$ for integers a, b. Assuming $b > 0$ we divide a by b, obtaining $a = q_1 b + r_1$ with $0 \leq r_1 < b$. If $r_1 > 0$ we continue: $b = q_2 r_1 + r_2, 0 \leq r_2 < r_1$ and so on $r_i = q_{i+2} r_{i+1} + r_{i+2}, 0 \leq r_{i+2} < r_{i+1}$ until we obtain a zero remainder, which will certainly happen sooner or later; at that point the algorithm stops. It is easy to check that the last nonzero remainder is the $\gcd(a, b)$ and using the equations in reverse order we easily obtain the sought solution. The algorithm amounts to the expansion of a/b as a continued fraction. The same algorithm holds in $k[X]$, for any field k.

Exercise 1.5. Prove that for coprime a, b Euclid's algorithm leads to an integral solution (m,n) of $aX + bY = 1$ after at most const. $\cdot \log\min(|a|,|b|) + 1$ steps. (Also, find a "best-possible" constant.)

Exercise 1.6. Prove that if a, b are coprime positive integers, for all sufficiently large integers r there exists a solution of $aX + bY = r$ in non-negative integers. (Also, prove that the largest r for which there are not such solutions is $(a - 1)(b - 1) - 1$.)

Exercise 1.7. Let A be an $r \times n$ matrix with entries in $\mathbb{Z}$ and let $\mathbf{v} \in \mathbb{Z}^r$. Prove that the equation $A\mathbf{x} = \mathbf{v}$ has a solution $\mathbf{x} \in \mathbb{Z}^n$ if and only if the congruence $A\mathbf{x} \equiv \mathbf{v} \bmod m$ has a solution for all positive integers m. (Hint: The image $A(\mathbb{Z}^n)$ is a subgroup of $\mathbb{Z}^r$. Use the theorem of elementary divisors to find a basis $\mathbf{b_i}$ of $\mathbb{Z}^r$ so that $\delta_i \mathbf{b_i}$ generate the subgroup...)

1.3. Diophantine Approximation

Already in basic cases, like the linear one, diophantine equations embody a principle of 'Diophantine Approximation': basically, this is *the theory of approximations of real numbers by rational numbers*. Of course every real number can be approximated by rationals with arbitrary accuracy, but the really 'good' approximations are those which are simple compared to the accuracy by which they approach the target number. For instance, the fraction $141/100$ approaches $\sqrt{2}$ by about $1/250$, whereas the simpler fraction $99/70$ produces an accuracy superior to $1/10.000$; other remarkable examples, like Archimedes' inequalities $233/71 > \pi > 22/7$, go back to antiquity.

To describe these excellent approximations, and to establish how good they can be, depending on the target number, is the fundamental problem originating this theory. For us this study will be most relevant, because, as will be clearer and clearer, such kind of information is fundamental for studying diophantine equations.

Below we shall see several examples of this link, at various levels of depth. The link itself arises as follows: any infinite sequence of distinct integer points on a (plane) curve cannot remain bounded. Hence these points have to approach some "point at infinity" in a projective model of the curve. In an affine model, the points at infinity correspond to the slopes of the asymptotic directions of the curve (note also that the number of such slopes is at most the degree of the curve). In conclusion, the ratio between the coordinates of a large integer point on the curve yields a rational approximation to one of these slopes.

For instance, in the next section we shall see the interesting case coming from quadratic equations; but the linear case already gives an instance. Writing the fundamental linear equation $ay + bx = 1$ as $(a/b) - (-x/y) = 1/by$ we see that the rational number $-x/y$ provides a 'good' approximation to a/b. The word 'good' is motivated by the fact that if p/q is any rational number $\neq a/b$, then

$$|(a/b) - (p/q)| = |aq - bp| / |bq| \geq 1/|bq|.$$

(In fact, note that $aq - bp$ is a nonzero integer and is thus ≥ 1 in absolute value.[1]) In other words, the accuracy by which p/q approaches a/b, *relative* to the denominator q, cannot exceed $1/|b|$ for any rational fraction; this bound is attained precisely with the solutions of the linear equation $ax + by = \pm 1$.

What can we say for a general real number $\xi \in \mathbb{R}$? Note that for a rational approximation to ξ, with a *prescribed* denominator $y \in \mathbb{Z}$, the best that we can generally say (see Remark below) is $\min_{x \in \mathbb{Z}} |\frac{x}{y} - \xi| \leq \frac{1}{2y}$. Nonetheless, especially for irrational numbers, for special choices of y we can achieve substantially better inequalities. To see this, we start with a very elegant and simple celebrated lemma.

Theorem 1.1 (Dirichlet Lemma). *Let $\xi \in \mathbb{R}$ and let $Q > 0$ be a positive integer. Then there exist $p, q \in \mathbb{Z}$, such that $(p, q) = 1$ and*

$$0 < q \leq Q, \qquad \left| \xi - \frac{p}{q} \right| \leq \frac{1}{q(Q+1)}. \qquad (1.2)$$

Proof. We first prove that there exist p, q, not necessarily coprime, satisfying (1.2). For this, let us consider the sequence of $Q+1$ real numbers $0, \{\xi\}, \ldots, \{Q\xi\}$ in $[0, 1)$, where $\{x\}$ denotes the fractional part of the real number x. If we divide the interval $[0, 1)$ into the $Q + 1$ intervals $[j/(Q + 1), (j + 1)/(Q + 1))$, $j = 0, \ldots, Q$, we see (by "Dirichlet's box principle") that either there exists a single number of the sequence in each interval or there exists some interval containing two numbers of the sequence.

In the first case there exists an integer $q, 0 < q \leq Q$ such that $Q/(Q+1) \leq \{q\xi\} < 1$, whence for a suitable integer p, $|q\xi - p| \leq 1/(Q+1)$, as required.

In the second case there exist integers r, s satisfying $0 \leq s < r \leq Q$ and $|\{r\xi\} - \{s\xi\}| < 1/(Q + 1)$, which again leads to the sought conclusion, with $q = r - s$.

---

[1] David Masser once expressed [58] this last deduction as the "fundamental theorem of transcendence"; indeed, it plays a role so crucial that it is difficult to overestimate it.

Now, even if p, q are not coprime, we can simply factor out $\gcd(p, q)$ from both; the conditions (1.2) will continue to hold, and this proves the conclusion.   □

Remark 1.3. A slightly simpler argument would replace $Q + 1$ with Q, an almost equally useful result.

Exercise 1.8. Similarly to Dirichlet Lemma, prove that if $\xi_1, \ldots, \xi_r$ are reals and Q is a given positive integer, there exist a positive integer $q \leq Q^r$ and integers $p_1, \ldots, p_r$ such that $|q\xi_i - p_i| < Q^{-1}$. (For $r = 1$ we find back *almost* the previous lemma.) (Hint: Consider the $Q^r + 1$ points $(\{t\xi_1\}, \ldots, \{t\xi_r\})$ in the unit cube, for $0 \leq t \leq Q^r$. Subdividing the unit cube in Q^r small cubes of side $1/Q$ yields two points in a same small cube...)

Exercise 1.9. Let $a_1 < a_2 < \ldots$ be the sequence of integers of the form $2^r 3^s$, arranged in increasing order. Prove that the ratio a_{n+1}/a_n tends to 1 as $n \to \infty$.

Exercise 1.10. Let $\xi \in \mathbb{R}$. Suppose that $w > 0$ is such that for every integer $Q \geq 1$ there exist integers p, q with $|p|, |q| \leq Q$ and $0 < |q\xi - p| \leq Q^{-w}$. Prove that $w \leq 1$. (Hint: Fix a large Q and find coprime p, q with the said property. Then define $X \geq Q$ by $|q\xi - p| = X^{-w}$. Choose now t, u with the property, for $2X$ in place of Q. Finally, eliminate ξ to estimate $|pu - qt|$...)

Exercise 1.11. Prove that there exists $\xi \in \mathbb{R}$ such that for every real number w and infinitely many pairs (p, q) of positive integers we have $0 < |q\xi - p| < q^{-w}$. (Compare with the previous exercise, and see also the next chapter, especially Remark 2.4. Hint: define ξ by a series of rational numbers, with suitably rapid convergence.)

When ξ is irrational, we can prove that there are infinitely many 'good approximations':

Corollary 1.2. *Let* $\xi \in \mathbb{R} \setminus \mathbb{Q}$. *Then there exist infinitely many* $p, q \in \mathbb{Z}$ *such that* $(p, q) = 1$ *and*

$$|q\xi - p| < q^{-1}. \tag{1.3}$$

Proof. Let $(p_1, q_1), \ldots, (p_n, q_n)$ be pairs as in (1.3) and consider

$$\varepsilon = \min_i |q_i\xi - p_i|,$$

where $\varepsilon > 0$ since $\xi \notin \mathbb{Q}$. Taking any $Q > \frac{1}{\varepsilon}$ in Dirichlet Lemma, we obtain a fraction $\frac{p}{q}$ in lowest terms with $0 < q \leq Q$ and

$$\left| \xi - \frac{p}{q} \right| \leq \frac{1}{qQ} < \frac{1}{q^2}.$$

Then $|q\xi - p| \leq \frac{1}{Q} < \varepsilon$, so the pair (p, q) satisfies (1.3) and is distinct from any of the previous ones.   □

Remark 1.4. (**Approximations in function fields**) The 'exponent' 2 attributed to q^{-1} in the approximations $|\xi - (p/q)| \leq q^{-2}$ comes from the *double freedom* in choosing p, q. One may see even more clearly this principle by looking at a function field version of Dirichlet Lemma and of this corollary. For this, let $\xi(t)$ be a power series in $k[[t]]$ (where k is a field) and look at 'approximations' of ξ by rational functions $p(t)/q(t) \in k(t)$, with respect to the topology of $k[[t]]$: namely, we want that $p(t)/q(t)$ has a Taylor series at the origin which coincides with $\xi(t)$ up to a 'large' order. If $p, q \in k[t]$ are restricted to have degree $\leq n$ (which is like bounding p, q in Dirichlet Lemma) we have $2n + 2$ free coefficients. Imposing the vanishing of the first N coefficients of $q(t)\xi(t) - p(t)$ gives a linear system which can be solved nontrivially as soon as $2n + 2 > N$. Thus we can achieve that $\deg p, \deg q \leq n$ and $\text{ord}_{t=0}(q\xi - p) > 2n$. This shows how the '2' appears. To go even closer in the analogy with the numerical case, let us write $q(t) = t^n q^*(1/t)$, $p(t) = t^n p^*(1/t)$, where p^*, q^* are also polynomials of degree $\leq n$ (and are 'large' in $k[[t]]$). Then $\text{ord}_{t=0}(\xi - (p^*(1/t)/q^*(1/t))) > n + \deg q^* \geq 2 \deg q^*$ while $\text{ord}_{t=0}(q^*(1/t)) = -\deg q^* \geq -n$. See the supplements below for an important example related to $\exp(t)$.

Remark 1.5. (**Good approximations are rare**) For a real ξ and positive integer y let us put $\mu(y) = \mu(\xi, y) := \min_{x \in \mathbb{Z}} |x - \xi y|$. We have noticed that $\mu(y) \leq 1/2$ and this cannot be improved if $\xi = n + \frac{1}{2}$, with $n \in \mathbb{Z}$, for every odd y. Also, for every ξ it is easy to see that $\mu(y) \geq 1/3$ for infinitely many y. To go further, fix an irrational ξ and a positive $\varepsilon < \frac{1}{2}$. One may prove (see next exercise or e.g. [25]) that the density in $[1, T]$ of the set of y such that $\mu(y) \leq \varepsilon$ tends to 2ε as $T \to \infty$. All of this shows in particular that the approximations as in the Corollary are very rare. Actually, one can prove that the number of corresponding denominators up to T is $\ll \log T$.

Exercise 1.12. Let ξ be irrational and let $0 < \epsilon < 1$. Prove that for $T \to \infty$, the number of positive integers $q \leq T$ such that the *fractional part* $\{q\xi\} \leq \epsilon$ is $\sim \epsilon T$. (Hint: use Dirichlet Lemma with $Q = T$ to approximate ξ very well with a rational and argue with residue classes modulo the denominator.)

Exercise 1.13. Prove that the number of rational approximations p/q to ξ in reduced terms and such that $|\xi - (p/q)| \leq q^{-2}$ and $q \leq T$ is $\ll \log T$. (Hint: consider the difference of approximations p/q, p'/q' with $q < q'$, and observe that $|pq' - p'q| \leq 2q'/q$. Fix then p/q and vary p'/q' among a few other approximations.)

Remark 1.6. (**Irrationality criterion**) When $\xi = a/b$ is rational, the corollary does not hold. On the contrary, there exists a $c = c(\xi) > 0$ such that every other rational $p/q \neq a/b$ ($b, q > 0$) satisfies $|\xi - \frac{p}{q}| \geq \frac{c}{q}$. In fact, $|\xi - \frac{p}{q}| = \frac{|aq-bp|}{bq} \geq \frac{1}{bq}$ and we can take $c = \frac{1}{b}$.

Therefore, to prove that a given number ξ is irrational it suffices to find, for every $\epsilon > 0$, a rational fraction $p/q \neq \xi$ such that $|\xi - (p/q)| \leq \epsilon q^{-1}$.

This principle for instance leads quickly to a proof of the irrationality of $e = 2.7182....$ Assuming $e \in \mathbb{Q}$, for every $\frac{p}{q} \neq e$ we would have $q|e - \frac{p}{q}| > c$ for some constant $c > 0$. Now let n be an integer such that $\frac{1}{n} < c$ and consider the fraction $\sum_{i=0}^{n} \frac{1}{i!} = \frac{p}{q} \neq e$, where $q = n!$. We have

$$c < q \left| e - \frac{p}{q} \right| = n! \sum_{j>0} \frac{1}{(n+j)!} < n! \sum_{k \geq 0} \frac{1}{(n+1)!} \frac{1}{(n+1)^k}$$

$$= \frac{1}{n+1} \frac{1}{1 - \frac{1}{n+1}} = \frac{1}{n},$$

a contradiction.

Exercise 1.14. Prove that $\sqrt{2}$ is irrational by constructing good rational approximations to it. For this consider *e.g.* the equality $a_n - b_n\sqrt{2} := (1 - \sqrt{2})^n$, $a_n, b_n \in \mathbb{Z}$.

1.4. Pell Equation

After this brief digression on diophantine approximations, we go back to equations in two variables, turning to the quadratic case. Geometrically, these equations represent conic sections: ellipses, parabolas, hyperbolas.

We shall leave aside the problem of rational points; we only recall that if there exists a single rational point P_0 on such a curve C, then we may parametrize all of them by projecting C from P_0 to any given rational line (*i.e.* a line defined by an equation with rational coefficients) not containing P_0. So, the problem of rational points is easy if we only know one such point. (See the last supplement below for an algorithm to detect such a possible point.)

For integral points the situation is subtler. As we shall show below, the general equation of degree 2 can be reduced to a few basic cases. The most difficult and interesting of these is the following special diophantine equation, known as *Pell Equation*:

$$X^2 - dY^2 = 1, \qquad 0 < d \in \mathbb{Z}, \tag{1.4}$$

where d is not a square. This represents a hyperbola. We shall look for integral solutions different from the *trivial* ones $(\pm 1, 0)$. Note that when $d = n^2$ is a perfect square, the equation has only trivial integer solutions $(x, y) \in \mathbb{Z}^2$: in fact, we have $(x + ny)(x - ny) = 1$, so $x \pm ny$ have both to divide 1 and thus must be equal to ± 1.

Remark 1.7. (Parametrization of rational points) One could consider a parametrization for rational points (as alluded above) with the purpose of picking the integer ones among them. Projecting the hyperbola from $P_0 := (1, 0)$ to the line $X = 0$ by $P \mapsto (0, t)$ leads to the parametrization (in terms of a parameter $t = p/q \in \mathbb{Q}$)

$$(x, y) = (1, 0) - \left(\frac{2t^2}{t^2 - d}, \frac{2t}{t^2 - d} \right) = (1, 0) - \frac{2p}{p^2 - dq^2}(p, q),$$

where p, q are arbitrary coprime integers. (Note that the inverse of this map is $t = y/(1 - x)$.)

If (x, y) is to be an integer point, then $(p^2 - dq^2)$ has to divide $2p$, since $(p, q) = 1$. Also, we have $(p^2 - dq^2, p) = (d, p)$ and so $p^2 - dq^2$ divides $2d$ and so is an integer h in a certain finite set depending only on d. In substance, we find other equations of the same type as (1.4), namely $p^2 - dq^2 = h$. In other words, if we try to find integer solutions of (1.4) via a parametrization of its rational solutions, we end up with a problem of the same nature.

Naturally this parametrization of rational points is very interesting and useful in many problems. It can be obtained in the same way for any conic, provided we have a rational point P_0 on it, to be used as the center of projection ($P_0 = (1,.0)$ in the example).

It is very useful to view the equation (1.4) in the quadratic field $\mathbb{Q}(\sqrt{d})$, over which it becomes

$$(X - \sqrt{d}Y)(X + \sqrt{d}Y) = 1,$$

so that by a linear change of variables it takes the usual form of a hyperbola $ZW = 1$. If $(x, y) \in \mathbb{Z}^2$ is a non-trivial solution, where we can assume $x, y > 0$, then $x = \sqrt{1 + y^2 d} > y\sqrt{d}$ and $\left| x - y\sqrt{d} \right| = \frac{1}{x+y\sqrt{d}} < \frac{1}{2y\sqrt{d}}$, whence

$$\left| \frac{x}{y} - \sqrt{d} \right| < \frac{1}{2\sqrt{d}y^2}.$$

We remark that the approximation of $\sqrt{d}$ so obtained by the rational x/y is particularly good, as we can especially appreciate taking into account the last section. In fact the error is inversely-proportional to the square of y, whereas for a general y we can only expect an error inversely-proportional to y itself (recall Remark 1.5).

This observation, simple as it is, is very important, because it deeply links the diophantine equation (1.4) with diophantine approximation.

In fact, coming back to the problem of finding integer solutions of (1.4), we shall use, reciprocally, diophantine approximation. We shall prove the following

Theorem 1.3. *For any $d \in \mathbb{Z}$, $d > 0$, not a perfect square, there are infinitely many solutions $(x, y) \in \mathbb{Z}^2$ of the equation $X^2 - dY^2 = 1$.*

Proof. We start by proving the existence of a non-trivial solution. Consider the fractions $\frac{x}{y}, x, y \in \mathbb{N}, y > 0$, satisfying

$$\left| \frac{x}{y} - \sqrt{d} \right| \leq \frac{1}{y^2}, \qquad y > 0.$$

There are infinitely many of them, by Corollary 1.2. For them we have $|x - y\sqrt{d}| \leqslant y^{-1} \leqslant 1$, so that $x \leqslant 1 + y\sqrt{d}$ and

$$|x^2 - dy^2| \leqslant \frac{\left|x + y\sqrt{d}\right|}{|y|} \leqslant \frac{2|y|\sqrt{d} + 1}{|y|} \leqslant 2\sqrt{d} + 1.$$

Since $|x^2 - dy^2|$ is a nonzero integer, this proves that there is a nonzero integer M in the interval $(-1 - 2\sqrt{d}, 1 + 2\sqrt{d})$ such that the equation

$$X^2 - dY^2 = M$$

has infinitely many integer solutions (x, y), with $y > 0$.

We are not yet quite done, since M may well be $\neq 1$. To obtain solutions of Pell Equation, the principle is the following: in the first place, for a solution p, q, we have $(p + q\sqrt{d})(p - q\sqrt{d}) = 1$, so a solution yields the invertible element $p + q\sqrt{d}$ in $\mathbb{Z}[\sqrt{d}]^*$; hence it is sensible to seek invertible elements. Now, we have infinitely many solutions of $(x + y\sqrt{d})(x - y\sqrt{d}) = M$, which may be viewed as a factorization of M in the ring $\mathbb{Z}[\sqrt{d}]$. If we find two equivalent factorizations, that is up to a unit factor, then we have found a nontrivial invertible element. In turn, a way for showing that not all such factorizations can be inequivalent is to observe that if $M = aa' = bb'$ where $a \equiv b$ (mod M), then $a = b + qM = b(1 + qb')$, whence $b|a$; by symmetry $a|b$, as wanted. Here is some more detail to put these ideas in practice; since there are infinitely many solutions in $\mathbb{Z} \times \mathbb{Z}$ to $X^2 - dY^2 = M \neq 0$ and since $\mathbb{Z}/M\mathbb{Z} \times \mathbb{Z}/M\mathbb{Z}$ is a finite set, there exist elements $x_1 + y_1\sqrt{d}$ and $x_2 + y_2\sqrt{d}$ in $\mathbb{Z}[\sqrt{d}]$ such that $x_1, x_2, y_1, y_2 > 0$ and

$$\begin{cases} (x_1 + y_1\sqrt{d})(x_1 - y_1\sqrt{d}) = M; \\ (x_2 + y_2\sqrt{d})(x_2 - y_2\sqrt{d}) = M; \\ x_1 \equiv x_2 \pmod{M}; \\ y_1 \equiv y_2 \pmod{M}. \end{cases}$$

We may further suppose that $\frac{x_1}{y_1} \neq \frac{x_2}{y_2}$; in fact, otherwise the identity $x^2 - dy^2 = y^2((x/y)^2 - d)$ implies $y_1 = y_2$, which we may exclude.

Let us now write $(x_1 + y_1\sqrt{d})(x_2 - y_2\sqrt{d}) = A + B\sqrt{d}$, where $A = x_1x_2 - dy_1y_2$, $B = x_2y_1 - x_1y_2$ are integers; the above congruences give $A \equiv x_1x_2 - dy_1y_2 \equiv x_1^2 - dy_1^2 \equiv 0$ (mod M) and $B \equiv x_2y_1 - x_1y_2 \equiv 0$ (mod M). Hence $A = MA'$, $B = MB'$ for integers A', B'. Moreover we can directly check that

$$A^2 - dB^2 = (x_1^2 - dy_1^2)(x_2^2 - dy_2^2) = M^2. \qquad (1.5)$$

This gives $A'^2 - dB'^2 = 1$ and we have only to verify that the solution so obtained is non-trivial, *i.e.* that $B' \neq 0$. However $B' = 0$ implies $x_1/y_1 = x_2/y_2$, which has been excluded.

We have thus seen that some nontrivial solutions (x, y) exists. To obtain infinitely many of them it suffices to define integers x_n, y_n by $x_n + y_n\sqrt{d} := (x + y\sqrt{d})^n$, that is

$$x_n = \frac{(x + y\sqrt{d})^n + (x - y\sqrt{d})^n}{2},$$

$$y_n = \frac{(x + y\sqrt{d})^n - (x - y\sqrt{d})^n}{2\sqrt{d}}. \tag{1.6}$$

We have $x_n - y_n\sqrt{d} = (x - y\sqrt{d})^n$ (either by direct verification or by applying the involution $\sqrt{d} \mapsto -\sqrt{d}$), whence $x_n^2 - dy_n^2 = ((x+y\sqrt{d})(x - y\sqrt{d}))^n = 1$. Note that the solutions so obtained are pairwise distinct because (x, y) is a nontrivial solution, so $x + y\sqrt{d} \neq \pm 1$ is not a root of unity. □

The link equations-approximations goes both ways; in the proof we have used approximations$\rightarrow$ equations; but as we have previously observed, the solutions so obtained give back some especially good approximations to $\sqrt{d}$, even better (up to a constant factor) than those predicted by Corollary 1.2; we have:

Corollary 1.4. *There are infinitely many fractions $\frac{x}{y} \in \mathbb{Q}$, x, y coprime integers, such that*

$$\left| \frac{x}{y} - \sqrt{d} \right| \leqslant \frac{1}{2\sqrt{d}y^2}.$$

Exercise 1.15. Prove that for a non square positive integer d and for any positive integer m there are infinitely many integer solutions (x, y) of $X^2 - dY^2 = 1$ such that $m|y$. (Hint: use (1.6) or consider $X^2 - m^2dY^2 = 1$.)

1.4.1. Structure of the solutions and units in quadratic fields

As we have seen, solutions to Pell Equation $X^2 - dY^2 = 1$ are strongly related to units, *i.e.* invertible elements, in $\mathbb{Z}[\sqrt{d}]$ or generally in the ring of integers of the (real) quadratic field $\mathbb{Q}(\sqrt{d})$, the two rings being equal when d is squarefree and $d \not\equiv 1 \pmod 4$.

Recall that, generally, *if k is a finite extension of $\mathbb{Q}$ with ring of integers $\mathcal{O}_k$, then $\xi \in \mathcal{O}_k^*$ if and only if its norm is $N_{\mathbb{Q}}^k(\xi) = \pm 1$.* In fact, if ξ is a unit, its norm $N_{\mathbb{Q}}^k(\xi) \in \mathbb{Q}$ must be actually in $\mathbb{Z}$ and must be a unit of $\mathbb{Z}$, *i.e.* $= \pm 1$. Viceversa, if $\xi \in \mathcal{O}_k$ has norm ± 1, ξ is invertible in $\mathcal{O}_k$, the inverse being $\pm$ the product of its other conjugates. In our context, this argument extends to the rings $\mathbb{Z}[\sqrt{d}]$. Thus an element $x + y\sqrt{d}$, $x, y \in \mathbb{Z}$, is a unit of $\mathbb{Z}[\sqrt{d}]$ if and only if

$$N_{\mathbb{Q}}^{\mathbb{Q}(\sqrt{d})}(x + y\sqrt{d}) = x^2 - dy^2 = \pm 1.$$

In particular, when (x, y) is a solution of Pell Equation, $x + y\sqrt{d}$ is a unit in $\mathbb{Z}[\sqrt{d}]$. By a similar argument, for squarefree d, units of $\mathcal{O}_{\mathbb{Q}(\sqrt{d})}$ correspond to integers x, y such that $x^2 - dy^2 = \pm 4$; when $d \equiv 2, 3$ (mod 4) it is easy to see that x, y must be even and we obtain a solution of Pell Equation on dividing by 4; when $d \equiv 1$ (mod 4), it is easy to see that we obtain a solution to Pell Equation by cubing $x + y\sqrt{d}$.

Exercise 1.16. Prove the last statement.

Let us take a closer look at $\mathbb{Z}[\sqrt{d}]^*$, by considering the homomorphism

$$\Phi: \begin{array}{ccc} \mathbb{Z}[\sqrt{d}]^* & \to & \mathbb{R} \\ \xi & \mapsto & \log|\xi|. \end{array}$$

Its kernel is $\{\pm 1\}$, while its image H is a *discrete subgroup* of the additive group of real numbers.[2] In fact, for any $a > 0$ the condition $-a < \log|\xi| < a$ becomes $e^{-a} < |\xi| < e^a$; but the same then holds for the conjugate $\xi' = \pm\frac{1}{\xi}$ of ξ, i.e. $e^{-a} < |\xi'| < e^a$. Since $\xi = r + s\sqrt{d}$ and $\xi' = r - s\sqrt{d}$ both belong to $\mathbb{Z}[\sqrt{d}]$, with $r, s \in \mathbb{Z}$, we have

$$|r| = \left|\frac{\xi + \xi'}{2}\right| < e^a; \qquad |s| = \left|\frac{\xi - \xi'}{2\sqrt{d}}\right| < \frac{e^a}{\sqrt{d}},$$

leading to a finite number of possibilities for r, s and thus for ξ.

To go on, observe that discrete subgroups of $\mathbb{R}$ are cyclic. In fact, if such a subgroup G is nonzero, let g be any element of $G \setminus \{0\}$ whose absolute value is minimal. Then every $h \in \mathbb{R}$ can be written as $h = mg + k$, where $m \in \mathbb{Z}$ and $|k| < |g|$; if $h \in G$ then $k \in G$, so by minimality we have $k = 0$ and $h \in < g >$. This implies that g generates G.[3]

In our case we thus obtain the exact sequence

$$1 \to \{\pm 1\} \to \mathbb{Z}[\sqrt{d}]^* \to \mathbb{Z} \to 1,$$

---

[2] This means it is discrete with the induced topology, namely the set $H \cap (-\delta, \delta)$ is finite for every $\delta > 0$.

[3] Later we shall see a generalization to discrete subgroups of $\mathbb{R}^n$.

where the surjectivity is granted by the existence of non-trivial solutions to Pell Equation. From this ($\mathbb{Z}$ is a free group) we obtain

$$\mathbb{Z}[\sqrt{d}]^* \cong \{\pm 1\} \times \mathbb{Z}. \tag{1.7}$$

All elements $r + s\sqrt{d}$ of $\mathbb{Z}[\sqrt{d}]^*$ projecting on the generators of the cyclic group are called *fundamental units* of $\mathbb{Z}[\sqrt{d}]$; the corresponding pairs (r, s) are called *fundamental solutions*: they are solutions of one of the Pell-type equations $r^2 - ds^2 = \pm 1$.[4] We may further normalize (keeping the same terminology) by requiring that r, s in a fundamental solution are > 0, which determines it uniquely. (Similarly for the units of the full ring of integers of $\mathbb{Q}(\sqrt{d})$.)

It is very striking that fundamental units appear in the Dirichlet formula for the class number of the quadratic fields $\mathbb{Q}(\sqrt{d})$. In a sense, the larger the fundamental solution is, the smaller the class number is (see the notes below). Gauss conjectured that for infinitely many real quadratic fields there is unique factorization for their rings of integers; this would correspond to 'very large' fundamental solutions.

By (1.7) we can write any solution (x, y) by using the formula $x + y\sqrt{d} = \pm(r + s\sqrt{d})^n$, where (r, s) is a fundamental solution and n is in $\mathbb{Z}$. From this it immediately follows that there are $\ll \log T$ solutions (x, y) of the Pell equation in a box $|x|, |y| \leq T$.

Remark 1.8. (**The equation** $X^2 - dY^2 = -1$) From the above it follows that this equation has an integer solution (x, y) if and only if the fundamental units of $\mathbb{Z}[\sqrt{d}]$ have norm -1. In this case it also follows that a fundamental solution of the usual Pell Equation is obtained by considering $(x + y\sqrt{d})^2 = x^2 + dy^2 + 2xy\sqrt{d}$. Simple complete criteria for solvability are not known. By considering the quadratic character of -1 it is clear that for the equation to be solvable every odd prime factor of d must be $\equiv 1 \pmod 4$, but this is not a sufficient condition; however it is sufficient if d is prime (see next exercises).

Remark 1.9. (**Automorphs of quadratic forms**) It is well known that the solutions of Pell Equation are related to the orthogonal group over $\mathbb{Z}$ of the quadratic form $X^2 - dY^2$. In fact, let $M = \begin{pmatrix} r & s \\ t & u \end{pmatrix}$, $r, s, t, u \in \mathbb{Z}$ be in the orthogonal group, so that ${}^t M P M = P$ where P is the matrix $\begin{pmatrix} 1 & 0 \\ 0 & -d \end{pmatrix}$. This gives $r^2 - dt^2 = 1$, $rs = dut$, $s^2 - du^2 = -d$. The first equation is just the Pell Equation. Given an integral solution (r, t) of it, it is easily seen that one must have $s = \pm td$, $u = \pm r$ and conversely. In other words, M is of the shape

$$M = \begin{pmatrix} r & \pm dt \\ t & \pm r \end{pmatrix}, \text{ where } r^2 - dt^2 = 1.$$ The structure of the orthogonal group then follows from the above theory of Pell Equation.

---

[4] The equation with the minus sign is sometimes called the *negative Pell Equation*.

Exercise 1.17. Prove that $X^2 - 305Y^2 = -1$ has no solutions in integers. (Observe that $305 = 5 \cdot 61$. Hint: note the solution $489^2 - 305 \cdot 28^2 = 1$. If the equation had a solution we would have $489 + 28\sqrt{305} = (x + y\sqrt{305})^2$...)

Exercise 1.18. Prove that if d is a prime number $\equiv 1 \pmod 4$ then the equation $X^2 - dY^2 = -1$ has an integer solution. (Hint: let $a^2 - db^2 = 1$ be the fundamental solution of the Pell Equation; from $(a-1)(a+1) = dy^2$ find $a \pm 1 = 2du^2, a \mp 1 = 2v^2$, so that $v^2 - du^2 = \mp 1$. The plus sign is excluded from minimality, whence the conclusion. See also the supplements.)

Remark 1.10. **(Solutions in {3}-integers)** Similarly to solutions in integers, one can consider solutions in rationals whose denominators are divisible only by primes from a fixed finite set S; these numbers are called S-integers and have appeared more and more naturally in the theory of diophantine equations. We shall meet them often in the sequel; for the moment let us limit ourselves to a very simple but interesting example, where S consists, say, of the single prime 3 and we again deal with the Pell Equation. On clearing denominators, our question leads to the integer solutions of

$$X^2 - dY^2 = 9^n.$$

Every solution $(x, y) \in \mathbb{Z}^2$ to this equation corresponds to an element $\alpha = x + y\sqrt{d} \in \mathbb{Z}[\sqrt{d}]$ with norm $N_{\mathbb{Q}}^{\mathbb{Q}(\sqrt{d})}(\alpha) = 9^n$. The shape of such elements depends on the behaviour of the ideal (3) in the ring $\mathbb{Z}[\sqrt{d}]$: it can remain prime, ramify or split. In the first case x, y must be divisible by 3^n and we find nothing new. In any case, on considering ideal factorization in the appropriate quadratic ring, it is not difficult to see that all of the solutions are expressed by $\alpha = \pm \gamma^r \gamma'^s \eta^t$ with γ, γ' generators for suitable powers of the prime ideals above 3 (possibly $\gamma = \gamma'$) and η a fundamental unit. Through the map $(x, y) \mapsto x + y\sqrt{d}$, the solutions give rise to a multiplicative group of rank 2 or 3, while for Pell equation we have rank 1. In the present case the number of solutions in a box $|x|, |y| \le T$ will behave like a constant times $\log^h T$, with $h = 2$ or 3.

1.4.2. Effective solution of Pell and related equations

We now present an effective procedure to determine a non-trivial solution to Pell Equation, for any given non-square $d > 0$, based on the existence proof given above. Following such existence proof, the first step is to construct effectively sufficiently many 'good' approximations for $\sqrt{d}$.

Let us then start with a 'good' approximation, *i.e.* corresponding to integers r, s with $|s\sqrt{d} - r| < s^{-1}$. To construct a new good approximation p/q, we take $Q = cs$ in Dirichlet's lemma, with $c = [2 + 2\sqrt{d}]$, to obtain coprime integers p, q such that $0 < q \le Q$ and $|q\sqrt{d} - p| \le (Q+1)^{-1}$.

Of course we have $|q\sqrt{d} - p| < q^{-1}$, and we contend that $q > s$, so we have found indeed a different good approximation, with larger denominator.

In fact, $|p - q\sqrt{d}| < Q^{-1} = (cs)^{-1}$. If we had $q \leqslant s$, we would also have $|p + q\sqrt{d}| \leqslant 1 + 2q\sqrt{d} \leqslant 1 + 2s\sqrt{d}$ and $1 \leqslant |p^2 - dq^2| < \frac{1 + 2s\sqrt{d}}{cs} \leqslant \frac{s(1 + 2\sqrt{d})}{s(1 + 2\sqrt{d})} = 1$, which is absurd. Thus $s < q \leq cs$.

Applying repeatedly this remark from a 'starting' good approximation (e.g., $s = 1, r = [\sqrt{d}]$), we can construct successively good approximations $\frac{p_1}{q_1}, \ldots, \frac{p_n}{q_n}$, with $q_{i-1} < q_i \leq c^i$ and any fixed n. For each of them we have $|p - q\sqrt{d}| < q^{-1}$, whence $|p^2 - dq^2| < c$. Let us choose $n > (2c + 1)c^2$; by the pigeon-hole principle, at least $\frac{n}{2c+1} > c^2$ of them are such that $p^2 - dq^2 = m$, for some fixed m with $|m| < c$. Since the set $(\mathbb{Z}/m)^2$ has $m^2 < c^2$ elements, at least two approximations will be such that $p \equiv p' \pmod{m}$ and $q \equiv q' \pmod{m}$. As we have seen these lead to a non-trivial solution to Pell Equation. Thus a solution can be found taking n good approximations, whose largest denominator is at most c^n, with $c = [2 + 2\sqrt{d}]$ and e.g. $n = 1 + c^2(2c + 1)$.

The theory of class numbers gives better estimates (see the notes to this chapter) but the above suffices for an algorithm which is effective in principle.

The equation $X^2 - dY^2 = m$

We shall now give an effective procedure for the integer solutions to the related equation $X^2 - dY^2 = m$, where d is a positive, non-square given integer and where m is a given nonzero integer. Just as the Pell Equation is related to units, this one is related to the equation $N(\xi) = m$, where ξ is an integer in $\mathbb{Q}(\sqrt{d})$ and $N = N_{\mathbb{Q}}^{\mathbb{Q}(\sqrt{d})}$ is the norm.

Basically, we shall see that if there exists some solution at all, then there is one whose coordinates can be effectively bounded in terms of some solution of the Pell Equation.

Let (x, y) be a solution, and suppose we are in possession of a non-trivial solution (p, q) to the equation $p^2 - dq^2 = e$, where $e = \pm 1$ (e.g. to Pell Equation) where we can assume $x, y, p, q \geqslant 0$. Let us consider $\alpha = x + y\sqrt{d}$ and $\xi = p + q\sqrt{d}$ in $\mathbb{Z}[\sqrt{d}]$. We shall denote by a dash conjugation in $\mathbb{Q}(\sqrt{d})$ over $\mathbb{Q}$, so the inverse of $e\xi$ is the conjugate $\xi' = p - q\sqrt{d}$. Let also $c = \sqrt{|m|\,\xi}$. Since $\xi > 1$, there exists a $k \in \mathbb{Z}$ such that

$$c\xi^{-1} \leqslant \alpha\xi^k < c.$$

Now, put $\beta = \alpha\xi^k = u + v\sqrt{d} \in \mathbb{Z}[\sqrt{d}]$. Its norm is $N(\beta) = \beta\beta' = N(\alpha\xi^k) = N(\alpha)N(\xi)^k = me^k$ and its conjugate is $\beta' = u - v\sqrt{d} = \frac{me^k}{\beta}$.

Thus,

$$|\beta| = \alpha \xi^k < c;$$

$$|\beta'| = \frac{|m|}{\beta} = \frac{|m|}{\alpha \xi^k} \leqslant \frac{|m|\,\xi}{c} = c.$$

From these inequalities we obtain the effective bounds

$$\begin{cases} |u| = \left|\frac{\beta+\beta'}{2}\right| \leqslant \frac{|\beta|+|\beta'|}{2} < c = \sqrt{|m|\,\xi}; \\ |v| = \left|\frac{\beta-\beta'}{2\sqrt{d}}\right| \leqslant \frac{|\beta|+|\beta'|}{2\sqrt{d}} < \frac{c}{\sqrt{d}} = \sqrt{\frac{|m|\xi}{d}}. \end{cases}$$

Hence it suffices that we compute $u^2 - dv^2$ for these values of u, v. The original equation has a solution if and only if we find m or em among these values.

Example. As an example, let us show that the equation $X^2 - 82Y^2 = 2$ has no solution. Since $9^2 - 82 \cdot 1^2 = -1$, we can take $\xi = 9 + \sqrt{82} < 19$. Applying the bound we just computed we obtain $|v| < \sqrt{\frac{2\xi}{82}} < \sqrt{\frac{2 \cdot 19}{82}} < 1$ and hence $v = 0$, a contradiction. We note that this equation cannot be shown to be unsolvable by congruence considerations (see next remark and related exercise).

Exercise 1.19. Prove that if x, y are integers then xy cannot divide $x^2 + y^2 + 1$ unless the ratio is 3. (Hint: the condition amounts to the equation $x^2 + y^2 + 1 = qxy$, i.e. $(2x - qy)^2 - dy^2 = -4$, where $d = q^2 - 4$. Use the unit $\xi := (q + \sqrt{d})/2$ of the ring of integers of $\mathbb{Q}(\sqrt{d})$, as in the above theory.)

We finally note that the above effective procedure gives also informations on the structure of the full set of solutions. Namely, we have

Proposition 1.5. *All the integral solutions of $x^2 - dy^2 = m$ are obtained from the formula $x + y\sqrt{d} = \alpha \xi^k$ where α belongs to a certain finite set (effectively computable in terms of ξ and m) and where $k \in \mathbb{Z}$.*

Remark 1.11. (Congruences) Let us consider again the equation $X^2 - dY^2 = m$. For every positive integer N we can test its solvability through the congruence $X^2 - dY^2 \equiv m \pmod{N}$, which plainly gives a necessary condition. It is important to note that these conditions, for varying N, are not generally sufficient for the existence of an integral solution: consider e.g. the case $X^2 - 82Y^2 = 2$ of the last example, when one may verify the solvability of all the associated congruences.[5] However sometimes a wise choice of N can

---

[5] On the contrary, it is a celebrated theorem by Hasse and Minkowski that for homogeneous quadratic equations the nontrivial solvability of all the congruences ensures nontrivial solvability in integers; see the last supplement below and [74].

show quickly that there is no integer solution. Here we shall briefly recall how to check with a finite amount of computation all the congruences for a given equation. First, by the Chinese Remainder Theorem it suffices to consider the cases when N is a prime power. Then the following proposition gives a complete effective criterion.

Proposition 1.6. *Let d and $m > 1$ be integers. Then:*

(i) *If p does not divide $2md$ the congruence $X^2 - dY^2 \equiv m \pmod{p^h}$ has a solution for every h.*

(ii) *If $p^A||2md$, the congruence $X^2 - dY^2 \equiv m \pmod{p^h}$ has a solution for each h if and only if it has a solution for $h = 4A + 5$.*

The proof will use a lemma, basically due to Hensel.

Lemma 1.7. *Let $f \in \mathbb{Z}[X_1, \ldots, X_n]$ and p be a prime. Assume that there exists $\underline{a} = (a_1, \ldots, a_n) \in \mathbb{Z}^n$ such that $p^c||\mathrm{GCD}\left(\frac{\partial f}{\partial x_1}(\underline{a}), \ldots, \frac{\partial f}{\partial x_n}(\underline{a})\right)$, $p^m||f(\underline{a})$, with $m > 2c$. Then for every h the congruence $f(\mathbf{X}) \equiv 0 \pmod{p^h}$ has a solution $(x_1, \ldots, x_n)$, with $x_i \equiv a_i \pmod{p^{c+1}}$ for $i = 1, \ldots, n$.*

Proof. We shall prove by induction on $h \geqslant m > 2c$ that there exist $\delta \in \mathbb{Z}$ and $\underline{b} \in \mathbb{Z}^n$ such that:

- $f(\underline{b}) = \delta p^h$;
- $b_i \equiv a_i \pmod{p^{c+1}}$ $\forall i = 1, \ldots, n$;
- $\frac{\partial f}{\partial x_i}(\underline{b}) = \xi_i p^c$ $\forall i = 1, \ldots, n$;
- $p \nmid (\xi_1, \ldots, \xi_n)$.

For $h = m$ we take $\underline{b} = \underline{a}$. Assuming that the assertion is true for h, let us verify it for $h + 1$. Let us take $\mu = h - c \geqslant m - c \geqslant c + 1$ and

$$\underline{b}' = \underline{b} + p^\mu \underline{q} \equiv \underline{b} \equiv \underline{a} \pmod{p^{c+1}},$$

with $\underline{q} \in \mathbb{Z}^n$ still to be determined. Applying Taylor's formula, we obtain

$$f(\underline{b}') \equiv f(\underline{b}) + \sum_{i=1}^n p^\mu q_i \frac{\partial f}{\partial x_i}(\underline{b}) \equiv p^h \left(\delta + \sum_{i=1}^n q_i \xi_i\right) \pmod{p^{2\mu}};$$

since p does not divide every ξ_i, $\underline{q}$ can be chosen in such a manner that $f(\underline{b}') \equiv 0 \pmod{p^{2\mu}}$, where $2\mu = 2h - 2c \geqslant h + (m - 2c) > h$ implies

$$f(\underline{b}') = \delta' p^{h+1}.$$

Moreover, since $\mu > c$, we have

$$\frac{\partial f}{\partial x_i}(\underline{b}') \equiv \frac{\partial f}{\partial x_i}(\underline{b}) \quad (\text{mod } p^{c+1}),$$

which implies $p^c \| \text{GCD}\left(\frac{\partial f}{\partial x_1}(\underline{b}'), \ldots, \frac{\partial f}{\partial x_n}(\underline{b}')\right)$. This completes the proof of the inductive step. □

Proof of Proposition 1.6. First, observe that for a prime $p \nmid 2d$ the congruence $X^2 - dY^2 \equiv m \pmod{p}$ has always a solution. This is because both maps $X \mapsto X^2$ and $Y \mapsto dY^2 + m$ assume $\frac{p+1}{2}$ values in $\mathbb{Z}/p\mathbb{Z}$, thus two such values coincide.

Now assume that p does not divide $2md$. Then, since there is a solution modulo p, it is readily checked using Lemma 1.7 that we have solutions modulo every power of p.

Let now $p | 2dm$. Assume that the congruence $X^2 - dY^2 \equiv m \pmod{p^{4A+5}}$ has a solution (x, y). If $p^{2A+2}|(2x, 2dy)$, then $p^{A+1}|(x, y)$ and $p^{2A+2}|m$, which is absurd. Thus we can apply Lemma 1.7.

This plainly proves the proposition. □

Remark 1.12. (Diophantine equations and congruences) We have recalled above that for a given equation the 'test of congruences' offers a necessary, but not generally sufficient, condition for solvability in integers.[6] So, although sometimes useful, this test has such a severe limitation. Actually, another limitation comes from the fact that, roughly speaking, *if an equation has some known integer solutions, then 'usually' congruences cannot prove that they are the only ones*: take for instance the equation $f(x, y) = 0$ (e.g. $f \in \mathbb{Q}[X, Y]$ absolutely irreducible). It is a celebrated theorem of Weil (Riemann hypothesis for curves over finite fields) that the congruence $f(x, y) \equiv 0 \pmod{p^m}$ has 'many' solutions (at least $p + O(\sqrt{p})$) for every large enough prime p; in particular, for prime-power moduli with large p the congruence test will not be able to restrict 'rigidly' (that is, to finitely many given points) the residue class of the possible integral solutions. Concerning the finitely many remaining 'small' primes, if there is a solution $(a, b) \in \mathbb{Z}^2$, the congruence has anyway (at least) the solution (a, b); and if this solution is nonsingular, the above (Hensel's) lifting procedure shows that there is a whole bunch of solutions in a p-adic neighborhood of (a, b). Hence we cannot draw any finiteness consequence about other possible integer solutions: they could exist, subject only to be congruent to (a, b) modulo finitely many prime powers. In other words, for these equations the test may work only if there are no (nonsingular) solutions at all.

Sometimes one can use conguence considerations for *moduli which depend on the possible solutions*; when this applies, it leads to subtler criteria for non-solvability. Here is an interesting example, with the 'elliptic' equation $y^2 = x^3 +$

---

[6] An exception occurs for instance with linear equations, as in Exercise 1.7.

7, to be solved in integers. If $(a, b) \in \mathbb{Z}^2$ is a solution, we have $b^2+1 = a^3+8 = (a+2)(a^2-2a+4)$. Note that a must be odd (as shown by a congruence modulo 4). Hence all primes dividing the left side must be $\equiv 1 \pmod 4$ (by Euler's criterion - see also the supplements). However $a^2 - 2a + 4 = (a - 1)^2 + 3 \equiv 3 \pmod 4$, a contradiction. Hence the equation has no integer solutions.[7] See the exercises below for another similar example.

Further limitations of congruence methods also come from the fact that they never apply if one seeks solutions over an *arbitrary* number field.

Exercise 1.20. Prove that the congruence $x^2-82y^2 \equiv 2 \pmod N$ has a solution for every integer N. (It has been proved above that the equation $x^2 - 82y^2 = 2$ has no integer solution.)

Exercise 1.21. Prove that the equation $x^7 - 1 = (y^5 - 1)(x - 1)$ has no integer solutions with $x \neq 1$. (Hint: observe that any prime factor $p \neq 7$ of $(x^7 - 1)/(x - 1)$ is $\equiv 1 \pmod 7$ - look at the order of x mod p. Deduce that every factor of $y^5 - 1$ is either exactly divisible by 7 or $\equiv 1 \pmod 7$, which is not possible in view of $y^5 - 1 = (y - 1)(y^4 + y^3 + y^2 + y + 1)$.)

Exercise 1.22. (Chevalley-Warning) Let $f \in \mathbb{F}_q[X_1, \ldots, X_n], q = p^r$, have degree $d < n$. Prove that the number N of solutions of $f(P) = 0, P \in \mathbb{F}_q^n$ satisfies $N \equiv 0 \pmod p$. Deduce that (E. Artin's question answered by Chevalley) *if there is a solution, there is another one.* (Hint: Observe that $1 - f(P)^{q-1}$ is congruent mod p to the characteristic function of the solutions. Then, sum over P and use $d < n$ and the fact that $\sum_{x \in \mathbb{F}_q} x^s$ equals -1 or 0 according as $s > 0$ is or is not divisible by $q - 1$. See [59] or [74].)

Exercise 1.23. Apply the previous result to show that every conic over a finite field has a rational point. Prove this result by a different method. (Hint: the suggested proof is immediate. A second one may be obtained as follows: we may suppose that the conic is defined by $ax^2+by^2+cz^2 = 0, abc \neq 0$. To find an affine point with $z = 1$, count the sets $\{ax^2\}$ and $\{-c - by^2\}$ for x, y varying over $\mathbb{F}_q$.)

1.5. The general case of degree 2

In this section we shall consider the general equation $Q(x, y) = 0$ of degree 2, in particular describing an effective procedure, due to Lagrange and Gauss, to test its solvability in integers. We write

$$Q(X,Y) = aX^2 + bXY + cY^2 + dX + eY + f, \quad a, b, c, d, e, f \in \mathbb{Z}. \quad (1.8)$$

We can assume Q to be nonzero and irreducible over $\mathbb{Q}$, since we know from Section 1.2 how to treat rational linear equations in integers. In particular, we assume that Q has no multiple factors.

---

[7] Colliot-Thélène has shown to me that this may be explained in terms of the so-called *Brauer-Manin* obstruction, on looking at the quaternion algebra $(-1, x + 2) = (-1, x^2 - 2x + 4)$ on the function field $\mathbb{Q}(x, y)$ of our curve.

Moreover, we can assume Q to be *absolutely* irreducible, *i.e.* irreducible over any extension field. In fact, in the first place if Q is reducible its factors may be assumed to have algebraic coefficients. (The factors are linear and at least one of them contains infinitely many among the algebraic points on the curve $Q = 0$.) Now, the two linear factors must be conjugate over $\mathbb{Q}$,[8] up to a constant factor. Then any rational point on the line defined by one factor must also lie on the conjugate line; in conclusion, if Q is reducible over $\overline{\mathbb{Q}}$ but irreducible over $\mathbb{Q}$, there is at most one rational point, *i.e.* the intersection of the lines.

Now let $\Phi(X, Y) = aX^2 + bXY + cY^2$ be the binary quadratic form associated to Q. We distinguish three cases, depending on the sign of the discriminant $\Delta = b^2 - 4ac$ of Φ.

• The elliptic case

When $\Delta < 0$, the curve $Q = 0$ is an ellipses and the polynomial Φ is irreducible over $\mathbb{R}$. In this case there exists an effective constant $\delta > 0$ such that

$$\delta |\mathbf{v}|^2 \le |\Phi(\mathbf{v})|, \qquad \forall\, \mathbf{v} \in \mathbb{R}^2.$$

(One may take $\delta = \frac{|\Delta|}{4(|a|+|c|)}$.) Now, if $\mathbf{v} = (x, y)$ is a solution we have $|\Phi(\mathbf{v})| = |-dx - ey - f| \ll |\mathbf{v}| + 1$, whence $|\mathbf{v}|$ is bounded and so (since it is an integral vector) there are only finitely many solutions. (Note that this conclusion does not hold over general number fields; in fact, if the field contains $\sqrt{\Delta}$ this case merges with the hyperbolic case, treated below.)

• The parabolic case

When $\Delta = 0$ the curve is a parabola; we can write $\Phi = AL^2$, where $L(X, Y) = rX + sY$, with $A, r, s \in \mathbb{Z}$ and $(r, s) = 1$. Since r and s are coprime integers, there exist integers t and u such that $ru - st = 1$. Now, let $M(X, Y) = tX + uY$. Since $\det \left(\begin{smallmatrix} r & t \\ s & u \end{smallmatrix}\right) = ru - ts = 1$, we have the linear automorphism of $\mathbb{Z}^2$ given by $(L, M) = (X, Y)\left(\begin{smallmatrix} r & t \\ s & u \end{smallmatrix}\right)$. This gives the new equivalent equation in integers

$$Q(X, Y) = AL^2 + BL + C - DM = 0.$$

Let $l_1, \ldots, l_h$ be the incongruent solutions to $Al^2 + Bl + C \equiv 0 \pmod{D}$. Then all solutions to $Q(X, Y) = 0$ correspond in (L, M)-coordinates to

$$L = l_i + kD; \qquad M = \frac{AL^2 + BL + C}{D} = M_i(k),$$

---

[8] That is, defined by equations with conjugate coefficients.

where $M_i(k)$ is a certain quadratic polynomial and k is an arbitrary integer. Note that if one solution does exist, then there are infinitely many of them.

● *The hyperbolic case*

When $\Delta > 0$ the curve $Q = 0$ is a hyperbola. Now we consider the change of variables

$$\begin{cases} T = 2aX + bY + d, \\ U = (b^2 - 4ac)Y + bd - 2ae, \end{cases}$$

and the constants

$$\alpha = \Delta = b^2 - 4ac, \qquad \beta = bd - 2ae, \qquad \gamma = 4af - d^2.$$

Then, by 'completing the squares', we see that there is a one-to-one correspondence between integer solutions (x, y) of (1.8) and integer solutions (t, u) of

$$U^2 - \alpha T^2 = \beta^2 + \alpha\gamma \qquad (1.9)$$

satisfying

$$\begin{cases} u \equiv \beta \pmod{\alpha}; \\ \alpha t \equiv b(u - \beta) + \alpha d \pmod{2a\alpha}. \end{cases}$$

Now observe that $\beta^2 + \alpha\gamma \neq 0$, since otherwise Q would be reducible (over $\mathbb{Q}(\sqrt{\alpha})$).
When $\alpha = \delta^2$ is a square, the left side of 1.9 factors as $(U + \delta T)(U - \delta T)$; for a solution (t, u), both factors $u \pm \delta t$ lie among the divisors of the right side, which is non-zero. Then the set of solutions is finite and effectively computable.

When α is not a square, from the previous theory, we know how to compute integer solutions (t, u) of 1.9: they correspond to those elements $u + t\sqrt{\alpha} \in \mathbb{Z}[\sqrt{\alpha}]$ of the form $u + t\sqrt{\alpha} = \lambda_i \xi^k$, where λ_i belongs to a certain computable finite set $\Lambda \subset \mathbb{Z}[\sqrt{\alpha}]$ and ξ has norm 1, *i.e.* it corresponds to a solution of the Pell Equation $U^2 - \alpha T^2 = 1$.

We have now to determine which solutions (t, u) of (1.9) satisfy the above system of congruences. Since we are assuming that $\alpha = b^2 - 4ac$ is not a square, we have $a \neq 0$. Note that, modulo the ideal $J = (2a\alpha)$, there are only finitely many powers of ξ, since $\mathbb{Z}[\sqrt{\alpha}]/J$ is a finite ring; moreover we can compute these powers for any given J. Thus we just need to verify the congruences on a certain finite set of integer solutions (t, u), which is an effective procedure.

Note that if there is one solution, then there are infinitely many of them: if a solution $u + t\sqrt{\alpha} = \lambda_i \xi^k$ satisfies the congruences and if $\xi^{k+r} \equiv \xi^k \pmod{J}$, with $r \neq 0$, then for every integer m the solution $u' + t'\sqrt{\alpha} = \lambda_i \xi^{k+mr}$ satisfies the same congruences. On the other hand such an r exists since there are only finitely many powers of ξ modulo J; also, if two powers ξ^u, ξ^v are congruent, then ξ^{u-v} is congruent to 1, since ξ is invertible in $\mathbb{Z}[\sqrt{\alpha}]$, so *a fortiori* ξ is invertible modulo J.

Similarly to the previous case, when $\Delta > 0$ is not a square the equation $Q(X, Y) = 0$ has an integral solution if and only if it has infinitely many ones.

Exercise 1.24. Consider the affine conic C of equation $X^2 + XY - 36Y^2 = 4$.

(i) Show that the point $(1, 1) \in C(\mathbb{F}_2)$ lifts to a point in $C(\mathbb{Z}_2)$ (namely show that for each integer m the congruence $X^2 + XY - 36Y^2 \equiv 4 \pmod{2^m}$ has a solution with $X \equiv Y \equiv 1 \pmod 2$).
(ii) Parametrize the rational points on C.
(iii) Describe the integral points on C with parametric formulas and show that there do not exist such points with coprime coordinates.

Remark 1.13. (Density of integral points and points at infinity) When there are infinitely many integral points on a curve C, one can measure their 'density', for instance by counting how many of them lie in a square box $|x|, |y| \leq T$ of side $2T$, where T is a parameter tending to infinity; let us denote by $N(T) = N_C(T)$ this function.

In the case of lines in the plane, plainly $N(T)$ can have the order of magnitude of T. As to irreducible conics, we have seen that in the parabolic case the integral points can be parametrized by polynomial maps; this easily leads to examples when $N(T) \asymp \sqrt{T}$.

In the hyperbolic case the integral points are either finite in number or may be parametrized by exponential functions; this leads to an estimate $N(T) \ll \log T$. (The case of the ellipse leads to a finite set of integral points over $\mathbb{Z}$ and is equivalent to the hyperbolic case over a suitable finite extension field.)

This different behaviour of the growth of $N(T)$ (polynomial versus logarithmic) has its geometric origin in the fact that lines and parabolas have only a single point at infinity whereas a hyperbola has two such points. (A similar phenomenon happens over number fields other than $\mathbb{Q}$.)

This is a first indication that the number of points at infinity strongly affects the distribution of integral points; in fact, we shall note when stating Siegel's theorem later that three points at infinity for a curve prevent it from containing infinitely many integral points, no matter the number field we work with.

More generally, roughly speaking, the 'largest' is the divisor at infinity for an affine variety, the more sparse are the integral points on it. This is made precise with conjectures which shall not be touched here. (See for instance [17].)

Supplements to Chapter 1

Two applications of Dirichlet Lemma

First application: Integer solutions of $a^2 + b^2 = p$

We shall provide a proof of the following celebrated result, stated by Fermat and proved by Euler.

Theorem 1.8. *Let p be a prime number $\equiv 1$ (mod 4). Then there exist integers a and b such that $a^2 + b^2 = p$.*

Almost all of the different proofs of this theorem begin with the following

Lemma 1.9. *There exists $x \in \mathbb{Z}$ such that $x^2 + 1 \equiv 0$ (mod p).*

Proof. Of course this is contained in modern undergraduate courses, but we reproduce the following argument of Euler. The distinct non-zero squares modulo p are just the $r_i := i^2$ for $1 \leq i \leq (p-1)/2$. To every r_i we can associate some r_j with $r_i r_j \equiv 1$ (mod p); the only squares that are 'coupled' with themselves are those r such that $r \equiv r^{-1}$ (mod p), thus $r \equiv \pm 1$ (mod p). Since $(p-1)/2$ is even, the number of such singletons has to be even, thus both 1 and -1 must be squares and there exists an $x \in \mathbb{Z}$ such that $x^2 \equiv -1$ (mod p), as wanted. $\square$

Proof of Theorem 1.8. We shall look for sums of squares $a^2 + b^2$ that are multiples of p and, at the same time, small. After lemma 1.9, we can look at the congruences

$$(qx)^2 + q^2 \equiv 0 \quad (\text{mod } p),$$

taking then the least remainder x_q of $\pm qx$ modulo p. Thus we look for small q with small $|x_q| = |qx - pm|$. For this, we choose $\xi = x/p$, in Dirichlet Lemma; choosing also $Q = [\sqrt{p}]$ we obtain integers m and $0 < q \leq Q$ such that

$$\frac{|x_q|}{pq} = \left| \frac{x}{p} - \frac{m}{q} \right| \leq \frac{1}{q(Q+1)}.$$

Multiplying by pq we have $|x_q| \leq \frac{p}{Q+1} < \sqrt{p}$ and $x_q^2 + q^2 \equiv 0$ (mod p) with $0 < x_q^2 + q^2 < p + Q^2 \leq 2p$. Since $x_q^2 + q^2$ is a multiple of p, it must equal p, concluding the proof. $\square$

Exercise 1.25. Let d be an integer. Prove the existence of a number l_d such that if the congruence $x^2 \equiv d$ (mod m) has a solution, then for some $h \in \mathbb{Z}$ with $0 < |h| < l_d$ we can write hm in the form $a^2 - db^2$, with integers a, b. Also, give an estimate for l_d. (Hint: use Dirichlet lemma as above, which is the case $d = -1$.)

Second application: A factorization algorithm

Algorithms for factoring a given integer N have often been of interest for mathematicians, since the times of Fermat, and Gauss even said that the problem of finding satisfactory algorithms was one of the most important of the whole Number Theory. In recent times, the topic gained new interest due to its connection with certain new cryptographic systems, whose safety depends on the high complexity required to factorize huge numbers. We won't get into this here. We shall, nonetheless, give an algorithm that, though far from the most efficient known ones, is easily described and implemented and is much better than the 'obvious' one. This last is based on the remark that if $N = ab$, with $2 \leqslant a \leqslant b$, then $a \leqslant \sqrt{N}$. Thus we can find a by dividing N by every number from 2 to $[\sqrt{N}]$. (Even dividing just by primes, provided we know them, the number of divisions would not decrease by much.)

The algorithm we are going to describe takes at most $O(N^{\frac{1}{3}})$ steps and the fact it works depends crucially on Dirichlet Lemma. We shall first describe it, then show that it works for every $N > 512$.

```
  1.   Cycle A from 2 to ∛N.
  2.   If A divides N, return A.
3.     End cycle A.
4.     Cycle P from 1 to 4∛N.
  5.   Cycle C from ⌈√4NP⌉ to √4NP + ⁶√N/(4√P).
    6.   Compute D = √(C² − 4NP).
      7.   If D is an integer, return GCD(N, C − D).
  8.     End cycle C.
9.     End cycle P.
10.    Return N.
```

Let us first compute how many elementary operations are needed. The cycle of A is $\leqslant \sqrt[3]{N}$ steps long. Considering steps $(6) - (7)$ as elementary operations, every cycle of C uses at most $\frac{\sqrt[6]{N}}{4\sqrt{P}} + 1$ operations; summing on the cycle of P we obtain at most $\sum_{P=1}^{4\sqrt[3]{N}} \left(\frac{\sqrt[6]{N}}{4\sqrt{P}} + 1 \right) \leqslant 4\sqrt[3]{N} + \frac{\sqrt[6]{N}}{4}[2\sqrt{P}]_0^{4\sqrt[3]{N}} = 4\sqrt[3]{N} + \frac{\sqrt[6]{N}}{4} 2\sqrt{4\sqrt[3]{N}} = 5\sqrt[3]{N}$ elementary operations, as wanted.

We will now show that the algorithm does work. The first cycle finds out all factors $\leqslant N^{\frac{1}{3}}$. If this cycle does not return any factor and N is not a prime, then $N = ab$, with $N^{\frac{1}{3}} < a \leqslant b < N^{\frac{2}{3}}$. Taking $Q = \left[\frac{b}{N^{\frac{1}{3}}}\right]$ in Dirichlet's Lemma, we obtain integers r and s such that $1 \leqslant s \leqslant Q$ and $|\frac{a}{b} - \frac{r}{s}| < \frac{1}{s(Q+1)}$. Now let $P = rs$, $C = as + br$, and $D = |as - br|$.

First, note that P and C satisfy the inequalities required in the algorithm. Note that $P > 0$: otherwise $r \leqslant 0$ and $|\frac{a}{b}| \leqslant \frac{1}{s(Q+1)}$, so that $Q+1 \leqslant s(Q+1) \leqslant \frac{b}{a} < \frac{b}{N^{\frac{1}{3}}} < Q+1$, which is absurd. Also, $D < \frac{b}{Q+1} < N^{\frac{1}{3}}$. And $r \leqslant s\frac{a}{b} + \frac{1}{Q}$, so that $P \leqslant s(s\frac{a}{b} + \frac{1}{Q}) \leqslant Q^2\frac{a}{b} + 1 = Q^2\frac{N}{b^2} + 1 \leqslant 4N^{\frac{1}{3}} + 1$. Since $C^2 = 4NP + D^2$, we have $C \leqslant \sqrt{4NP} + \frac{D^2}{2\sqrt{4NP}} \leqslant \sqrt{4NP} + \frac{N^{\frac{1}{6}}}{4\sqrt{P}}$.

Now we have to show that steps $(4) - (9)$ will return a non-trivial factor for every $N > 512$. Since $(C - D)$ is either $2as$ or $2br$, the $\gcd(N, C - D)$ will be either $a(b, 2s)$ or $b(a, 2r)$, i.e. one of a, b, and $ab = N$. In the last case N divides $(C - D)$, so that $N \leqslant C - D \leqslant 2C \leqslant 2\sqrt{4NP} \leqslant 8N^{\frac{2}{3}}$, which implies $N \leqslant 2^9 = 512$.

The idea for the algorithm is to multiply $N = ab$ by a small number $P = rs$, in such a manner that their factors as and br have the same magnitude. The principle is that their sum C will then be near $\sqrt{4NP}$, nearer than one factor is near to the other; to find their sum one thus should look near $\sqrt{4NP}$. In fact in the above proof we looked for r and s such that $\frac{r}{s} \approx \frac{a}{b}$, i.e. $as \approx rb$. Now, $NP = (as)(rb)$ is a decomposition into almost equal factors.

A cyclotomic solution of certain Pell equations

Let $p \neq 2$ be a prime number and consider the cyclotomic field $\mathbb{Q}(\zeta_p)$. It is well known that it is a normal extension of $\mathbb{Q}$, with cyclic Galois group $G \cong \mathbb{F}_p^*$, and that its ring of integers is $\mathbb{Z}[\zeta_p]$. We construct two polynomials in $\mathbb{Z}[\zeta_p][X]$, namely $\Phi_R(X) = \prod_R (X - \zeta_p^r)$ and $\Phi_N(X) = \prod_N (X - \zeta_p^n)$, where $R = \mathbb{F}_p^{*2}$ is the set of all nonzero quadratic residues modulo p and $N = \mathbb{F}_p^* \setminus R$ is the set of all quadratic nonresidues. Clearly we have $\Phi_R(X)\Phi_N(X) = \Phi(X) = X^{p-1} + \ldots + X + 1$.

If we let G act on the (coefficients of the) polynomials, we see that any σ corresponding to some $m \in \mathbb{F}_p^*$ either fixes both Φ_R and Φ_N ($m \in R$) or exchanges them ($m \in N$). Hence the two polynomials have their coefficient in the unique quadratic subfield F of $\mathbb{Q}(\zeta_p)$, fixed by the subgroup of G corresponding to R, and they are conjugate there; it is well known that $F = \mathbb{Q}(\sqrt{\varepsilon p})$, where $\varepsilon = \pm 1$ according as $p \equiv \pm 1 \pmod 4$ (as can be seen either by considering ramification or by explicit evaluation of the square of a Gauss sum). Then, since the coefficients of both Φ_R, Φ_N are algebraic integers we have $\Phi_R(X) = \frac{1}{2}(A(X) + \sqrt{\varepsilon p}B(X))$ and $\Phi_N(X) = \frac{1}{2}(A(X) - \sqrt{\varepsilon p}B(X))$, with $A, B \in \mathbb{Z}[X]$.

From this we obtain $4\Phi(X) = A^2(X) - \varepsilon p B^2(X)$, a result due to Gauss (*Disquisitiones Arithmeticae*). We now put $X = 1$ and denote $A = A(1)$ and $B = B(1)$, so that A and B are integers with $4p = A^2 - \varepsilon p B^2$; this implies that p divides A, i.e. $A = pC$. Then $B^2 - \varepsilon p C^2 = -4\varepsilon$.

Assume $p \equiv 1 \pmod 4$. Then $\varepsilon = 1$ and, consequently, $C \neq 0$. This means that $\omega = \frac{1}{2}(B + C\sqrt{p})$ is a non-trivial unit of the ring of integers of $F = \mathbb{Q}(\sqrt{p})$; note that it can be explicitly expressed in terms of ζ_p. Also, the norm is $N(\omega) = \frac{1}{4}(B^2 - pC^2) = -1$, a property that for other quadratic fields is not verified by any unit. Finally, we have $B \equiv C \pmod 2$; this implies that $\omega^3 = t + u\sqrt{p}$, where $t, u \in \mathbb{Z}$ satisfy $t^2 - pu^2 = -1$; Pell Equation can be solved by taking ω^6.

This can be used to give an alternative (more explicit) argument in Exercise 1.18. Also, it yields an 'explicit' solution of Pell Equation for $d = p$; in particular, this can be used to estimate the magnitude of a fundamental solution for $p \equiv 1 \pmod 4$.

Remark 1.14. (i) It may be shown that B, C are both even (resp. odd) if and only if $p \equiv 1 \pmod 8$ (resp. $p \equiv 5 \pmod 8$). This may be proved for instance by inspection of the action of a Frobenius element at a prime lying above 2 in the full cyclotomic field. Another relevant issue in this direction is to consider the quotient ring $\mathcal{O}/2\mathcal{O}$, where $\mathcal{O}$ is the ring of integers of $\mathbb{Q}(\zeta_p)$. This is either $\mathbb{F}_2^2$ or $\mathbb{F}_4$ according to the splitting of the prime 2. This structure essentially yields what asserted concerning ω^3.)

(ii) The formula may be also motivated as follows: it is an easy exercise to show that for l, m prime to p, the ratio $\mu_{l,m} := (\zeta^l - 1)/(\zeta^m - 1)$ (where $\zeta = \zeta_p$) is a unit of the ring of integers of $\mathbb{Q}(\zeta)$. Let now l be a quadratic non-residue; taking the norm of $\mu_{l,1}$ to the quadratic subfield F of $\mathbb{Q}(\zeta)$ we clearly obtain a unit of $\mathcal{O}_F$ and it is easy to recover in this way (exercise) the previous results.

(iii) It is known that the solution of the Pell Equation obtained in this way is essentially the h-th power of a fundamental unit, where h is the class-number of F.

Remark 1.15.   (A Pell Equation over number fields). We have already remarked how the solvability of Pell Equation yields the structure of units of the ring of integers of a quadratic field. More generally, a celebrated theorem of Dirichlet predicts the structure of units in the ring of integers $\mathcal{O}_K$, for an arbitrary number field K; this result says that if K has r real embeddings and $2s$ complex nonreal embeddings, the group of units in $\mathcal{O}_K$ has rank $r + s - 1$. (See Chapter 3 for precise statements in the more general case of S-units.)

This result of Dirichlet helps also to clarify the issue of a Pell Equation over K, i.e. an equation $X^2 - dY^2 = 1$, where $d \in \mathcal{O}_K$ is fixed and nonzero and where the variables X, Y are supposed to run in $\mathcal{O}_K$. For instance, an easy analysis, which we omit here, allows one to derive from Dirichlet's theorem that *the Pell Equation over $\mathcal{O}_K$ has infinitely many solutions unless K is a totally real field and d is negative in all embeddings of K in $\mathbb{R}$.* In turn, by a known result of Hilbert-Landau on real fields, the said condition on d amounts to the fact that $-d$ is a sum of squares of elements of K.

A Pell Equation in polynomials

Let us consider a polynomial version of Pell Equation:

$$x^2(t) - \Delta(t)y^2(t) = 1,$$

where $x, y, \Delta \in k[t]$, k being a field, assumed here to be algebraically closed of characteristic $\neq 2$. We view Δ as fixed (and x, y as polynomial unknowns), non-constant with even degree 2δ; we also assume that it has no multiple factors (note that we can absorb every square in y).[9]

We consider the affine curve $\mathcal{C}$ over k, defined in $\mathbb{A}^2$ by $u^2 = \Delta(t)$, of genus $\delta - 1$. This curve is nonsingular (but its projective closure in $\mathbb{P}_2$ is singular for

---

[9] Actually, whereas in the classical case of integers to absorb the mentioned square does not lead to new restrictions to solvability, for this polynomial case the square factors lead to quite relevant issues, concerning *generalized Jacobians*, on which we cannot pause here. (For more, see *e.g.* the author's forthcoming survey paper *Unlikely Intersections and Pell's equations in polynomials*, to appear in: Trends in Contemporary Mathematics, V. Ancona, E. Strickland (eds.), Springer-INdAM Series, Vol. 8, 2014.)
Nevertheless, here we shall absorb the square factors inside y for the sake of simplification.

$\delta > 1$); the normalization $\tilde{C}$ of its projective closure has two points at infinity, denoted ∞_+, ∞_-.

For a solution (x, y), considering the functions in $k(C)$ given by $\varphi_\pm = x \pm uy$, we may write the equation as $\varphi_+ \varphi_- = 1$. Since both functions are regular on the affine curve and their product is 1, none of them can vanish at some affine point, *i.e.* their divisor (on $\tilde{C}$) is supported at infinity, whence $\operatorname{div}(\varphi_+) = m_+ \infty_+ - m_- \infty_-$. Since the divisor has degree zero, we have $\operatorname{div}(\varphi_+) = m(\infty_+ - \infty_-)$. The integer m equals $\pm \deg x(t)$ and determines the function up to a constant factor, and the equation says that this constant has to be ± 1. The solution is trivial if and only if $m = 0$, so a necessary condition for non-trivial solvability is that the divisor $m(\infty_+ - \infty_-)$ is principal for some $m \neq 0$. The same arguments show that this condition is also sufficient.

When the curve has genus 0, Δ has degree 2 and there always exists a function ψ whose divisor is $\infty_+ - \infty_-$. Necessarily $\psi = r(t) + us(t)$ for polynomials r, s (of degrees 1, 0) and $\psi^* := r - us$ has divisor $-\operatorname{div}(\psi)$ (since the involution $u \mapsto -u$ exchanges ∞_+ and ∞_-). Hence $r^2 - \Delta s^2$ is constant, which can be assumed to be 1. Now all the solutions of Pell Equation will be given by $\varphi_+ = \pm \psi^m$, in full analogy with the arithmetical case. This genus zero case is obtained by taking for instance $\Delta(t) = t^2 - 1$; with this choice it is an easy matter to see that the solutions lead and are related to the Tchebychev polynomials, which appear in so many mathematical topics.[10] We also remark that the general case of genus zero reduces to this choice of Δ by a suitable linear variable change $t \mapsto at + b$.

When the curve has genus 1 (*i.e.* $\deg \Delta = 4$), it is a celebrated fact that $\tilde{C}$ can be equipped with a group law, inherited from $\operatorname{Pic}_0(\tilde{C})$, where the origin can be any prescribed point. The fact that $m(\infty_+ - \infty_-)$ is the divisor of a non-constant function says that $\infty_+ - \infty_-$ is torsion in $\operatorname{Pic}_0(\tilde{C})$, of order dividing m. These torsion divisors are quite special; there exists anyway (at least for curves defined over $\overline{\mathbb{Q}}$) an algorithm to determine whether a given divisor is torsion or not. (Though elementary, this is not obvious; see *e.g.* [83].)

Let us study this interesting genus 1 case in a little more detail. By a variable change $t \mapsto at + b$ ($a \in k^*, b \in k$) one can suppose that $\Delta(t) = t^4 + c_2 t^2 + c_3 t + c_4$. The condition that $\infty_+ - \infty_-$ is torsion of order dividing a given integer m may be translated into an algebraic relation (dependent on m, but *a priori* trivial) among the coefficients c_i of Δ.

For actual computations, it is perhaps convenient to perform a further variable change; for instance, by the substitution $x = t^2 - u$, $y = \sqrt{2}(t(x + (c_2/2)) + (c_3/4))$ one obtains an elliptic curve E defined by a Weierstrass equation $y^2 = x^3 + (c_2/2)x^2 - c_4 x - (c_2 c_4/2) + (c_3^2/8)$. One of the points $\infty_\pm$ goes to the single point $\infty \in E$ at infinity on E while the other one goes to $(-c_2/2, \sqrt{2}c_3/4)$.

Let us take a special instance, over $\mathbb{C}$, by setting $c_2 = 6, c_3 = 2\sqrt{2}, c_4 = 0$. We obtain the curve $y^2 = x^3 + 3x^2 + 1$ with the point $(-3, 1)$. This point is not

---

[10] The Tchebyshev polynomial $T_d(t)$ of degree d satisfies $T_d(t + t^{-1}) = t^d + t^{-d}$.

a torsion point[11], as can be seen *e.g.* by applying the Lutz-Nagell Theorem (see [83]) after doubling the point. Hence for the corresponding $\Delta(t) = t^4 + 6t^2 + 2\sqrt{2}t$ the polynomial Pell Equation has only the trivial solutions over $\mathbb{C}$.

In particular, for every integer m, the alluded algebraic relation among the c_i is not trivial, for otherwise $\infty_+ - \infty_-$ would always be torsion, no matter the coefficients, which is not the case in view of the example just given. If the coefficients (as in the example) do not satisfy any of these infinitely many algebraic relations, for varying m, there are no non-trivial solutions to the Pell Equation, and conversely.

Reversing the procedure of the example, we may also start from a given curve E in Weierstrass form $y^2 = f(x)$ with a monic cubic polynomial $f \in k[x]$ with no multiple roots. We may then select a finite point $P = (a, b)$ on E and construct a corresponding quartic model $C : u^2 = \Delta(t)$ of E, so that the points ∞ and P go at infinity: for this we can argue explicitly on setting $t = (y + b)/(x - a), u = x - (t^2/2) + c$ for a suitable constant c. The image of ∞ on E will be a point ∞_+ at infinity on C and the image of P will be the other point at infinity ∞_-. If we choose P of exact order m on E, then $\infty_+ - \infty_-$ will have exact order m in $\mathrm{Pic}(C)$, *i.e.* the minimal solution of the Pell Equation will have $x(t)$ of degree m.

So, over $\mathbb{C}$ or $\overline{\mathbb{Q}}$, we can find examples with no solutions and also examples with minimal solution of any given degree.[12]

When the curve has larger genus, things are similar but the actual calculations get nasty: to check non-trivial solvability, we have to determine whether $\infty_+ - \infty_-$ is torsion on the Jacobian variety associated to the curve.

When k is a finite field (or its algebraic closure), every algebraic point on the Jacobian is torsion, so the Pell Equation always has nontrivial solutions. Very interesting problems arise if we restrict k to be *e.g.* a given number field. The question now overlaps with the celebrated deep question of the field of definition of torsion points; for elliptic curves in turn this leads to the problem of rational points on modular curves, but we cannot enter into this here (see [83] for references).

In any case, if there is any solution, as in the arithmetical case and in the case of genus zero, every other solution is obtained by considering $\pm$ the powers φ_+^n.

Exercise 1.26. Adapt the method of this chapter to prove that for a finite field k the Pell Equation in polynomials always has non-trivial solutions.

Padé Approximations to $\exp(x)$ and celebrated irrationalities

We shall now prove that e^t is irrational for every non-zero rational number t. In doing so we will use the so-called Hermite-Padé approximations to the exponential function, as in Remark 1.4.

For every natural number n there exist polynomials $P_n, Q_n \in \mathbb{Z}[X]$ of degree $\leq n$ and such that the rational function $\frac{P_n(x)}{Q_n(x)}$ approximates $\exp(x)$ (in the

---

[11] We are now implicitly working with ∞ as the origin

[12] One may generalize even to several variables, but probably in these cases the degree of a minimal solution is bounded, unless there is a substitution reducing everything to the case of a single variable.

natural topology of $\mathbb{Q}[[x]])$ with an error of the order of $O(x^{2n+1})$. This existence follows immediately from linear algebra, but we are going to determine explicitly the relevant polynomials using the simple idea of taking $n + 1$ derivatives of the equation $Q_n(x)e^x - P_n(x) = O(x^{2n+1})$. We shall use the notation $D := d/dx$, $I = $ identity.

Note that for every polynomial $A(x)$ we have $D(A(x)e^x) = (D+I)(A(x))e^x$, so we obtain $(I + D)^{n+1}(Q_n(x))e^x = O(x^n)$, whence $(I + D)^{n+1}(Q_n(x)) = O(x^n)$. The degree of the left side is $\le n$, so $(I + D)^{n+1}(Q_n(x)) = cx^n$, where c may be any constant. (All of this also proves that P_n, Q_n are uniquely determined up to a constant factor.)

Taking $c = 1$ we have $Q_n(x) = (I+D)^{-n-1}x^n$. The differential operator may be expanded with the binomial theorem, to obtain $Q_n(x) = \sum_{j=0}^{n} \binom{-n-1}{j} D^j x^n = \sum_{j=0}^{n}(-1)^j \binom{n+j}{j} \frac{n!}{(n-j)!} x^{n-j} \in \mathbb{Z}[x]$. Observing that $P_n(x)e^{-x} - Q_n(x) = O(x^{2n+1})$ one can similarly determine $P_n(x)$, finding $P_n(x) = Q_n(-x)$.

Now we shall see how these good functional approximations behave under specialization. Let s be any positive integer and consider the power series $R_n(x) := Q_n(x)e^x - P_n(x)$. As above we obtain $D^{n+1}R_n(x) = (I + D)^{n+1}(Q_n(x))e^x = x^n e^x$, from which by integration we get the explicit formula $R_n(x) = x^{2n+1}\sum_{m \ge 0} \frac{(m+n)!}{(m+2n+1)!} \frac{x^m}{m!}$. Note that the term $\frac{(m+n)!}{(m+2n+1)!}$ is bounded by $(\frac{1}{n})^{n+1}$. Let us specialize x at s; then we obtain integers $q_n := Q_n(s)$ and $p_n := P_n(s)$ such that $q_n e^s - p_n = R_n(s)$. This is positive and bounded by $s^{2n+2}(\frac{1}{n})^{n+1}\sum_{m \ge 0}\frac{s^m}{m!} = (\frac{s^2}{n})^{n+1}e^s \le c_1(\frac{c_2}{n})^{n+1}$, where c_1 and c_2 are numbers depending only on s. As n goes to infinity, we have $0 < q_n e^s - p_n \le c_1(\frac{c_2}{n})^{n+1} \to 0$. In other words, the good approximations $\frac{P_n(x)}{Q_n(x)}$ to e^x become by specialization good approximations $\frac{p_n}{q_n}$ to e^s. If e^s were rational, say $e^s = \frac{a}{b}$, then $|q_n e^s - p_n| \ne 0$ would be bounded below by $\frac{1}{b}$, a contradiction for large n. This proves that e^s is irrational for every non-zero integer s. Now, for a rational number $t = \frac{s}{r}$, the equality $(e^t)^r = e^s$ shows that e^t is also irrational.

Similar arguments prove that $\exp(t)$ does not lie in $\mathbb{Q}(\sqrt{-1})$ for nonzero $t = s/r \in \mathbb{Q}(\sqrt{-1})$, which implies the irrationality of π. Let us sketch some details. As above we reduce to the case $t = s \in \mathbb{Z}[\sqrt{-1}]$. Assuming $\exp(s) \in \mathbb{Q}(\sqrt{-1})$, we specialize the 'approximation form' $Q_n(x)e^x - P_n(x)$ to $x = s$; the difference with the case $s \in \mathbb{Z}$ is that now we cannot conclude (by positivity) that this is nonzero. However we can deduce as above that, for large n, $|Q_n(s)e^s - P_n(s)|$ is so small that it must vanish. But then we repeat this with the derivative $(Q_n(x) + Q'_n(x))e^x - P'_n(x)$. The argument will produce the sought contradiction unless we are in the 'bad' case when they both vanish at s, which implies that the determinant $Q_n P'_n - P_n(Q_n + Q'_n)$ vanishes at s. However this cannot happen, because the determinant is a nonzero constant multiple of x^{2n}: in fact, in the first place it has order $\ge 2n$ at $x = 0$, as follows on eliminating e^x from the two approximation forms; moreover, it has degree $\le 2n$ and it cannot vanish identically, for otherwise $Q_n(x)e^x$ and $P_n(x)$ would be linearly dependent functions; this concludes the proof. This last 'Wronskian' argument is common in diophantine approximation; for instance, it will appear again in Thue's proof (see next chapter) and in the last chapter.

Exercise 1.27. Quantify how good are the approximations to the number e (or more generally to e^t) so obtained.

Exercise 1.28. (i) Prove the stated series formula for $R_n(x)$. (Hint: integrate $n+1$ times the Taylor series for $x^n e^x$ and take into account that $R_n(x)$ vanishes at the origin at order at least $2n+1$.)

(ii) Prove by a direct asymptotic argument the nonvanishing of $Q_n(s)e^s - P_n(s)$ for large enough n and fixed s. (Hint: writing the above series as $R_n(x) = x^{2n+1} \frac{n!}{(2n+1)!} \sum_{m \geq 0} c_{m,n} \frac{x^m}{m!}$, prove that, for fixed $m \geq 0$, $\lim_{n \to \infty} c_{m,n} = 2^{-m}$. Then deduce, for fixed x_0, the asymptotic formula $R_n(x_0) \sim x^{2n+1} \frac{n!}{(2n+1)!} e^{\frac{x_0}{2}}$.)

Exercise 1.29. In the second part of the above proof, replace $\sqrt{-1}$ with $\sqrt{-d}$, d a positive integer, and modify the arguments to obtain a proof that $\exp(t)$ does not lie in $\mathbb{Q}[\sqrt{-d}]$ for nonzero $t \in \mathbb{Q}(\sqrt{-d})$.

In turn, deduce that π^2 is irrational.

Rational points on conics

Although in these lecture notes we are mainly concerned with integral points, we add here a simple effective algorithm to describe the rational points on conics in $\mathbb{P}_2$ (they are defined by quadratic homogeneous equations in three variables), which is a desirable addendum to the above analysis of integral points on conics in $\mathbb{A}^2$. Also, this will be used in Ch. II, Supplements, for an algorithm concerning integral points on curves of genus 0.

Let the conic, supposed irreducible, be expressed by the equation $aX^2 + bY^2 = cZ^2$, where $a, b, c \in \mathbb{Q}^*$. (This expression may be always obtained after a suitable linear change of coordinates; i.e., one has to 'complete the squares'.) Assuming we are in possession of a rational point P_0 on the conic, we may project the conic from P_0 to any rational line not containing P_0, to parametrize all the rational points; a familiar example is the parametrization $(2t/(1+t^2), (1-t^2)/(1+t^2))$ for the (affine) circle $X^2 + Y^2 = 1$, obtained by projecting from $(-1, 0)$ to the line $X = 0$. We leave to the interested reader the task of writing formulas for the general case.

Thus our problem is to decide about the possible existence of a single rational point and in the affirmative case to calculate some such point for a given curve. The argument below, following [74], goes back to Legendre. We may first clear denominators to assume $a, b, c \in \mathbb{Z}$; then on multiplying by c and absorbing c^2 by the substitution $Z \to Z/c$, we may further assume $c = 1$ and $a, b \in \mathbb{Z}$. For $|a| + |b| = 2$ we may clearly calculate a rational point unless $a = b = -1$, when there is none; for $|a| + |b| > 2$ we use a descent procedure. By symmetry we can assume $|b| \leq |a|$ (so $|a| > 1$); we may further suppose a, b squarefree, by absorbing square factors in the variables. We must decide whether $x^2 a = z^2 - by^2$ for suitable integers x, y, z not all zero (when b is not a square this amounts to a being a norm from $\mathbb{Q}(\sqrt{b})$ to $\mathbb{Q}$). If there is a solution, we may assume $\gcd(x, y, z) = 1$. We contend that $\gcd(a, y) = 1$. For otherwise letting p be a common prime divisor, p would divide z so p^2 would divide ax^2, hence p would also divide x (since a is squarefree), a contradiction. Hence b is a quadratic residue of a and we may write $b \equiv u^2 \pmod{a}$, where $|u| \leq |a|/2$. This means that $u^2 - b = aq$ where $|q| \leq ((a^2/4) + |a|)/|a| < |a|$, because

$|a| \geq 2$. Multiplying we find $a^2 x^2 q = (z^2 - by^2)(u^2 - b) = v^2 - bw^2$ for suitable v, w (e.g. $v = zu + by, w = z + uy$). This reduces the problem from the pair $\{a, b\}$ to the pair $\{q, b\}$. Since $|a| + |b| > |q| + |b|$ this provides a descent and an associated algorithm. Note that this method not only gives an algorithm to check the existence of a rational point, but it also allows to calculate such a point if there are any. (Another existence algorithm is based on congruences: contrary to what happens for integral points, for rational points they provide a necessary and sufficient condition, in virtue of a celebrated theorem of Hasse and Minkowski - see [26] or [74]. Such result is valid for general homogeneous quadratic equations; for conics in $\mathbb{P}_2$ it may be proved by descent, following the argument just given.)

Exercise 1.30. Decide whether $5X^2 - 11Y^2 = 26Z^2$ has a rational point.

Exercise 1.31. Following the above proof, show that the projective conic $aX^2 + bY^2 = cZ^2$ $(a, b, c \in \mathbb{Z} \setminus \{0\})$ has a rational point if and only if it has a real point and the corresponding congruence is solvable in coprime integers, for all moduli.

Prove that in fact it suffices to test the solvability of the congruences for all prime-power moduli, excluding the condition on the real point. (Actually, one can alternatively exclude the test for an arbitrary prime at our choice, but the present argument does not contain this subtler fact. In more compact language: the conic has a rational point if and only if it has a point in all but one fields $\mathbb{R}$ and $\mathbb{Q}_p$, p a prime number. See [26] or [74] for proofs.

Exercise 1.32. Prove that an odd prime number p is a sum of two squares of rationals if and only if it is congruent to 1 modulo 4. (Hint: consider the conic defined by $x^2 + y^2 - pz^2$ and apply the result of the previous exercise. See also the first supplement above for another proof. It is further to be noted that the solvability in rationals here itself implies the solvability in integers. One way to prove this implication, following Aubry and Davenport-Cassels, is as follows: take a rational solution (x_0, y_0), and pick an integral point (u, v) nearest to it; take then the line through these two points and intersect it with the circle $X^2 + Y^2 = p$. It turns out that if (x_0, y_0) is not integral, the other intersection is also rational, with a smaller denominator; see [74].)

Exercise 1.33. Prove an analogous statement with the field $\mathbb{Q}$ replaced by a field $\mathbb{F}_q(t)$. (Hint: a proof entirely analogous to the above one works. It also works, with suitable assumptions, for a field $\mathbb{Q}(t)$.)

Exercise 1.34. Write a formula parametrizing the rational points on the conic $aX^2 + bY^2 = cZ^2$, in terms of a rational point P_0 on it.

A theorem of Fermat

The diophantine equations considered in these lecture notes mainly concern integral points on curves. The theory of rational points, already for curves of genus one, is a sophisticated topic by itself and therefore we shall not practically make any attempt to describe it. Just for the sake of example we give here a brief proof of one of Fermat's statements: *there do not exist right angled triangles with integer sides and square area.* This corresponds in fact to a problem of rational points on a curve of genus 1.

Before this, we recall a celebrated formula for Pythagorean triples, namely integer solutions of $X^2 + Y^2 = Z^2$. On dividing by Z^2 we see that we are concerned with rational points on the unit circle. They may be described with the method of projection (or in an arithmetical way, as in [49]). The result is as follows: *If l, m, n are positive coprime integers such that $n^2 = m^2 + l^2$ then one among l, m, say l, is even, and there exist positive coprime integers x, y, z such that $n = x^2 + y^2$, $l = 2xy$, $m = x^2 - y^2$.* It also follows that x, y have opposite parity.

Suppose now that a right angled triangle has square area. On dividing by a gcd we can assume that its sides are coprime and that it has minimal area among such triangles.

By the above, the sides are therefore expressed by $x^2 - y^2$, $2xy$, $x^2 + y^2$ (with x, y coprime and of opposite parity). Hence the area is $(x^2 - y^2)xy$. Since the three factors are positive, pairwise coprime and their product is a square, they must all be squares, so $x = t^2$, $y = u^2$, $x^2 - y^2 = v^2$, whence

$$t^4 - u^4 = v^2,$$

where t, u, v are pairwise coprime and t, u have opposite parity. If t were even, the sum $v^2 + u^4$ would be a multiple of 4, so u, v would be even, which is not the case. Hence t is odd and u is even.
On factoring we obtain $(t^2 + u^2)(t^2 - u^2) = v^2$. The two factors on the left are coprime, because they are odd and t, u are coprime. Hence they are both squares, so there are positive integers z, w such that

$$t^2 + u^2 = z^2, \qquad t^2 - u^2 = w^2.$$

These two equations give two Pythagorean triples so, since t is odd,

$$t = a^2 - b^2, \quad u = 2ab, \qquad t = r^2 + s^2, \quad u = 2rs,$$

where $(a, b) = (r, s) = 1$. This also yields $ab = rs$. If $d := (a, r)$ we have $a = d\alpha$, $r = d\rho$ with coprime α, ρ, whence $\alpha b = \rho s$. Hence α divides s, $s = \alpha\sigma$, and $b = \rho\sigma$. Also, note that $(d, \sigma) = 1$.
We further find $a^2 - b^2 = r^2 + s^2$ whence $d^2\alpha^2 - \rho^2\sigma^2 = d^2\rho^2 + \alpha^2\sigma^2$.

Note that either α, ρ or d, σ have opposite parity: otherwise they would all be odd (e.g. if α, ρ are even, a, b would be even) and also a, b would be odd, and t even, which is not the case. Say that α, ρ have opposite parity, the other case being similar. The above equations give $d^2(\alpha^2 - \rho^2) = \sigma^2(\alpha^2 + \rho^2)$, whence, since d^2, σ^2 are coprime, d^2 divides $\alpha^2 + \rho^2$: $\alpha^2 + \rho^2 = hd^2$. Then $\alpha^2 - \rho^2 = h\sigma^2$ and multiplying we obtain $\alpha^4 - \rho^4 = (hd\sigma)^2$. This gives another solution of our original equation, again in integers > 0, and with the same parity-conditions. But $t = r^2 + s^2 = d^2\rho^2 + \alpha^2\sigma^2 > \alpha$, whence the involved numbers are still > 0 but strictly smaller. This is not possible and the contradiction proved the stated impossibility.

This proof is similar to Fermat's original argument. (See also [59].) It corresponds to a descent on the curve of genus 1 defined by $y^2 = x^4 - 1$.

Exercise 1.35. Prove the stated formulas for Pythagorean triples. (Hint: use the already mentioned method of projection, or observe that $\left(\frac{l}{2}\right)^2 = \frac{(n+m)}{2}\frac{(n-m)}{2}$, where the factors on the right are coprime.)

Notes to Chapter 1

Algorithms for factoring a polynomial over $\mathbb{Z}$ may be also obtained by looking at the reduction modulo p for several primes p. Note however that there exist irreducible polynomials which become reducible modulo every prime.

As is well known, the linear equation $aX + bY = 1$ can be solved by expanding a/b as a continued fraction (which is equivalent to Euclid's algorithm); this again fits with the view that there is a correspondence between solutions and good approximations for a/b. The linear equation in several variables can be easily reduced to the case of two variables, with appropriate conditions on the coefficients which ensure solvability.

When ξ is a real irrational, the result of Corollary 1.2 can be refined to prove the existence of infinitely many pairs (p, q) such that $|q\xi - p| \leqslant 1/\sqrt{5}q$. This conclusion is optimal for some ξ, e.g. $\frac{1+\sqrt{5}}{2}$.

It is not difficult to show that an inequality as (1.3) is *essentially* optimal for *almost all* real numbers (in Lebesgue measure), in the sense that for almost all ξ the inequality $|q\xi - p| < q^{-1-\epsilon}$ has only finitely many integer solutions, for every positive ϵ. Intuitively, this result appears natural; in fact, for integers q having N (decimal) digits, such an approximation p/q yields about $(2+\epsilon)N$ digits of ξ. But in the choice for p, q we dispose only of $2N$ digits, yielding a *gain of information*, which can be only rarely possible. However it is usually extremely difficult to prove that a given real number ξ has this finiteness property. The celebrated Roth Theorem, which we shall recall below, states that all the (irrational) algebraic numbers have this 'bad approximation' property. (See [25] for all of this.)

Rational approximations for function as described in Remark 1.4 are classically called *Padé approximations*. They can be 'explicitly' written down for some classes of functions. In some cases specialization of a Padé approximation for a function $\xi(t)$ at suitable points $t = t_0$ produces good numerical rational approximations for the values $\xi(t_0)$. This idea has been used by Hermite to prove the irrationality of the powers of the exponential function, as in the above supplement on $\exp x$; see also the last chapter.

Pell Equation appeared already in antiquity, for instance in India in VII Century a.D. (see [93]). It was proposed by Fermat as a problem to English mathematicians, as Brounker and Wallis. Later, Euler erroneously attributed it to Pell.

It turned out that already the Indians had a method of solution (so called "cyclic-method"), and also Brounker and Wallis found a similar one. (See [42] and [93].) However the first rigorous existence proof was given by Lagrange.

Efficient algorithms of solution of Pell Equation are obtained by expanding $\sqrt{d}$ as a continued fraction (necessarily periodical). In fact, continued fractions produce the 'best' rational approximations to a real number. (See [25].) The link between solutions of Pell Equations and approximations to $\sqrt{d}$ was noted already by Euler and Lagrange (see [93]).

Dirichlet's celebrated class number-formulas among other things relate the fundamental solution of Pell Equation (for squarefree $d > 1$) with the class number $h(d)$ of the quadratic field $\mathbb{Q}(\sqrt{d})$. Roughly speaking, one expects the minimal solution to be 'large' when the ring of integers is 'near' to unique factorization. (It is conjectured that there are infinitely many d such that it has unique factorization.) Dirichlet's formulas yield $2h(d) \log |\epsilon| = d^{1/2} L_d(1)$, where ϵ is a fundamental unit > 1 and L_d is the Dirichlet series associated to $\mathbb{Q}(\sqrt{d})$. (See [61].) This formula leads to the estimate $|\epsilon| \leq \exp(d^{1/2}(1 + \log d^{1/2})/2)$, which itself provides an algorithm to calculate ϵ for a given d.

As sketched above, the theory of Pell Equation is essentially the theory of units in quadratic rings. For a general number field k with ring of integers $\mathcal{O}_k$, the group $\mathcal{O}_k^*$ was studied by Dirichlet, who proved that it is the product of a finite group of roots of unity times a free abelian group of rank $r_1 + r_2 - 1$, where r_1 (resp. $2r_2$) is the number of real (resp. complex) embeddings of k (see e.g. [61] for proofs). The proof shows that some set of generators can be calculated effectively for a given number field.

This result can be rephrased by saying that $\mathbb{G}_m(\mathcal{O}_k)$, i.e. the set of integral points (over k) for the algebraic group $\mathbb{G}_m$, forms a finitely generated group, of rank $r_1 + r_2 - 1$. We shall recall a more general result for 'S-units' in the next chapters.

The equation $N_{\mathbb{Q}}^k(\xi) = \mu$ in the unknown $\xi \in k$, where k is a number field, $N_{\mathbb{Q}}^k$ is the norm and $\mu \in \mathbb{Q}^*$ is fixed, can be treated effectively by using the mentioned result by Dirichlet (see [23]). Similarly to the quadratic case, all the solutions lie in finitely many families $\alpha \eta_1^{a_1} \cdots \eta_r^{a_r}$, where α runs through a finite set depending on μ, the η_i are generators for $\mathcal{O}_k^*$ and the a_i run through $\mathbb{Z}$. In fact, if ξ, η are solutions and $\xi \equiv \eta$ (mod μ) then $\xi \eta^{-1} \in \mathcal{O}_k^*$. This result also allows to transform certain algebraic diophantine equations into exponential diophantine equations.

Concerning quadratic equations and congruences, it is known that, for a given $d > 0$ the 'test' of congruences, for deciding whether $X^2 + dY^2 = m$ has an integral solution, gives a necessary and sufficient condition (*i.e.* there is a *local-global principle*) only for finitely many positive integers d (so-called *numeri idonei*). However their complete list is still unknown (just conjectured). See [26].

The effective solution of the general diophantine quadratic equation in two variables admits an important application to an effective algorithm to establish whether two given binary quadratic forms over $\mathbb{Z}$ are equivalent over $\mathbb{Z}$ (*i.e.* there is a unimodular linear integral transformation sending a form to the other). This was well known to Lagrange and to Gauss. See [26] for proofs and generalizations to higher dimensional quadratic forms.

For some factorization algorithms see [34].

For the cyclotomic solution of Pell Equation, known to Gauss, see also [35]. There is also a *modular* solution of Pell Equation, due to Kronecker. See [82].

For the Pell Equation in polynomials, see [5] for a discussion, an application and for references.

The proof of irrationality of $\exp(t)$ is taken from [78]. It is very instructive, because it exploits Hermite's principle that (under appropriate conditions, satisfied here) functional approximations give numerical approximations by specialization. See [7] or [78] for general developements of these ideas.

The above algorithm for detecting whether a conic has a rational point is due in essence to Legendre. Different, but still elementary, algorithms over number fields more general than $\mathbb{Q}$ may be obtained with a little algebraic number theory. (See *e.g.* [97].) Further, different algorithms may be obtained by using the Hasse local-global principle. (See also [26] for the higher dimensional case.)

Chapter 2
Thue's equations and rational approximations

In this chapter we continue to deal only with solutions of diophantine equations in the classical sense, namely in integers of $\mathbb{Z}$. We shall present a proof of the theorem of Thue, showing finiteness for equations $f(X, Y) = c$, where $c \neq 0$ is constant and f is a homogeneous polynomial satisfying some natural necessary conditions. The proof will almost immediately follow from Thue's celebrated result in diophantine approximation, providing a new, deeper, example of the fundamental link between these theories. To better illustrate the basic principles of Thue's quite intricate, though elementary, proof, we shall limit ourselves to the original result, and only briefly recall the important sharpenings due to subsequent authors. Prior to details, we shall also present the main points of this argument. Finally, in the 'Supplements' we shall present some applications to the finiteness of integral points on other curves, a short proof of a theorem of Runge and a brief discussion of a function-field Thue Equation.

2.1. Thue Equations

We have seen that certain quadratic equations like $X^2 - dY^2 = c$ may have infinitely many integral solutions, a fact known since centuries. On the other hand in the XIX Century it had been noted that several diophantine equations of the same type, but of higher degree, like $X^3 - 2Y^3 = 1$, have only finitely many integral solutions; moreover nobody could produce infinitely many integral solutions for any such equation.

At the beginning of XX Century A. Thue made a most significant advance, by studying diophantine equations of the general shape $f(X, Y) = c$ for a *homogeneous* polynomial f with integer coefficients. He proved the following theorem:

Theorem 2.1 (Thue, 1909). *Let $f \in \mathbb{Z}[X, Y]$ be homogeneous, not a constant times a power of a linear or quadratic polynomial, and let c be* *a nonzero integer. Then the equation $f(X, Y) = c$ has only finitely many integer solutions.*

This result, which includes as a special case the 'Pell-like' equations in higher degree $X^d - aY^d = m$, was impressively general for Thue's time, when diophantine equations were treated with *ad hoc* methods, and no finiteness conclusion for whole families of arbitrarily large degrees was known. It is also to be remarked that the single case of the theorem is almost always far from trivial (and probably often as difficult as the general statement). Thue's method, which we shall explain in full detail, was also epoch-making.

Remark 2.1.

(i) Observe that the condition on f cannot be omitted from the statement, in view of the structure of solutions of linear diophantine equations and Pell Equation, explained in Chapter 1.

(ii) Observe that in proving the theorem one may assume that f is irreducible (over $\mathbb{Q}$) of degree $d \geq 3$. In fact, write a factorization $f = bf_1^{a_1} \cdots f_r^{a_r}$ where $b \in \mathbb{Z}$ and where the f_i are pairwise non-proportional irreducible homogeneous polynomials over $\mathbb{Z}$. Then the equality $f(x, y) = c$ with integers x, y plainly leads to finitely many systems of the shape $f_i(x, y) = c_i, i = 1, \ldots, r$, for suitable divisors c_i of c. Now, if $i \geq 2$ each system has only finitely many (complex) solutions by Bezout Theorem, because the curves $f_i(X, Y) = c_i, i = 1, 2$ have no common component (for instance they have no common points at infinity). Hence we may suppose $i = 1$. But then the assumption amounts to saying that f has degree ≥ 3.

In view of this remark (ii), for the rest of this discussion we assume that f is a homogeneous polynomial in $\mathbb{Z}[X, Y]$, irreducible over $\mathbb{Q}$ and of degree $d \geq 3$. We shall refer to the equation

$$f(X, Y) = c \tag{2.1}$$

as a *Thue Equation*.

Remark 2.2. (A probabilistic argument) We may ask: why do we expect Thue's theorem to be true, while *e.g.* Pell Equation has infinitely many integer solutions? Beyond the proof, we shall see several 'heuristic' reasons for that; for instance, here is a plausible 'probabilistic' argument for the Thue Equation $X^3 - 2Y^3 = c$.

We can estimate heuristically the probability that a given number N is of both shapes x^3 and $c + 2y^3$. The 'density' of cubes around N is roughly $1/3N^{2/3}$, while the density of numbers of the shape $c + 2y^3$ around N is $1/6N^{2/3}$. Hence, assuming that 'being a cube' and of the form 'c plus twice a cube' are in some sense independent events, the probability that a random integer N is of both types is proportional to $N^{-4/3}$. The series $\sum_N N^{-4/3}$ converges, so the number of solutions we expect is finite.

Note that for a Pell Equation $X^2 - dY^2 = 1$ a similar argument suggests $\sim (N^{-1/2})^2 = N^{-1}$ for the probability that $N = x^2 = 1 + dy^2$, so the number of solutions up to T should be around $\approx \sum_{N=1}^{T} N^{-1} \sim \log(T)$, which indeed is what happens! However such arguments, while sometimes useful for forming some sort of idea, cannot be trusted generally; for instance, they would predict only finitely many solutions for the diophantine equation $Y^2 = X^3 + X^2$, which however can be parametrized by $X = t^2 - 1, Y = t^3 - t$.

A simple application

A nice and simple application of Thue's Theorem 2.1 is to *integral values of rational functions on* $\mathbb{Q}$. Namely, given a rational function $r(t) \in \mathbb{Q}(t)$ we ask: do there exist infinitely many rational values $t = t_0 \in \mathbb{Q}$ such that $r(t_0) \in \mathbb{Z}$? If we require $t_0 \in \mathbb{Z}$, it is not difficult to show that there are only finitely many ones unless $r(t)$ is a polynomial, but for $t_0 \in \mathbb{Q}$ the question is more delicate. In the exercises below we suggest an algorithm for answering this question for any given $r(t)$, depending on Thue's theorem. (The algorithm only anwers the question, but does not produce the possible points t_0. The thing will be discussed again in the Supplements, in connection with integral points on curves of genus 0.)

Exercise 2.1. Let $r(t) \in \mathbb{Q}(t)$. Prove that if $r(t_0) \in \mathbb{Z}$ for infinitely many $t_0 \in \mathbb{Z}$ then $r \in \mathbb{Q}[t]$. Also, assuming $r \in \mathbb{Q}[t]$, produce an algorithm to decide whether the set of such t_0 is nonempty and observe that this happens if and only if this set is infinite (in which case it is easily parametrized).

Exercise 2.2. Let $f, g \in \mathbb{Z}[X, Y]$ be coprime homogeneous polynomials. Prove that for coprime $p, q \in \mathbb{Z}$, the $\gcd(f(p,q), g(p,q))$ divides a fixed computable integer $D_{f,g} \neq 0$ (independent of p, q). (Hint: Taking the resultant with respect to X we obtain an equation $A(X, Y)f(X, Y) + B(X, Y)g(X, Y) = dY^m$ for suitable $A, B \in \mathbb{Z}[X, Y], d \neq 0, m \in \mathbb{N}$. Similarly with X, Y interchanged. Now specialize at (p, q).)

Exercise 2.3. Given $r \in \mathbb{Q}(t)$, use Exercise 2.2, Theorem 2.1 and the effective results of the previous chapter to produce an algorithm for deciding whether the set $\{t_0 \in \mathbb{Q} \mid r(t_0) \in \mathbb{Z}\}$ is infinite. (Hint: write $r(p/q) = f(p,q)/g(p,q)$ for coprime homogeneous f, g of the same degree. One finds that if $r(t)$ has three or more poles in $\mathbb{P}_1(\overline{\mathbb{Q}})$ there are only finitely many t_0; otherwise the question reduces to integral points on lines or conics. See also the supplement below on integral points on curves of genus zero.)

Exercise 2.4. Decide whether $t/(2t^2 - 41)$ assumes infinitely many integral values as t runs through $\mathbb{Q}$.

Remark 2.3. For rational functions of two or more variables, things are distinctly deeper. For instance, to describe the set of integral values (at integral or rational points) of a rational function of x, y amounts to describe the integral points on a suitable open subvariety of a blow-up of $\mathbb{P}_2$ at a corresponding set of points, and this kind of problem can easily be extremely difficult. Here we do not pause further on it.

For the sake of amusing example, we only recall the result of Exercise 1.19 which states that the only integral value of $(x^2 + y^2 + 1)/xy$, for integers x, y, is the value 3. A similar instance comes from the following exercise.

Exercise 2.5. (i) Prove that every integral value of the rational function $(x^2 + y^2)/(xy + 1)$ at an integral point $(x, y) \in \mathbb{Z}^2$ is a perfect square. Prove that each square actually appears and that the corresponding set of (x, y) is Zariski dense in $\mathbb{A}^2$. (Hint: use the quadratic equation $x^2 - qxy + y^2 = q$ and apply the theory developed above, based on Pell's equation; alternatively, note that if (x, y) is a solution then also $(x, qx - y)$ is a solution for the same q, and use a descent argument.)

(ii) Discuss the same problem, but allowing x, y to be rational numbers. For instance: Do there exist non-square integral values (at a rational point)? Do there exist infinitely many such values? Do there exist prime integral values? (Hint: the appearance of a q among such values amounts to the quadratic homogeneous equation $x^2 - qxy + y^2 - qz^2 = 0$ in three variables; this may be rewritten as $(2x - qy)^2 + (4 - q^2)y^2 - 4qz^2 = 0$, and by the local-global principle it depends only on certain local conditions on q and $q^2 - 4$. For instance, it turns out that for (nontrivial) solvability each odd prime dividing q exactly must be 1 modulo 4. As to the questions, one finds e.g. that $q = 65$ is a non-square integral value and that $q = 4801$ is a suitable prime value.) It is less easy to prove that there are infinitely many non-square integral values at rational points. The following argument is due to R. Heath-Brown. Let $n > 1$ be an odd integer and set $x = (n^4 - n^2 + 16)/(2n)$, $y = 2/n$. Then $(x^2 + y^2)/(xy + 1) = q$ with $q = (n^4 - 2n^2 + 17)/4$. Then q is a non-square. As remarked by Heath-Brown, this would yield infinitely many prime values provided one assumes well-known conjectures on prime values of polynomials.

More generally, Heath-Brown sets $x = u/v$, $y = 2/v$ so that $u^2 + 4 = 2qu + qv^2$. If we then put $w = u - q$ we need $q^2 + qv^2 - (w^2 + 4) = 0$. This has a solution $q = (-v2 + r)/2$ if $r^2 = v^4 + 4w^2 + 16$. In turn, this may be solved by setting $r - 2w = c$ where c is a constant dividing some integral value $v^4 + 16$.

As to a general criterion, applying the local-global principle as suggested above proves that an integral value q occurs if $q = r^2 + s^2$ with integers r, s such that the system $r^2 + s^2 - t^2 + 2u^2 = -2$, $r^2 + s^2 - v^2 - 2w^2 = 2$ has an integral solution. This also suggests suitable asymptotic formulae for the distribution of integral values, seemingly very difficult to prove at the moment, even with sophisticated techniques.

Relations with Diophantine Approximation

Similarly to the treatment of Pell Equation, the starting point of Thue's method is to observe that an integral solution to $f(x, y) = c$ provides a 'very good' rational approximation x/y to some root of $f(X, 1)$. It is clear that x/y must go near to *some* root, but if the roots are distinct it can go near to *only one* of them; this implies that in fact it must go *very near* to it. To put this on a more solid ground, we write

$$f(X, Y) = a_0 \prod_{i=1}^{d} (X - \xi_i Y) \in \mathbb{Z}[X, Y], \qquad (2.2)$$

where $a_0 \neq 0$ is an integer and where the ξ_i are distinct algebraic numbers (recall that f is supposed to be irreducible over $\mathbb{Q}$). Defining $\eta = \eta_f :=$ $\min_{i \neq j} |\xi_i - \xi_j| > 0$ as the minimum distance between distinct roots, we prove the following simple

Proposition 2.2. *If $f(x, y) = c$ for integers x, $y \neq 0$, there exists a root ξ of $f(X, 1)$ with $|\xi - \frac{x}{y}| \leq B|y|^{-d}$, where $B = |c|(2/\eta)^{d-1}$ depends only on f and c.*

Proof. We have $a_0 \prod_{i=1}^{d} (x - \xi_i y) = c$. Let ξ be one of the ξ_i such that $|x - \xi y| =: \mu$ is minimum. Certainly $\mu^d \leq |c|/|a_0| \leq |c|$.

Suppose first that $|y| \leq 2\mu/\eta$. Then

$$\mu|y|^{d-1} \leq \mu^d (2/\eta)^{d-1} \leq |c|(2/\eta)^{d-1},$$

whence $\mu/|y| \leq |y|^{-d}|c|(2/\eta)^{d-1} \leq B|y|^{-d}$, proving the conclusion in this case.

In the remaining case we have $|y| > 2\mu/\eta$. Hence if $\xi_j \neq \xi$ we have $|x - \xi_j y| = |x - \xi y + (\xi - \xi_j)y| \geq |\xi - \xi_j||y| - \mu \geq \eta|y| - \mu \geq \eta|y|/2$.

From this we deduce that $|c| = |a_0| \prod_{i=1}^{d} |x - \xi_i y| \geq \mu(\eta|y|/2)^{d-1}$ and the conclusion again follows. $\qquad\square$

In Section 1.3 we have seen that all real irrational numbers ξ admit rational approximations x/y so that $|\xi - \frac{x}{y}| \leq y^{-2}$. However this proposition shows that for solutions of the Thue equation (where c is fixed) with $d \geq 3$ we have much better approximations if $|y|$ is large. In the Notes to the preceding chapter we have remarked that only a set of measure 0 of real numbers ξ can admit infinitely many approximations such that $|\xi - \frac{x}{y}| \leq y^{-2-\epsilon}$ for an $\epsilon = \epsilon(\xi) > 0$ (see also Exercise 2.6 below). Hence it is reasonable to expect that approximations as good as those coming from the integer solutions of the Thue Equation are finite in number (here we have another heuristic argument for Theorem 2.1): this is what Thue could prove, deducing therefore Theorem 2.1.

In the sequel we shall explain in detail his proof, but we shall now pause for several considerations on the rational approximations to algebraic numbers.

2.2. Rational approximations to algebraic numbers

As we have already seen in the last chapter, the rational numbers are 'badly' approximable with rationals: see for instance Remark 1.6. To go further, take a quadratic irrational like, $e.g.$, $\xi = \sqrt{d}$. For every $p + q\sqrt{d} \in \mathbb{Z}[\sqrt{d}]$ with $p, q > 0$ we have $(p + q\sqrt{d})(p - q\sqrt{d}) = p^2 - dq^2 \neq 0$, so $|p - q\sqrt{d}| \geqslant \frac{1}{p+q\sqrt{d}}$. Moreover, when $|p - q\sqrt{d}|$ is, say, ≤ 1, we have $p + q\sqrt{d} \leqslant 1 + 2q\sqrt{d} \leqslant 3q\sqrt{d}$. Hence we obtain $|\sqrt{d} - \frac{p}{q}| = \frac{1}{q}|p - q\sqrt{d}| \geqslant \frac{1}{q}\frac{1}{3q\sqrt{d}} = \frac{1}{3\sqrt{d}q^2}$. Therefore the approximations to $\sqrt{d}$ may improve what predicted by Corollary 1.2 at most by a constant factor. In particular, for every $\varepsilon > 0$ there are only finitely many approximation $\frac{p}{q}$ such that $\left|\xi - \frac{p}{q}\right| < q^{-2-\varepsilon}$.

The same conclusions hold for any quadratic irrational in place of $\sqrt{d}$, with similar calculations.

For algebraic numbers of higher degree the situation is far more subtle, as we shall see. However a simple idea is sufficient to realize that in some sense no algebraic number can be approximated 'too well' by means of rationals. This is due to Liouville:

Theorem 2.3 (Liouville 1844). *Let ξ be an algebraic number of degree d. Then there exists a (computable) number $c = c(\xi) > 0$ such that, for every $p, q \in \mathbb{Z}$ with $q > 0$ and $\frac{p}{q} \neq \xi$, $\left|\xi - \frac{p}{q}\right| \geq \frac{c}{q^d}$.*

Proof. Let $f(X) = a_0 X^d + \cdots + a_d \in \mathbb{Z}[X]$ be a minimal polynomial of ξ, where $a_0 > 0$. The idea is very simple: if p/q is a rational very near to ξ, then $f(p/q)$ is very near to 0. However it cannot be too near (unless it vanishes), because it is a rational number with denominator (dividing) q^d. To translate this argument in formulas, observe that if $f(p/q) \neq 0$, with $p, q \in \mathbb{Z}$ and $q > 0$ as usual, we have

$$|f(p/q)| = \frac{|a_0 p^d + \ldots + a_d q^d|}{q^d} \geqslant \frac{1}{q^d},$$

because the numerator is a nonzero integer and must be therefore ≥ 1 in absolute value. (This simple deduction is at the basis of the whole theory!) We may plainly suppose that ξ is real (because $|\xi - (p/q)| \geq |\Im\xi|$) so the Mean Value Theorem yields

$$|f(p/q)| = |f(\xi) - f(p/q)| = \left|\xi - \frac{p}{q}\right| \cdot |f'(\alpha)|,$$

where α lies between ξ and $\frac{p}{q}$.

Now, if $|\xi - \frac{p}{q}| > 1$ the conclusion is true with $c = 1$; otherwise $\alpha \in I := [\xi - 1, \xi + 1]$, whence $|f'(\alpha)| \leq \max_{t \in I} |f'(t)| = M$, say, where M depends only on f and can be easily estimated in terms of the coefficients. Finally, combining the last two displayed inequalities we obtain the conclusion with $c := \min(1, 1/M)$. □

Remark 2.4. (**Some transcendental numbers**) This theorem gave the first proof, even 'constructive', of the existence of transcendental numbers. Actually, any real number ξ which for every integer m admits a rational approximation p_m/q_m ($p_m, q_m \in \mathbb{Z}, q_m > 1$) such that $0 < |\xi - \frac{p_m}{q_m}| \leq q_m^{-m}$, must necessarily be transcendental: if ξ was algebraic of degree d then by Theorem 2.3 we would have $cq_m^{-d} \leq q_m^{-m}$, whence $2^{m-d} \leq q_m^{m-d} \leq 1/c$ for all m, a contradiction.

Such numbers are named *Liouville numbers* and it is easy to 'construct' examples of them. A classical one is $\xi = \sum_{n>0} 10^{-n!}$: the partial sum $\sum_{0<n\leqslant m} 10^{-n!}$ has $q_m = 10^{m!}$ as a possible denominator and approximates ξ with an error less than $10^{1-(m+1)!} < 10^{-mm!} = q_m^{-m}$.

Liouville numbers however have null Lebesgue measure, as is very easy to prove (see next exercise). Thus trying to apply this theorem is not a good way to decide whether a number arising independently (*i.e.* a number not constructed *ad hoc*) is transcendental or not, because we do not expect a 'random' number to be a Liouville number. For instance e and π are known not to be of this type (but they are known to be transcendental: see *e.g.* [7]).

Exercise 2.6. Prove that the set of real numbers ξ such that for some $\varepsilon > 0$ there exist infinitely many rationals p/q with $|\xi - (p/q)| < q^{-2-\varepsilon}$ has Lebesgue measure 0. (Hint: Note that we can use the same ε for all ξ. Then restrict to $\xi \in (0, 1)$ and consider the intervals with endpoints $(p/q) \pm q^{-2-\varepsilon}$ for large q. See also the notes to the previous chapter.)

Exercise 2.7. Prove that e is not a Liouville number. (Hint: Follow the irrationality proof of e^n given in the supplements to Chapter 1.)

Note that Liouville Theorem is essentially best possible for rational numbers ($d = 1$): if $\xi = r/s$ is a rational fraction with r, s coprime integers, there are infinitely many integer pairs (p, q) so that $qr - ps = 1$, which implies $|\xi - (p/q)| = (sq)^{-1}$. The same holds for quadratic irrationals ($d = 2$) in view of Corollary 1.2. For higher degree, the result is no more best possible; in fact Thue, making a most significant advance, proved the following:

Theorem 2.4 (Thue, 1909). *Let ξ be an algebraic real number of degree $d \geq 3$. For every $\varepsilon > 0$ there is a number $\gamma = \gamma(\xi, \varepsilon) > 0$ such that for all $p, q \in \mathbb{Z}, q > 0$,*

$$\left| \xi - \frac{p}{q} \right| > \frac{\gamma}{q^{1+\frac{d}{2}+\varepsilon}}. \tag{2.3}$$

Equivalently, we may clearly rephrase this result by saying that *for every* $\varepsilon > 0$ *there are only finitely many rationals* $\frac{p}{q}$, *with* $q > 0$, *such that*

$$\left| \xi - \frac{p}{q} \right| < \frac{1}{q^{1+\frac{d}{2}+\varepsilon}}. \tag{2.4}$$

Exercise 2.8. Prove that in fact the two statements are equivalent.

Note that for $d \geq 3$ we have $1 + (d/2) < d$, so the conclusion improves on Liouville's, leaving aside the computability of γ. Concerning this (important) issue, as we shall see in the next sections, Thue's very subtle (elementary) proof is unfortunately ineffective, in the sense that it does not allow to compute neither a possible number γ in (2.3), nor the finitely many fractions satisfying (2.4). (However it yields an upper bound for their number: see Remark 2.11 below.)

Theorem 2.4 implies Theorem 2.1

Beyond its intrinsic interest, Theorem 2.4 has also an immediate application to the proof of Theorem 2.1. To prove this implication, let $f(X, Y) \in \mathbb{Z}[X, Y]$ be a homogeneous polynomial of degree $d \geq 3$, irreducible over $\mathbb{Q}$ (we have noted in Remark 2.1 (ii) that it suffices to prove Theorem 2.1 with these assumptions). Then, if (x, y) ($y \neq 0$) is an integer solution to $f(X, Y) = c$, by Proposition 2.2 we obtain an approximation $|\xi - (x/y)| \leq B|y|^{-d}$ for some root ξ of $f(X) := f(X, 1)$, where B is defined in that Proposition and depends only on f, c. Assuming Theorem 2.4 for $\varepsilon = 1/4$, say, we also have $|\xi - (x/y)| > \gamma(\xi)|y|^{-1-(d/2)-(1/4)}$, for a number $\gamma(\xi) = \gamma(\xi, 1/4) > 0$. Comparison of estimates leads to $(\min_{f(\xi,1)=0} \gamma(\xi))|y|^{(d/2)-(5/4)} \leq B$. Since there are only finitely many roots ξ to consider, the left side is $\gg_f |y|^{1/4}$ (since $d/2 \geq 3/2$), whence we get a bound for $|y|$ and the finiteness of integer solutions.

Remark 2.5.   Note that the very proof of Liouville Theorem prevents any such finiteness implication. In fact, the proof of Liouville's lower bound was obtained by allowing the homogeneized polynomial $f(X, Y) := Y^d f(X/Y)$ to take at the integral point (p, q) the minimum possible positive value 1 (in absolute value); we cannot thus hope to extract from this argument the fact that $|f(p, q)|$ is usually larger.

Exercise 2.9. Apply Theorem 2.4 and Proposition 2.2 to prove that *if* $f(X, Y) \in \mathbb{Z}[X, Y]$ *is a homogeneous polynomial of degree* $d \geq 3$, *irreducible over* $\mathbb{Q}$, *and if* $g(X, Y) \in \mathbb{Z}[X, Y]$ *has degree* $< (d/2) - 1$, *then the diophantine equation* $f(X, Y) = g(X, Y)$ *has only finitely many integral solutions.*

Exponent of approximation

After Liouville's and Thue's results, a natural question for a real number ξ concerns the so-called *exponent* of approximation $e(\xi)$: we define it as *the infimum of the real numbers δ such that the inequality $|\xi - \frac{p}{q}| < q^{-\delta}$ has only finitely many solutions in rationals p/q.*

So, $e(\xi) = 1$ for $\xi \in \mathbb{Q}$, $e(\xi) \geq 2$ for all $\xi \in \mathbb{R} \setminus \mathbb{Q}$ (by Corollary 1.2) and $e(\xi) \leq 2$ for almost all real numbers (as we have remarked above, see Ex.2.6).

For an algebraic ξ of degree $d := [\mathbb{Q}(\xi) : \mathbb{Q}]$, Liouville's result implies that $e(\xi) \leq d$ while Thue's Theorem implies $e(\xi) \leq 1 + (d/2)$.

Thue's method and result were subsequently refined by several authors: C.L. Siegel proved $e(\xi) \leq 2\sqrt{d}$, A.O. Gelfond and F. Dyson proved $e(\xi) \leq \sqrt{2d}$.[1] Finally, in 1955 K. F. Roth proved $e(\xi) = 2$ for all irrational algebraic numbers ξ. We restate explicitly his result in equivalent homogeneous form as:

Theorem 2.5 (Roth 1955*). *Let ξ be an algebraic number and let $\varepsilon > 0$. Then there are only finitely many rationals p/q, $p, q \in \mathbb{Z}, q > 0$, such that $|q\xi - p| < q^{-1-\varepsilon}$.*

These theorems have also been extended to approximations by numbers in a given number field, and even with respect to p-adic absolute values. We shall recall some statements in this direction in the next chapter.

For transcendental numbers the exponent is known only in a few cases. For instance, it is known to be 2 (*i.e.*, the expected value) for the values $\exp(t)$ of the exponential function at rationals $t \in \mathbb{Q}$. By definition, the exponent is ∞ precisely for Liouville numbers.

Exercise 2.10. Define an exponent of approximation over $k[[t]]$ and prove that (for char$(k) = 0$) the exponent of $\exp(t)$ is 2. (Hint: Prove by differentiation that if $p, q \in k[t]$ have degree $\leq n$ and are not both 0 then $\text{ord}_{t=0}(q(t)\exp(t) - p(t)) \leq 2n + 1$.)

Exercise 2.11. Apply Theorem 2.5 and Proposition 2.2 to prove that *if $f(X,Y) \in \mathbb{Z}[X, Y]$ is a homogeneous polynomial of degree $d \geq 3$, irreducible over $\mathbb{Q}$, and if $g(X, Y) \in \mathbb{Z}[X, Y]$ has degree $< d - 2$, then the diophantine equation $f(X, Y) = g(X, Y)$ has only finitely many integral solutions.*

---

[1] For other proofs of Thue's theorem (in improved form) see, *e.g.*, Dickson's book [38] or Mordell's book [59].

2.3. Thue's method and later developements

2.3.1. A rough sketch of Thue's proof

In order to clarify Thue's elementary but intricate argument, we wish to present now with little detail the main points and motivations for Thue's actual proof, which will be given with full detail in the next section. Below we isolate and briefly describe the individual steps of the proof.

A gap principle

A fundamental point is a so-called *gap-principle* for rational approximations to a given number; roughly this may be expressed by saying that *good rational approximations to a same number ξ yield big gaps between their denominators*. More precisely, if we have two very good approximations $|\xi - (p/q)| \leq q^{-\nu}$ and $|\xi - (r/s)| \leq s^{-\mu}$ with $\mu, \nu > 1$, $0 < q^\nu \leq s^\mu$, and $\frac{p}{q} \neq \frac{r}{s}$, by the triangle inequality we have

$$\left|\frac{p}{q} - \frac{r}{s}\right| \leq \left|\frac{p}{q} - \xi\right| + \left|\xi - \frac{r}{s}\right| \leq \frac{1}{q^\nu} + \frac{1}{s^\mu} \leq \frac{2}{q^\nu}.$$

On the other hand, since $\frac{p}{q} \neq \frac{r}{s}$, we have $\left|\frac{p}{q} - \frac{r}{s}\right| = \frac{|ps-rq|}{qs} \geq \frac{1}{qs}$ (note that we have used once more the fact that "a positive integer is ≥ 1"). By comparison we obtain the 'gap'

$$s \geq \frac{q^{\nu-1}}{2}.$$

Note that we are assuming *a priori* $s \geq q^{\nu/\mu}$; however the displayed inequality is stronger if, roughly, $\nu > 1 + (\nu/\mu)$, *i.e.*, $\nu > \mu/(\mu - 1)$.

Exercise 2.12. Suppose that for a fixed $m > 2$ we have a sequence of distinct approximations $\{p_n/q_n\}$ to ξ, with $q_1 < q_2 < \ldots$ and $\left|\xi - \frac{p_i}{q_i}\right| < \frac{1}{q_i^m}$. Prove that $q_i > c^{(m-1)^i}$ for some $c > 1$. (Hint: apply the above with $\mu = \nu = m$.)

At this point the argument is as follows: starting from an 'excellent' approximation to the algebraic number ξ (large enough and so good to satisfy (2.4)) one constructs a whole sequence of 'relatively good' approximations (with *exponent* still > 1 but not as good as the previous one). Moreover these approximations have not too large gaps. Because of this, another possible 'excellent' approximation would have a 'small gap' with respect to some element of the sequence; this however would contradict the *gap principle*. Note that to start the construction we need a large excellent approximation, which could or could not exist; so what we are really proving is that *two* large excellent approximations cannot exist. This leads to the mentioned ineffectivity.

We shall now briefly describe Thue's construction of the said sequence and give a bit of detail for the conclusion.

Construction of new approximations from a given one

Let p/q be the alluded excellent approximation (chosen so that q is large enough), so $|\xi - (p/q)| \le q^{-\mu}$ for a $\mu > 1 + (d/2)$. Thue constructed, for each integer n, polynomials $P_n(X)$, $Q_n(X) \in \mathbb{Z}[X]$ of degree $\le n$ so that $P_n(X) - \xi Q_n(X)$ has a zero of 'large' order M ($\sim 2n/d$) at $X = \xi$. Moreover the coefficients of these polynomials are not too large ($\ll B^n$).

Now, putting $X = p/q$, we obtain that $P_n(p/q) - \xi Q_n(p/q)$ is very small, because of the large order zero at $X = \xi$ and because p/q is very near to ξ. This means that $p_n/q_n := P_n(p/q)/Q_n(p/q)$ is a good approximation to ξ.

It turns out that the construction allows $|\xi - (p_n/q_n)| \ll q_n^{-\nu}$ for some $\nu > 1 + (2/d)$ and $q_{n+1} \ll q_n^{1+\lambda}$ for a very small $\lambda = \lambda_q > 0$. Also, for large enough q, $p_n/q_n \to \xi$, so $q_n \to \infty$.

Remark 2.6.   Thue succeeded in an explicit construction only in a few special cases. After some unsuccessful attempts, he constructed P_n, Q_n indirectly, using some linear algebra with upper bounds for some integer solution. (See Siegel's lemma in the next section.) Note that we have $2n + 2$ coefficients at disposal; if we want them to be rational, each vanishing condition at ξ gives d linear conditions defined over $\mathbb{Q}$ (because $d = [\mathbb{Q}(\xi) : \mathbb{Q}]$). So in principle we can achieve multiplicity $M \ge (2n + 1)/d$; actually, to achieve also good bounds for the coefficients, it is necessary to impose only a multiplicity (slightly) smaller than $(2n + 1)/d$.

Conclusion of the proof

Now, fix another excellent approximation r/s, so $|\xi - (r/s)| \le s^{-\mu}$ and let n be the maximum index so that $q_n \le s^{\mu/\nu}$; then $s^{\mu/\nu} < q_{n+1} \ll q_n^{1+\lambda}$. By the gap principle (with q_n in place of q) if $p_n/q_n \ne r/s$, we have $s \gg q_n^{\nu-1}$, whence $q_n^{\nu-1} \ll q_n^{\nu(1+\lambda)/\mu}$. It follows that $\mu \le \nu(1+\lambda)/(\nu - 1) + o(1)$. However $\nu > 1 + (2/d)$ so $\mu < (1 + (d/2))(1 + \lambda) + o(1)$, which is not true for λ very near 0 and for a small enough '$o(1)$' term (both conditions being ensured for large enough q).

A crucial difficulty

Note that the required contradiction goes through an apparently small point: namely, to apply the gap principle we need $p_n/q_n \ne r/s$. Actually, to verify this point is a real obstacle, and was a major difficulty in the subsequent refinements of Thue's theorem. Thue overcame this difficulty by differentiating the polynomials P_n, Q_n: he thus obtained 'independent' polynomials with similar properties. In fact, the polynomial $(P_n^{(j)}(X) - \xi Q_n^{(j)}(X))/j!$ has still multiplicity $\geq M - j$ at $X = \xi$, which suffices if j is not too large; also, the coefficients are still integers and the division by $j!$ ensures they do not increase too much compared with the original ones. One can now specialize at p/q and repeat the previous pattern. Of course, *a priori* the original difficulty can persist. However using a 'Wronskian' (we shall see this in detail) one can prove that for a suitable j (not too large) this cannot occur: for otherwise the Wronskian $P_n' Q_n - P_n Q_n'$ would have a zero of high order at p/q, which would force the coefficients to be too large. (And here there is also an alternative argument relying on degrees.)

Precursors of Thue's method

This proof method was completely new, but was inspired by discoveries of previous authors.

Very interesting notes by Masser [57] relate Thue's method with the recursion method of Newton for approximating zeros of differentiable functions; this numerical procedure and its later refinements (by Halley and others) with derivatives of higher order might have suggested the construction of the auxiliary function.

Also, some kind of gap principle was implicit in the existing irrationality and transcendency proofs known at Thue's time; Hermite had proved the transcendency of e and Lindemann that of π (by similar, but technically more involved ideas). Also, Hermite proofs worked by first constructing *functional* approximations for $\exp(x)$ by means of rational (or algebraic) functions, and then specializing these functions at rational points r, in order to obtain good rational approximations for $\exp(r)$. (See the relevant supplement to Ch. I.) A similar technique was used by Thue in dealing with approximations to roots of rational numbers, by using Padé approximations for $(1 + x)^{1/r}$. Then Thue realized that this explicit constructions would not work generally and thus constructed indirectly the approximating polynomials. This idea of achieving numerical approximations from functional ones is clearly present (though in an entirely new way) in the above pattern of Thue's proof, on specializing $P_n(X) - \xi Q_n(X)$ at $X = p/q$.

2.3.2. A reformulation and some later refinements

It has been recognized that for several purposes it is convenient and simpler to reformulate and axiomatize the above pattern as follows.

Step 1: Construction of auxiliary polynomials. Note that the above conditions for the polynomials P_n, Q_n can be rephrased by saying that the polynomial $F_n(X, Y) := P_n(X) - Y Q_n(X)$ *has* $\deg_X F_n \le n$, $\deg_Y F_n \le 1$ *and a zero at* (ξ, ξ) *of multiplicity* $\ge M$ *with respect to* X. (The number M will be chosen $\approx (2 - \lambda)n/d$, for a sufficiently small $\lambda > 0$.)

We thus want to construct (for each $n \in \mathbb{N}$) $F_n(X, Y) \in \mathbb{Z}[X, Y]$ with these properties and with 'not too large' coefficients. This will be achieved by Siegel's lemma in the next section: linear algebra + estimates for an integral solution.

Step 2: Upper bound for $F_n(p/q, r/s)$. In the previous sketch we had compared $p_n/q_n := P_n(p/q)/Q_n(p/q)$ with r/s (this is implicit in the gap principle). This corresponds to an upper bound for $F_n(p/q, r/s)$. This value will be small because of the large order zero of F_n at (ξ, ξ) and the fact that $p/q, r/s$ are both very near ξ. Also, of course it will be important now that F_n has not too large coefficients.

Naturally, to take the maximum advantage we must choose n so that the contributions of: (i) the zero-multiplicity $M \approx (2 - \lambda)n/d$ with respect to X, and of: (ii) the zero-multiplicity 1 with respect to Y, are 'nearly' equal. This amounts to $q^M \approx s$, *i.e.* $q^{(2-\lambda)n} \approx s^d$, so we shall choose $n \approx \frac{d \log s}{(2-\lambda) \log q}$.

Step 3: Lower bound for $F_n(p/q, r/s)$. This represents the second half of the gap principle. In the present setting, we note that $F_n(p/q, r/s)$ is a rational number with denominator dividing $q^n s$. **If** it is nonzero, we shall have $|F_n(p/q, r/s)| \ge (q^n s)^{-1}$. The calculations show that if p/q and r/s are sufficiently good approximations to ξ (and if q, s are large enough) then this will contradict Step 2, concluding the argument.

Step 4: Non vanishing of $F_n(p/q, r/s)$. To ensure Step 3, we must prove that $F_n(p/q, r/s) \ne 0$; this corresponds to the above mentioned 'crucial difficulty'. We shall achieve this point, however, not quite for F_n itself, but for a suitable derivative of it (of not too large order) with respect to X. We shall prove that some polynomial $(h!)^{-1} \partial^h F_n(X, Y)/\partial X^h$ evaluated at $(p/q, r/s)$ still satisfies a good *Upper bound* but does not vanish. Then the same steps as above will be applied to $(h!)^{-1} \partial^h F_n(X, Y)/\partial X^h$ in place of F_n.

As alluded above, the required nonvanishing will follow from an analysis of the Wronskian $P_n' Q_n - P_n Q_n'$, which cannot have too high multiplicity at p/q due to the smallness of its coefficients. Here an alternative approach involves the degree of the Wronskian rather than its coefficients.

At the light of this reformulation, we also note that Liouville's simple result follows this pattern, but uses a polynomial in a single variable; this 'rigidity' prevents to go beyond the exponent d.

Improvements of Thue's result

As recalled in the previous section, after Thue several authors obtained sharpened results. Siegel obtained the exponent $2\sqrt{d}$ (in place of Thue's $1 + (d/2)$) by allowing the polynomial $F_n(X, Y)$ to have arbitrary degree in Y as well. The additional freedom gained by the higher degree in Y allows to impose a larger vanishing with respect to X, which leads to the improvement. (See also [59].) Here we also see the advantage of the new formulation: Siegel's step would not be clearly visible with the previous pattern of constructing rational approximations. We may say that we are now constructing approximations to ξ by algebraic numbers, possibly of degree > 1. After Siegel, Gelfond and Dyson independently obtained the exponent $\sqrt{2d}$ by a similar method, but increasing the zero-multiplicity also with respect to Y. (See [47].)

Dyson's approach is particularly interesting and original, because the nonvanishing of Step 4 is obtained not by considering the magnitude of the coefficients, but by an algebraic analysis, through the so-called Dyson's lemma: this involves *degrees* rather than *coefficients*; it bounds the number of zeros with certain multiplicity conditions for a polynomial in two variables (and in a sense generalizes the fact that a polynomial in one variable cannot have more zeros than its degree). This approach has been subsequently extended to several dimensions and also used by Bombieri for an effective analysis. (We shall give below a simple example of the idea. For further informations see the notes to this chapter.)

Finally, Roth obtained his final result, *i.e.* Theorem 2.5. His proof proceeds along the same steps as above, but uses polynomials in several variables $X_1, \ldots, X_r$, evaluated at rational approximations $p_1/q_1, \ldots, p_r/q_r$ with rapidly increasing denominators. This approach had been already suggested by Siegel, but until Roth's 1955 proof it was not clear how to overcome the fundamental difficulty coming from Step 4. Roth resolved this point in the celebrated *Roth's lemma* by an intricate analysis of Wronskians in several variables. (See [17] or [25].)

Remark 2.7.   All of these improvements suffer from the same ineffectivity as Thue's Theorem, because they proceed along the same lines: use of some excellent approximations to show that there cannot be others; however there is no information on these first ones, whose existence is purely hypothetical. However we shall see that the proofs may in principle become effective provided one is in possession of sufficiently good approximations to start with (just one may be sufficient for Thue's Theorem).

2.4. Proof of Thue's Approximation Theorem

2.4.1. Preliminaries

As above, we let ξ be a fixed algebraic number of degree $d \geq 3$. We fix an $\varepsilon > 0$ and for the whole section we shall call *excellent* any rational approximation p/q $(p, q \in \mathbb{Z}, q > 0)$ satisfying (2.4), namely

$$\frac{p}{q} \text{ is excellent (with respect to } \varepsilon) \text{ if } (p, q) = 1 \text{ and } \left| \xi - \frac{p}{q} \right| \leq \frac{1}{q^{1 + \frac{d}{2} + \varepsilon}}.$$

$$(2.5)$$

We have to prove that there are only finitely many excellent approximations.

We start by introducing a little terminology.

Differential operators

We shall use the differential operators $D_j = \frac{1}{j!} \frac{\partial^j}{\partial X^j}$ instead of the usual $D^j = \frac{\partial^j}{\partial X^j}$. This will be useful because D_j does not increase too much the coefficients, compared to D^j. Note that at the same time D_j maps $\mathbb{Z}[X]$ to itself. We shall also put $D = D_1$ and $P' := DP$.

Norms of polynomials

For the rest of this chapter only, we define the norm $\|P\|$ of a polynomial $P(X) = \sum_{i=0}^{n} a_i X^i \in \mathbb{Z}[X]$ as

$$\|P\| = \max_{0 \leq i \leq n} |a_i|.$$

The following properties, valid for every $A, B \in \mathbb{Z}[X]$ and every integer $j \geq 0$ are very easily verified and their proof is left as an exercise for the interested reader.

$$\|A + B\| \leq \|A\| + \|B\|, \quad \|AB\| \leq (\deg(A) + 1) \|A\| \|B\|; \quad (2.6)$$

$$\|D_j A\| \leq \binom{\deg A}{j} \|A\| \leq 2^{\deg(A)} \|A\|. \quad (2.7)$$

Exercise 2.13. Prove the above properties.

Further conventions

We have fixed an $\varepsilon > 0$ and we now define $\mu = 1 + \frac{d}{2} + \varepsilon > 1 + \frac{d}{2}$. Then, fix also (as is clearly possible) a positive rational $\lambda < \frac{1}{2}$ such that

$$\delta := \left(1 + \frac{2\varepsilon}{d}\right)(1 - \lambda) - 1 > 0. \quad (2.8)$$

(We may fix *e.g.* a rational ε and set $\lambda = \varepsilon/(2 + 2\varepsilon)$.) In the sequel we shall denote by $B_1, B_2, \ldots$ numbers > 1 depending only on $\xi, \varepsilon, \lambda$. An easy inspection of the arguments below would show that they can be explicitly estimated in terms of the said quantities, but we shall not pursue in this task.

Now, since ξ is an algebraic number of degree d, every positive power of ξ can be written as

$$\xi^r = c_{r,0} + c_{r,1}\xi + \cdots + c_{r,d-1}\xi^{d-1}, \qquad c_{rs} \in \mathbb{Q}. \qquad (2.9)$$

Proposition 2.6. *There exists an integer $b > 0$ depending only on ξ such that $b^r c_{rs} \in \mathbb{Z}$ for all $r, s \geq 0$. Moreover, $|c_{rs}| \leqslant B_1^r$.*[2]

Proof. We let $b > 0$ be a common denominator for the c_{ds} and we define $B_1 := 1 + \max_{r,s \leq d} |c_{rs}|$ (this is legitimate because B_1 depends only on ξ). The inequality of the proposition is trivially true for $0 \leq r \leq d$. For $r \geq d$ we have

$$\xi^{r+1} = c_{r0}\xi + c_{r1}\xi^2 + \ldots + c_{r(d-2)}\xi^{d-1}$$

$$+ c_{r(d-1)}(c_{d0} + \ldots + c_{d(d-1)}\xi^{d-1})$$

$$= c_{r(d-1)}c_{d0} + (c_{r(d-1)}c_{d1} + c_{r0})\xi + \ldots$$

$$+ (c_{r(d-1)}c_{d(d-1)} + c_{r(d-2)})\xi^{d-1}.$$

This inductively shows that $\max_s |c_{rs}| \leq B_1^r$ and also that b^r is a common denominator for the c_{rs}, as wanted. $\qquad\square$

At the light of the previous section, let us review the main steps of the proof:

0. We suppose to be given excellent approximations u, v to ξ, with sufficiently large denominators (see (2.24) below). We shall actually also require that v has a much larger denominator than u (see (2.26)). These conditions can certainly be satisfied if there are infinitely many excellent approximations.

Depending on u, v we shall:

1. Construct a polynomial $F(X, Y)$ with not too large integer coefficients and such that $D_j F(\xi, \xi) = 0$ for 'several' j.

2. Prove an upper bound for $|D_i(F(u, v))|$.

---

[2] For notational convenience we shall often write, also in the sequel, c_{rs} in place of $c_{r,s}$. Also, here B_1 can be chosen as $B_1 := 1 + \max_{0 \leq s < d} |c_{d,s}|$.

3. Prove a lower bound for $|D_i(F(u, v))|$, when $D_i F(u, v) \neq 0$.

4. Prove that $D_i F(u, v) \neq 0$ for a suitably small i and obtain a contradiction.

2.4.2. Construction of polynomials F_n

For Step 1 we shall actually construct a whole sequence of polynomials F_n; our F will be F_n for a suitable n depending on u, v.

We shall construct F_n of the shape $F_n(X, Y) = P_n(X) + Y Q_n(X) \in \mathbb{Z}[X,Y]$, with $\deg_X F_n = \max(\deg P_n, \deg Q_n) \leq n$, such that $D_j F_n(\xi,\xi) = 0$ for $0 \leq j < m$. We shall soon specify m as well.

We have $2n + 2$ free coefficients of P_n, Q_n at our disposal, which we view as variables; each condition $D_j F_n(\xi, \xi) = 0$ corresponds to a single linear condition in these variables. However this linear condition is defined over $\mathbb{Q}(\xi)$, while we want the solution (i.e. the coefficients of F_n) to be over $\mathbb{Q}$ (and actually in $\mathbb{Z}$). To get equations defined over $\mathbb{Q}$ we simply use a linear basis for $\mathbb{Q}(\xi)/\mathbb{Q}$, e.g. the basis $1, \xi, \ldots, \xi^{d-1}$, to transform each original equation into d new linear equations, this time over $\mathbb{Q}$. Since we have m conditions $D_j F_n(\xi, \xi) = 0$ we get a total of md homogeneous linear equations (over $\mathbb{Q}$) in $2n+2$ variables. Thus the condition $md < 2n+2$ ensures a non-trivial solution (P_n, Q_n). Actually, to obtain good bounds for the integral solutions we shall choose an m slightly smaller than $(2n + 2)/d$.

Let us now see more precisely what kind of linear system we obtain. We write $P_n(X) = x_0 + \ldots + x_n X^n$, $Q_n(X) = y_0 + \ldots + y_n X^n$. Then we have:

$$D_j F_n(\xi, \xi) = D_j P_n(\xi) + \xi D_j Q_n(\xi)$$

$$= \sum_{s=j}^{n} x_s \binom{s}{j} \xi^{s-j} + \sum_{s=j}^{n} y_s \binom{s}{j} \xi^{s-j+1}$$

$$= \sum_{s=j}^{n} x_s \sum_{l=0}^{d-1} \binom{s}{j} c_{s-j,l} \xi^l + \sum_{s=j}^{n} y_s \sum_{l=0}^{d-1} \binom{s}{j} c_{s-j+1,l} \xi^l$$

$$= \sum_{l=0}^{d-1} \xi^l \sum_{s=j}^{n} \left(x_s \binom{s}{j} c_{s-j,l} + y_s \binom{s}{j} c_{s-j+1,l} \right)$$

$$= L_{n,0}^{(j)}(\underline{x}) + \xi L_{n,1}^{(j)}(\underline{x}) + \cdots + \xi^{d-1} L_{n,d-1}^{(j)}(\underline{x}),$$

where $\underline{x} := (x_0, \ldots, x_n, y_0, \ldots, y_n)$ and where

$$L_{n,l}^{(j)}(\underline{x}) := \sum_{s=j}^{n} \left(x_s \binom{s}{j} c_{s-j,l} + y_s \binom{s}{j} c_{s-j+1,l} \right)$$

are linear forms. Here we have used equations (2.9).

In view of Proposition 2.6 the linear forms $b^{n+1}L_{n,l}^{(j)}$ have integral coefficients. Also, note that by the same proposition the coefficients of $L_{n,l}^{(j)}$ are bounded in absolute value by $2^n B_1^{n+1}$, so the linear forms $b^{n+1}L_{n,l}^{(j)}$ have integral coefficients bounded in absolute value by $b^{n+1}2^n B_1^{n+1} \leq B_2^n$, where we may choose for example $B_2 := (2bB_1)^2$; B_2 indeed depends only on ξ, as in our convention.

Our vanishing conditions amount to $L_{n,l}^{(j)}(\underline{x}) = 0$ for $j = 0, 1, \ldots, m-1$ and $l = 0, 1, \ldots, d-1$. To fulfill them with integral and not too large x_h, y_k we shall use the following celebrated *Siegel lemma*; there are several versions of it (see the notes) but we shall need just the simplest:

Lemma 2.7 (Siegel). *For $i = 1, \ldots, N$, $j = 1, \ldots, M$, where $N > M$, let a_{ij} be integers of absolute value at most $A \geq 1$. Then there exist integers $t_1, \ldots, t_N$, not all zero, such that*

$$|t_i| \leq (NA)^{\frac{M}{N-M}}, \qquad \sum_{i=1}^{N} a_{ij}t_i = 0, \quad j = 1, \ldots, M.$$

Proof. The principle of the proof is to consider the image of a finite N-dimensional 'cube' of integers by the map given by the M linear forms expressing the equations. Since the dimension of the target space is $M < N$, for a suitably large cube the map will not be injective, so by difference we shall obtain a 'small' integral vector in the kernel.

To carry out this program, let $T = \left[(NA)^{\frac{M}{N-M}}\right] \geq 1$, $I_T = \{0, 1, \ldots, T\}$ and consider the integral vectors $\underline{x} = (x_1, \ldots, x_N) \in I_T^N \subset \mathbb{Z}^N$; there are $(T+1)^N$ of them.

Consider the map $L: \mathbb{Z}^N \to \mathbb{Z}^M$ defined by $L(\underline{x}) = \left(\sum_{i=1}^{N} a_{ij}x_i\right)_{j=1,\ldots,M}$. To bound the values of this map, define $S_+^{(j)} = \sum_{i=1}^{N} \max\{0, a_{ij}\}$ and $S_-^{(j)} = \sum_{i=1}^{N} \min\{0, a_{ij}\}$. Then $S_+^{(j)} - S_-^{(j)} = \sum_{i=1}^{N} |a_{ij}| \leq NA$ and $S_-^{(j)}T \leq \sum_{i=1}^{N} a_{ij}x_i \leq S_+^{(j)}T$ for $\underline{x} \in I_T^N$.[3]

The interval $\left[S_-^{(j)}T, S_+^{(j)}T\right]$ contains $S_+^{(j)}T - S_-^{(j)}T + 1 \leq NAT + 1$ integers, thus the image of I_T^N under L consists of at most $(NAT+1)^M$ points. By definition of T, we have $T + 1 > (NA)^{\frac{M}{N-M}}$, so that $(T+1)^{N-M} > (NA)^M$ and

$$\#\left(I_T^N\right) = (T+1)^N > (NA(T+1))^M \geq (NAT+1)^M \geq \#\left(L(I_T^N)\right).$$

---

[3] Replacing $S_+ - S_-$ by NA would be equally useful, but would't lead to such a 'clean' result.

This implies that there exist two distinct vectors $\underline{x}', \underline{x}'' \in I_T^N$ such that $L(\underline{x}') = L(\underline{x}'')$. Let $\underline{t} = (t_1, \ldots, t_N) := \underline{x}' - \underline{x}'' \neq 0$. Then $L(\underline{t}) = \left(\left(\sum_i a_{ij} t_i \right)_j \right)_{j=1}^M = 0$, with $|t_i| \leqslant T \leqslant (NA)^{\frac{M}{N-M}}$ for $i = 1, \ldots, N$, as wanted. $\square$

Exercise 2.14. Prove that the exponent $M/(N - M)$ of A in the conclusion cannot be improved. (Hint: take for instance $M = 2n - 2$, $N = 2n$, and consider the system $t_{i+1} = At_i$, $t_{i+n+1} = At_{i+n}$, for $i = 1, \ldots, n - 1$.)

We now apply this lemma to the md linear forms $b^{n+1} L_{n,l}^{(j)}$, for $j = 0, 1, \ldots, m - 1$, $l = 0, 1, \ldots, d - 1$, in the $2n + 2$ variables x_h, y_k, $h, k = 0, 1, \ldots, n$; we have already remarked that these forms have integral coefficients bounded in absolute value by B_2^n.

For notational simplicity we shall consider only the values of n such that $(2n + 2)(1 - \lambda)$ is an integral multiple of d; recall that we chose λ to be rational, so these values of n exist and form a whole arithmetic progression. Also, we define

$$m := \frac{(2n + 2)(1 - \lambda)}{d}. \tag{2.10}$$

For n lying in the said progression, the values of m so obtained will also lie in a certain progression, both progressions depending only on λ and d. With these conventions, we apply Siegel Lemma 2.7 with $N := 2n + 2$, $M = md = N(1 - \lambda) < N$, and $A := B_2^n \geq 1$. Note that $N - M = \lambda N$, hence $M/(N - M) = (1 - \lambda)/\lambda$.

The conclusion of the lemma delivers a nonzero vector of coefficients $(\underline{x}, \underline{y})$, with

$$|x_h|, |y_k| \leqslant (NA)^{\frac{M}{N-M}} = \left((2n + 2) B_2^n \right)^{\frac{1-\lambda}{\lambda}} \leqslant (4B_2)^{\frac{n}{\lambda}} \leqslant B_3^n,$$

where $B_3 := (4B_2)^{1/\lambda}$ depends only on ξ, λ. This bound means that our polynomials P_n, Q_n satisfy

$$\|P_n\|, \|Q_n\| \leqslant B_3^n \tag{2.11}$$

Let now $u = \frac{p}{q}$ and $v = \frac{r}{s}$, $p, q, r, s \in \mathbb{Z}$, $(p, q) = (r, s) = 1$, be two excellent approximations to ξ, with $0 < q < s$, i.e.

$$|\xi - u| \leqslant \frac{1}{q^\mu}, \qquad |\xi - v| \leqslant \frac{1}{s^\mu}, \qquad 0 < q < s. \tag{2.12}$$

We shall obtain a contradiction provided q, s satisfy suitable inequalities, which shall be specified later (see (2.24) and (2.26)). This will imply the finiteness of the set of excellent approximations to ξ.

2.4.3. Upper bound for $|D_j F_n(u, v)|$

We let F_n be the polynomials constructed in 2.4.2. We shall now perform Step 2 in our program by proving an upper bound for $|D_j F_n(u, v)|$.

We have $F_n(X, Y) = P_n(X) + Y Q_n(X) \in \mathbb{Z}[X, Y]$. Recall that $F_n(X, \xi)$ has a zero of order $\geq m$ at $X = \xi$ (and lies in $\mathbb{Z}[\xi][X]$). Hence we may write $F_n(X, \xi) = (X - \xi)^m R_n(X)$ for a suitable polynomial R_n. (It lies in $\mathbb{Z}[\xi][X]$ but we won't need this.) Therefore $F_n(X, Y) = F_n(X, \xi) + (Y - \xi) Q_n(X) = (X - \xi)^m R_n(X) + (Y - \xi) Q_n(X)$. Then, for $j = 0, 1, \ldots, m - 1$,

$$D_j F_n(X, Y) = (X - \xi)^{m-j} R_{n,j}(X) + (Y - \xi) D_j Q_n(X), \qquad (2.13)$$

where $R_{n,j}(X)$ is a certain polynomial (see the next exercise for an explicit formula).

We want to bound $\|R_{n,j}\|$ and $\|D_j Q_n\|$.

As to this last quantity, inequalities (2.7) and (2.11) immediately yield $\|D_j Q_n\| \leq (2B_3)^n$.

As to the first one, observe that, by (2.13), we have $D_j F_n(X, \xi) = (X - \xi)^{m-j} R_{n,j}(X)$. Then, thinking of $\mathbb{C}[X]$ as embedded in $\mathbb{C}[[X]]$ and writing the formal series

$$\frac{1}{(\xi - X)^s} = \frac{\xi^{-s}}{\left(1 - \frac{X}{\xi}\right)^s}$$

$$= \sum_{r \geq 0} \binom{-s}{r} \xi^{-r-s} X^r = \sum_{r \geq 0} (-1)^r \binom{s+r-1}{r} \xi^{-r-s} X^r$$

we obtain $R_{n,j}(X) \in \mathbb{C}[X] \subset \mathbb{C}[[X]]$ as the formal product of $D_j F_n(X, \xi)$ and $\frac{1}{(X-\xi)^{m-j}}$. Since $R_{n,j}$ has degree at most $n - m$, we need only consider the first $n - m$ powers of X arising from the product. The coefficients of the formal series, for $s = m - j$ and $r \leq n - m$, can be immediately bounded by $2^n \max(1, |1/\xi|)^n$. Also, the coefficients of $D_j F_n(X, \xi) = D_j P_n(X) + \xi D_j Q_n(X)$ are bounded by $\|D_j P_n\| + |\xi| \|D_j Q_n\| \leq 2^n (1 + |\xi|) B_3^n$ (again by inequalities (2.7) and (2.11)).

All of this implies $\|R_{n,j}\| \leq (n + 1) 2^{2n} \max(1, |1/\xi|)^n (1 + |\xi|) B_3^n$. Together with the previous estimate on $D_j Q_n$, we thus have

$$\|R_{n,j}\|, \|D_j Q_n\| \leq B_4^n, \qquad j = 0, 1, \ldots, m - 1, \qquad (2.14)$$

where we can choose $B_4 = 16 \max(1, |1/\xi|)(1 + |\xi|) B_3$, which agrees with our conventions on dependencies of the numbers B_t.

We can now give the required upper bound for $|D_j F_n(u, v)|$, when u, v are excellent approximations as in (2.12); this last fact also implies

$|u|$, $|v| \leqslant 1 + |\xi|$, whence we have:

$$|D_j F_n(u, v)| = |(u - \xi)^{m-j} R_{n,j}(u) + (v - \xi) D_j Q_n(u)|$$
$$\leqslant |u - \xi|^{m-j} (n+1) \|R_{n,j}\| (1+|\xi|)^n + |v-\xi| (n+1) \|D_j Q_n\| (1+|\xi|)^n$$
$$\leqslant (|u - \xi|^{m-j} + |v - \xi|)(n + 1)(1 + |\xi|)^n B_4^n$$
$$\leqslant (q^{-\mu(m-j)} + s^{-\mu}) B_5^n, \tag{2.15}$$

where we may choose $B_5 = 2(1 + |\xi|) B_4$.

Remark 2.8. In these calculations we needed to estimate the norm of the factor $R_{n,j}(X)$ of $D_j F_n(X, \xi)$, in terms of the norm of $D_j F_n(X, \xi)$. We have done this directly by the series expansion, but we could have used the so-called *Gelfond's inequality*, which says roughly that $\|fg\| \gg \|f\| \|g\|$, where the implied constant depends only on $\deg(fg)$ (one may take it $2^{-\deg(fg)}$). See the exercises below, next chapter (section on Mahler's measure) and the notes for references.

Exercise 2.15. Prove the formula $R_{n,j}(X) = \sum_{l=0}^{j} \binom{m}{l}(X - \xi)^{j-l} D_{j-l} R_n(X)$.

Exercise 2.16. Let $F(x) = x^d + a_1 x^{d-1} + \ldots + a_d = (x - \xi_1) \cdots (x - \xi_d) \in \mathbb{C}[x]$. Also, set $a_0 := 1$, $A = A_F := \max_{i=0}^{d}(|a_i|)$ ($= \|F\|$ in the above notation) and $B = B_F := 1 + |a_1| + \ldots + |a_d|$.

(i) Prove that $A \leq B \leq \prod_{i=1}^{d}(1 + |\xi_i|)$.
(ii) Let $0 \leq r \leq d$. Prove that there exists a point z_0 on the unit circle such that $\prod_{i=1}^{r} |z_0 - \xi_i| \geq 1$. (Hint: consider for instance $\int_0^1 \prod_{i=1}^{r}(e^{2\pi i\theta} - \xi_i) e^{2\pi \sqrt{-1}\theta} d\theta$.)

Suppose now that $\xi_1, \ldots, \xi_r$ are exactly those ξ_i which have absolute value ≤ 2.

(iii) Prove that for each z on the unit circle we have

$$\prod_{i=r+1}^{d} |z - \xi_i| \geq 3^{r-d} \prod_{i=r+1}^{d}(1 + |\xi_i|).$$

(iv) Prove that, for z_0 as in (ii), $(d+1)A \geq B \geq |F(z_0)| \geq 3^{-d} \prod_{i=1}^{d}(1+|\xi_i|)$.

Exercise 2.17. Using *e.g.* the results of the previous exercise, prove (a version of) Gelfond's inequality: *For complex polynomials f, g of degrees m, n we have, for $d = m + n$, $(d + 1)A_f A_g \geq A_{fg} \geq 3^{-d}(d + 1)^{-2} A_f A_g$.*

2.4.4. Lower bound for $|D_i F_n(u, v)|$.

This step is very easy. Note that $D_i F_n(X, Y)$ has integral coefficients and degree $\leq n - i$ in X and ≤ 1 in Y. Hence $D_i F_n(u, v)$ is a rational number whose denominator is $\leq q^{n-i} s$. We deduce that

Either $D_i F_n(u, v) = 0$ or $|D_i F_n(u, v)| \geq (q^{n-i} s)^{-1}$ (2.16)

2.4.5. An upper bound for the multiplicity at (u, v)

In order to benefit from (2.16) we now deal with the more delicate Step 4 (non vanishing); we estimate the least value of j such that $D_j F_n(u, v) \neq 0$. Let then h be this value (possibly 0 or ∞) so we have $D_h F_n(u, v) \neq 0$ and $D_j F_n(u, v) = 0$ for every $j = 0, 1, \ldots, h - 1$, i.e.

$$P_n^{(j)}(u) + v Q_n^{(j)}(u) = 0, \qquad j = 0, 1, \ldots, h - 1.$$

Eliminating v from any pair of these equations we obtain

$$\left(P_n^{(j)} Q_n^{(i)} - Q_n^{(j)} P_n^{(i)} \right)(u) = 0, \qquad i, j = 0, 1, \ldots, h - 1. \tag{2.17}$$

For $i = 0$, $j = 1$ this says that the so-called *Wronskian* $W = W_{P_n, Q_n} :=$ $P_n Q_n' - P_n' Q_n \in \mathbb{Z}[X]$ of P_n and Q_n vanishes at $X = u$. More generally, the Leibnitz rule $D^j(AB) = \sum_{i=0}^{j} \binom{j}{i} A^{(j-i)} B^{(i)}$ yields

$$W^{(j)}(u) = \left(P_n Q_n' - Q_n P_n' \right)^{(j)}(u)$$

$$= \sum_{i=0}^{j} \binom{j}{i} \left(P_n^{(j-i)} Q_n^{(i+1)} - Q_n^{(j-i)} P_n^{(i+1)} \right).$$

So (2.17) in fact implies that

$$W^{(j)}(u) = 0, \qquad j = 0, 1 \ldots, h - 2. \tag{2.18}$$

To exploit this important information (which is empty for $h < 2$) we first need to show that W does not vanish identically. Now, it is a general fact that $W_{P,Q} = 0$ if and only if P and Q are linearly dependent polynomials (over the constant field of the derivation): in fact, this equivalence is clear if $Q = 0$, whereas if $Q \neq 0$, P and Q are linearly dependent if and only if P/Q is a constant, i.e. if and only if its derivative, which is just W/Q^2, is zero.

Note now that our actual polynomials P_n, Q_n are not linearly dependent. In fact, assume the contrary; then $F_n(X, \xi) = P_n(X) + \xi Q_n(X)$ would be a constant multiple of P_n or Q_n. On the other hand $F_n(X, \xi)$ is nonzero (because $\xi \notin \mathbb{Q}$ and $P_n, Q_n \in \mathbb{Q}[X]$ are not both zero) and by construction has a zero of multiplicity at least m at $X = \xi$. But since we are assuming that P_n, Q_n are linearly dependent, this would be true also for both P_n and Q_n. Now, these polynomials have rational coefficients, so they would have a zero of multiplicity $\geq m$ also at each of the d conjugates of ξ over $\mathbb{Q}$. But by (2.10) we have $md = (2n + 2)(1 - \lambda)$ and since we have chosen $\lambda < 1/2$ we have $md > n + 1$, forcing P_n, Q_n to vanish, since both have degree $\leq n$. This is a contradiction and therefore P_n, Q_n are linearly independent and $W \neq 0$.

To conclude the argument recall (2.18), which says that $(X - u)^{h-1} = \frac{1}{q^{h-1}}(qX - p)^{h-1}$ divides $W(X)$. Now, $W(X)$ has integral coefficients and $(qX - p)^{h-1}$ is a primitive polynomial (because p, q are supposed to be coprime); hence by Gauss Lemma $(qX - p)^{h-1}$ divides $W(X)$ in $\mathbb{Z}[X]$. This implies that q^{h-1} divides the leading coefficient of W so in particular $\|W\| \geq q^{h-1}$ because W is not identically zero.

On the other hand, by (2.11) and the properties of the norm, we have

$$\|W\| \leq \|P_n Q'_n\| + \|Q_n P'_n\| \leq 2n(n+1)B_3^{2n} \leq B_6^n,$$

where we may choose $B_6 = 16B_3^2$. Hence we deduce

$$h \leq 1 + \frac{n \log B_6}{\log q}. \tag{2.19}$$

Remark 2.9. (An alternative approach) Note that this bound for h is obtained by arithmetical considerations and heavily depends on the magnitude of q. As mentioned above, another approach relying this time only on the involved degrees was found by Dyson. Such argument is entirely algebraic, and it eliminates the dependence on q from this piece of the proof; it was realized by Bombieri (see [15] and [16]) that this fact is very important in effectivity questions. Let us see a simple version of this approach for the present context. It is inspired by notes of D. Masser, as quoted in [16].

Recall that by our very construction we have $j! D_j F_n(\xi, \xi) = P_n^{(j)}(\xi) + \xi Q_n^{(j)}(\xi) = 0$ for $j = 0, \ldots, m-1$. Taking any pair of these equations and eliminating as in the above arguments, we find $W^{(j)}(\xi) = 0$ for $j = 0, \ldots, m - 2$. Also, since W has rational coefficients the same vanishing must occur at each of the d conjugates of ξ over $\mathbb{Q}$.

At this point, collecting our informations on $W \in \mathbb{Z}[X]$, we conclude that: (i) W does not vanish identically; (ii) it has degree $\leq 2n - 1$; (iii) it has a zero of order $\geq h - 1$ at $u \in \mathbb{Q}$; (iv) it has a zero of order $\geq m - 1$ at each conjugate of ξ. Note that since u is rational, u is different from each of such conjugates. Therefore we infer that $d(m - 1) + (h - 1) \leq 2n - 1$, i.e., $h \leq 2n - dm + d$. Recalling $dm = (2n + 2)(1 - \lambda) \geq 2n(1 - \lambda)$ we thus find

$$h \leq 2\lambda n + d. \tag{2.20}$$

After the proof we shall observe in Remark 2.10 how this may be used in place of (2.19) to conclude.

Exercise 2.18. For a field k and for rational functions $P_1, \ldots, P_h \in k(X)$ (or more generally for elements of a differential field) define the Wronskian $W_{P_1,\ldots,P_h} = \det(P_i^{(j-1)})_{i,j=1,\ldots,h}$. Here $P^{(j)}$ denotes the usual j-th derivative. Prove that W vanishes identically if and only if the P_i are linearly dependent over the constant field of the derivation (which is k in characteristic 0). (Hint: reduce to the case $P_1 = 1$ and use induction on h.)

2.4.6. Conclusions

To keep the contributions of u, v in the upper bound (2.15) 'nearly' equal we impose that $q^m \approx s$. More precisely, recall that m was restricted to lie in a certain arithmetical progression depending only on λ and d. We let τ be the difference of such progression and we define m as the largest integer in the progression such that $q^m \leq s$. Hence, m will satisfy

$$\frac{\log s}{\log q} - \tau \leq m \leq \frac{\log s}{\log q}. \tag{2.21}$$

We now determine n from equation (2.10); we note that n will be actually an integer since m lies in the said progression. Now, recall that $D_h F_n(u, v) \neq 0$. Then, by (2.15),

$$|D_h F_n(u, v)| \leq (q^{-\mu(m-h)} + s^{-\mu}) B_5^n \leq 2q^{-\mu(m-h)} B_5^n.$$

On the other hand, comparing with (2.16) yields $(q^{n-h} s)^{-1} \leq 2q^{-\mu(m-h)} B_5^n$, i.e., taking logarithms,

$$\mu m - \mu h + h - n \leq \frac{\log s}{\log q} + \frac{\log 2 + n \log B_5}{\log q} \tag{2.22}$$
$$\leq m + \tau + \frac{\log 2 + n \log B_5}{\log q},$$

where the last inequality follows from (2.21). Inserting the estimate (2.19) for h this gives

$$(\mu - 1)m - n \leq (\mu - 1)h + \tau + \frac{\log 2 + n \log B_5}{\log q} \tag{2.23}$$
$$\leq \mu \left(1 + \frac{n \log B_6}{\log q}\right) + \tau + \frac{\log 2 + n \log B_5}{\log q}.$$

Now, we have $\mu = 1 + (d/2) + \varepsilon$ while m is given by (2.10), so $m > 2n(1 - \lambda)/d$. Hence, (2.23) yields

$$\left(\left(1 + \frac{2\varepsilon}{d}\right)(1 - \lambda) - 1\right) n \leq \mu \left(1 + \frac{n \log B_6}{\log q}\right) + \tau + \frac{\log 2 + n \log B_5}{\log q},$$

whence, recalling (2.8), i.e. $\delta := ((1 + \frac{2\varepsilon}{d})(1 - \lambda) - 1)$,

$$\delta n \leq \mu(1 + \frac{n \log B_6}{\log q}) + \tau + \frac{\log 2 + n \log B_5}{\log q}.$$

Recall we had chosen λ so small that $\delta > 0$. Let us now suppose that

$$\log q > \mu \frac{4\log(2B_5B_6)}{\delta}. \tag{2.24}$$

This is certainly possible if there are infinitely many excellent approximations, since B_5, B_6, δ depend only of ξ, ε, λ (so ultimately only on ξ, ε). Then the last displayed equation yields

$$\delta n \le \mu + \frac{\delta}{4}n + \tau + \frac{\delta}{4}n \le \mu + \tau + \frac{\delta}{2}n,$$

whence

$$n \le \frac{2(\mu + \tau)}{\delta}.$$

However comparison with (2.21) produces now a bound for s, *i.e.*

$$\frac{\log s}{\log q} \le \tau + m \le \tau + n \le \tau + \frac{2(\mu + \tau)}{\delta}. \tag{2.25}$$

In other words, if

$$\log s > \log q \left(\tau + \frac{2(\mu + \tau)}{\delta} \right) \tag{2.26}$$

we obtain a contradiction. In particular this proves the finiteness of the set of excellent approximations to ξ. □

Remark 2.10. (Conclusion with the alternative approach) Let us see how the algebraic approach for Step 4 outlined in Remark 2.9 may equally lead to the sought conclusion. Inserting in (2.22) the 'alternative' estimate (2.20) in place of (2.19) we arrive at

$$(\mu - 1)m - n \le (\mu - 1)h + \tau + \frac{\log 2 + n \log B_5}{\log q}$$

$$\le \mu(2\lambda n + d) + \tau + \frac{\log 2 + n \log B_5}{\log q}.$$

in place of (2.23). Using again $\mu = 1 + (d/2) + \varepsilon$ and $m > 2n(1 - \lambda)/d$ we obtain

$$\left(\left(1 + \frac{2\varepsilon}{d}\right)(1 - \lambda) - 1 - 2\mu\lambda \right) n \le \tau + \frac{\log 2 + n \log B_5}{\log q}.$$

We may now plainly suppose that $\lambda > 0$ has been chosen so small that $\delta' := (1 + \frac{2\varepsilon}{d})(1 - \lambda) - 1 - 2\mu\lambda > 0$, and now the proof can be concluded as before, by arguing with a sufficiently large q, e.g. $\log q > 2\frac{\log(2B_5)}{\delta'}$.

Remark 2.11. (Estimating the number of solutions) Let us note how the above proof allows us to estimate effectively the number of excellent approximations (although it does not allow to find them! See especially next remark).

This is not completely obvious, and it involves again a gap principle. What we have proved above is that *if p/q is excellent and if q is so large to satisfy* (2.24) *then any other excellent approximation r/s with $s > q$ satisfies* (2.25). In other words, *there exist numbers C_1, C_2 computable in terms of the original data (i.e. ξ, μ) such that if $q > C_1$ and if $s > q$ then $\log s / \log q < C_2$.* Now, the problem is that we do not know anything about the possible solutions p/q satisfying $q > C_1$. We have no means to estimate the minimal such $q > C_1$ and so, although we know that sufficiently large solutions do not exist, this is not enough to control *a priori* the 'intermediate' ones, whose number could grow with q. An escape to this difficulty comes from the gap principle, as explained at the beginning of the sketch of Thue's ideas. We have seen that for distinct excellent approximations $p/q, r/s$, with $s > q$, in fact we must have $s > q^{\mu-1}/2$; hence, if q is large enough, since $\mu - 1 > 1$, we have $\log s / \log q > c_3$, where $c_3 > 1$ is computable and dependent only on ξ, μ. Applying this inequality to consecutive pairs of excellent approximations p_i/q_i with $q_i > C_1$, ordered according to their denominators, we find that $\log q_{i+1} > c_3 \log q_i$, concluding that the number of intermediate solutions is at most $1 + \log C_2 / \log c_3$, providing the required bound.

In the final chapter we shall prove a sharp estimate of the number of solutions of a Thue Equation, with rather different, though not unrelated, methods; see especially Corollary 5.11.

Remark 2.12. (Effectiveness) We have already noted that this proof-pattern leads to ineffectivity, since we need at least *two* excellent approximations (moreover with sufficiently large denominators) to obtain the sought contradiction. However, in principle, we would have effectiveness if we were in possession of an excellent approximation p/q satisfying the lower bound (2.24) above; in fact we have seen that this leads to the explicit upper bound (2.25) (in terms of q) for the denominator of any other excellent approximation r/s. Unfortunately, the required magnitude for q is so large that no example is known when this is verified, and it is indeed very likely that no example actually exists. However, by means of this principle and with an explicit construction of the polynomials P_n, Q_n (using Padé approximations for $\sqrt[3]{1+x}$), A. Baker [B] proved for instance that we have the effective lower bound $|\sqrt[3]{2} - \frac{p}{q}| \geqslant 10^{-6} q^{-2.955}$ for every rational p/q. By means of completely different results coming from his advances in techniques of transcendental number theory for linear forms in logarithms, A. Baker also obtained around 1970 fully general effective bounds for the solutions of the Thue equation (see [7]). Reversing the above link 'equations-approximations', this yielded an effective exponent of approximation improving on Liouville, *i.e.* $< d$. Later Bombieri obtained some effective results by an entirely different method, nearer to Thue's one, but using Dyson's algebraic approach for Step 4 (nonvanishing) instead of the arguments of Thue and Siegel, depending on the magnitude of q; see [12]. Still later Bombieri introduced other important devices to obtain a full effective treatment (see [15]). We finally remark that the effective *exponents* obtained by Baker and Bombieri are just a little smaller than Liouville's and very far from Roth's and even from Thue's (moreover they also depend on the individual number to be approximated, not just on its degree). In this superficial discussion we omit reference to subsequent technical improvements relying on the said principles.

Another description of the method

A posteriori we can read the steps of the method still in another way as follows:

Step 1: Interpolation. We construct F as above, *i.e.* $F_j(\xi, \xi) = 0$ for several j.

Step 2: Extrapolation. Comparing Upper and Lower bounds we get $F_j(u, v) = 0$ for several j.

Step 3: Zero estimates. There are too many zeros (either for reasons of height, as in the first approach, or for reasons of degree, as in the alternative approach). So we have a contradiction.

This pattern resembles many proofs in transcendental number theory (see [7]).

Supplements to Chapter 2

Finiteness of integral points on certain curves

We have noted how Thue's and Roth's results in Diophantine Approximations lead to finiteness results for the integral points on curves defined by equations $f(X, Y) = g(X, Y)$, where f is homogeneous, say without multiple factors, and g has 'small' degree with respect to f. (Recall *e.g.* Theorem 2.1 and Exercise 2.11.) General as they are, such results do not cover the case of arbitrary (plane) curves. This general case was investigated by Siegel, who gave a complete geometric classification of affine curves which may have infinitely many integral points.

Before recalling a statement of Siegel's result, we add a little precision and facts concerning the present notion of integral points and the associated relevant concept of divisor at infinity. We speak of 'integral point' for an (affine) algebraic variety $\mathcal{X}$ only when $\mathcal{X}$ is embedded in some affine space $\mathbb{A}^n$; in this case we just mean a point of $\mathcal{X}$ with coordinates in $\mathbb{Z}$.[4] We denote by $\mathcal{X}(\mathbb{Z})$ the set of integral points. Also, we denote by $\tilde{\mathcal{X}}$ the projective closure of $\mathcal{X}$ in $\mathbb{P}_n$ and we define the *divisor at infinity* as the sum of the irreducible components of $\tilde{\mathcal{X}} \setminus \mathcal{X}$. Thus, when $\mathcal{X} = \mathcal{C}$ is a curve the divisor at infinity is a (formal) sum of distinct points.

Siegel's Theorem.* *If an affine irreducible curve $\mathcal{C}$ has infinitely many integral points, then it has genus zero and at most two points at infinity.*

---

[4] See the Notes for more general notions. Also, we stress that in these first two chapters we are working essentially only over $\mathbb{Z}$ and 'integral points' are meant in this classical sense. In later chapters we shall work with integral points over rings of S-integers in number fields, as is customary (and advantageous) in modern times.

Remark 2.13.

  (i) Note that Pell Equation represents an affine hyperbola with two points at infinity. Thus the result is in a way best-possible.

 (ii) It is important to note that by taking an integral ring extension we may replace the curve with a normalization of it; in doing this the number of integral points will not decrease, but the number of points at infinity can possibly increase (if the projective curve $\tilde{C}$ is singular there). Thus the number 'two' may be referred to as the number of points at infinity in a desingularization.

(iii) We further remark that it does not matter whether 'irreducible' is referred to $\mathbb{Q}$ or any extension field. In fact, suppose that C is irreducible over $\mathbb{Q}$. Since C has infinitely many points over $\overline{\mathbb{Q}}$ we may assume that the components are defined over $\overline{\mathbb{Q}}$ and conjugate over $\mathbb{Q}$. Then any rational point on C, being fixed by conjugation, must lie in the intersection of the components. Thus, if there are two or more components over $\overline{\mathbb{Q}}$, there are only finitely many rational points (which moreover can be computed).

This theorem of Siegel actually holds in general form for S-integers of any number field. The known proofs require some geometrical tools in addition to results in Diophantine Approximation, and lie beyond the scope of the present lecture notes. (See *e.g.* [77] for a modern presentation of Siegel's proof.) However we shall now see some examples of curves, not of Thue's type, for which nevertheless Thue's Theorem 2.1 leads in an elementary way to a finiteness proof for the integral points.

We start with the so-called *Double Pell Equation*, namely

$$Y^2 - aX^2 = r, \qquad Z^2 - bX^2 = s, \tag{2.27}$$

where a, b, r, s are fixed nonzero integers, to be solved simultaneously in integers. We also impose the harmless condition that $as \neq br$ (if $as = br$ the problem boils down to the case of a single equation). We have seen in Section 1.5 that for suitable values of a, b, r, s one (or both) of these two Pell's type equations may have individually infinitely many integral solutions; namely, it may happen that *e.g.* $ax^2 + r$ is a square infinitely often. However, as we shall now prove, *for any a, b, r, s such that $as \neq br$, it may happen only finitely many times that $ax^2 + r$ **and** $bx^2 + s$ are simultaneously squares*. In other words, we have

Theorem 2.8. *For any fixed nonzero integers a, b, r, s with $as \neq br$ the simultaneous equations (2.27) have only finitely many integer solutions.*

Before proving this, let us pause for a few geometrical considerations (where we assume some simple facts from the geometry of curves of genus 1; see [83]).

It is easy to see that if $as \neq br$ and $abrs \neq 0$ the equations (2.27) define a nonsingular irreducible curve C in affine 3-space $\mathbb{A}^3$, which has four points A_1, A_2, A_3, A_4 at infinity (*i.e.* in its projective closure $\tilde{C}$ in $\mathbb{P}_3$), and genus 1. One may also check that (for $i \neq j$) $A_i - A_j$ has order 2 in the Picard group of $\tilde{C}$, *i.e.* $2(A_i - A_j)$ is the divisor of some rational function on C.

As to the integral points on $\mathcal{C}$, the above stated Siegel's theorem directly leads to their finiteness, actually in two ways (genus or number of points at infinity). On the other hand, **it is not possible by a purely algebraic argument to recover this finiteness result from Thue's theorems proved above**. By this we mean, roughly, that this system of equations (2.27) cannot be *directly* reduced to a Thue Equation by any type of polynomial substitution. Let us prove this fact: the existence of such a substitution would yield in particular a regular non-constant map ψ from $\mathcal{C}$ to some Thue's curve $\mathcal{C}'$: $f(X, Y) = c$, (f homogeneous, without multiple factors, of degree ≥ 3, c nonzero constant). Now, it is easy to see (exercise below) that such a curve is nonsingular (even at infinity) and irreducible, so by a well-known formula it has genus $(d-1)(d-2)/2$. Hence, by the Hurwitz genus formula, if there is a non-constant rational map ψ from $\mathcal{C}$ to $\mathcal{C}'$ we must have $d = 3$ and the map must be unramified, even at ∞. If this is the case the Thue curve has three points B_1, B_2, B_3 at infinity. Their inverse images by ψ must lie among the four points at infinity of $\mathcal{C}$, because ψ is supposed to be regular on $\mathcal{C}$. Since ψ is unramified we have $\#\psi^{-1}(P) = \deg \psi$ for each P, hence $\deg \psi = 1$. Therefore ψ extends to an isomorphism of the projective closures of our curves, hence also the differences $B_i - B_j$ must have order two in the Picard group, namely $2(B_i - B_j)$ would be the divisor of a function on $\mathcal{C}'$. Now we see that this is not true. It is readily checked that the tangent to $\mathcal{C}'$ in B_i has a triple intersection with $\mathcal{C}'$ in B_i. Also, the line at infinity intersects $\mathcal{C}'$ in B_1, B_2, B_3. All of this leads to the linear equivalence $3B_i \sim B_1 + B_2 + B_3$, whence $3(B_i - B_j) \sim 0$ for all i, j. Now, if we had $2B_2 \sim 2B_1$, we would obtain $B_2 \sim B_1$. This is however impossible since the curve $\mathcal{C}'$ has genus 1 and $B_1 \neq B_2$.

In spite of this negative conclusion, we shall reduce the problem to a Thue equation, by means of an elementary procedure, which at bottom amounts to taking an unramified cover.

Proof of Theorem 2.8. The equations (2.27) yield (as $- br)X^2 - sY^2 + rZ^2 = 0$. We view this as defining a nonsingular conic $\mathcal{Q}$ in $\mathbb{P}_2$. Each integral solution to our system produces a rational point on $\mathcal{Q}$, so we may certainly suppose that $\mathcal{Q}$ has a rational point P_0. We may then start by using the method of projection, as in Remark 1.7, to parametrize the rational points on $\mathcal{Q}$: for instance, if $X(P_0) \neq 0$, we may project $\mathcal{Q}$ from P_0 to the line $\mathcal{L}_0 : X = 0$. Explicitly, if $R = (0 : t : u)$ is a point on $\mathcal{L}_0$, the line $\mathcal{L} := RP_0$ meets $\mathcal{Q}$ in P_0 and another point $P := \phi(R)$. On the one hand P will be projected to R from P_0. On the other hand, to recover P from R, observe that $\mathcal{L}$ will be defined by an equation $lX + uY - tZ = 0$, where $l = l(t, u)$ is a certain linear form in t, u, with coefficients depending on P_0. Intersecting $\mathcal{L}$ with $\mathcal{Q}$ and noting that $\mathcal{L} \cap \mathcal{Q} = \{P_0, P\}$, expresses $P = \phi(R)$ in the form $P = (\phi_1(t, u) : \phi_2(t, u) : \phi_3(t, u))$ for suitable quadratic forms ϕ_1, ϕ_2, ϕ_3 with rational coefficients. Also, the ϕ_i have no common factor (over $\mathbb{C}$): in fact, in the first place a possible common factor would be linear (if it had degree 2 the ϕ_i would divide each other and ϕ would be constant). But then on dividing out this factor we see that ϕ would be defined by three coprime linear forms, which parametrize a line; so $\mathcal{Q}$ would be reducible, a contradiction. Since the ϕ_i are defined up to a constant factor, we may also assume that their coefficients are integers.

Take now an integral point $(x_1, y_1, z_1) \in \mathcal{C}$. This produces a rational point $P = (x_1 : y_1 : z_1)$ on $\mathcal{Q}$ and thus a rational point $R := (0 : t_1 : u_1) \in \mathcal{L}_0$, where we may assume that t_1, u_1 are coprime integers. By the above we have $(x_1 : y_1 : z_1) = (\phi_1(t_1, u_1) : \phi_2(t_1, u_1) : \phi_3(t_1, u_1))$. Now, the $\gcd(x_1, y_1, z_1)$ divides $\gcd(r, s)$ and so has a finite number of possibilities. Let $d = \gcd(\phi_i(t_1, u_1))$. Since the ϕ_i have no common (identical) factor and since t_1, u_1 are coprime, it is easily seen that d divides a fixed integer. (Look e.g. at the resultant, of any two of the ϕ_i, with respect to both t, u; see also Exercise 2.2.) So also this gcd has only finitely many possibilities (that is, independently of x_1, y_1, z_1, t_1, u_1). But then we must have equations $x_1 = \rho\phi_1(t_1, u_1), y_1 = \rho\phi_2(t_1, u_1), z_1 = \rho\phi_3(t_1, u_1)$, where ρ takes only finitely many rational values as x_1, y_1, z_1 vary; hence for our purposes we may suppose that ρ is fixed. Substituting these formulas into the first of equations (2.27) we obtain a Thue equation $F(t_1, u_1) = r$ where $F = \rho^2(\phi_2^2 - a\phi_1^2)$ is a homogeneous polynomial of degree 4 with rational coefficients. We shall now prove that F has no multiple factors (even over $\mathbb{C}$). This may be checked directly, but we can argue as follows: put $\lambda = t/u$. If F has a square factor then $f(\lambda) := F/u^4$ is the product of a polynomial of degree ≤ 2 and the square of a linear one. Then the curve $v^2 = f(\lambda)$ has a rational component. But this component is birational to $\mathcal{C}$: in fact, the above calculations show that we may parametrize $\mathcal{C}$ by $x = \rho\sqrt{r}\phi_1(\lambda, 1)/v, y = \rho\sqrt{r}\phi_2(\lambda, 1)/v, z = \rho\sqrt{r}\phi_3(\lambda, 1)/v$. Since $\mathcal{C}$ has genus 1 we have a contradiction.

Hence we may apply Thue's Theorem 2.1 (here it would suffice that F be not a square) and obtain our finiteness conclusion.

Exercise 2.19. Prove directly that F is not a square η^2, as follows: then η would divide $W_\partial := (\partial\phi_2)\phi_1 - (\partial\phi_1)\phi_2$, where ∂ is either derivation with respect to t or to u. But, from the fact that $(\phi_1 : \phi_2 : \phi_3)$ lies on $\mathcal{Q}$, show that ϕ_3 divides W_∂ as well and that W_∂ would then vanish, so ϕ_1/ϕ_2 would be constant, which is impossible.

Remark 2.14. Here is a viewpoint on this proof: Let $\mathcal{C}'$ be the Thue curve of degree 4 (and genus 3) defined in $\mathbb{A}^2$ by $\phi_2(t, u)^2 - a\phi_1(t, u)^2 = r$. We have a map $\mathcal{C}' \to \mathcal{C}$: $(t, u) \mapsto (\phi_1(t, u), \phi_2(t, u), \phi_3(t, u))$. That this map goes indeed to $\mathcal{C}$ follows from the fact that $(\phi_1(t, u) : \phi_2(t, u) : \phi_3(t, u)) \in \mathcal{Q}$ and clearly it maps finite points to finite ones. This map corresponds to a function field inclusion $\mathbb{C}(\mathcal{C}) \subset \mathbb{C}(\mathcal{C}')$ and we may also describe it as follows. Writing $\lambda = t/u$, the above mentioned projection shows that λ parametrizes $\mathcal{Q}$, so $\mathbb{C}(\mathcal{Q}) = \mathbb{C}(x/z, y/z) = \mathbb{C}(\lambda)$. We may write $\phi_2(t, u)^2 - a\phi_1(t, u)^2 = u^4 f(\lambda)$, say, where f has degree 4 and has no repeated roots. This shows that $\mathbb{C}(\mathcal{C}') = \mathbb{C}(\lambda, f(\lambda)^{1/4})$. Also, we have $\mathbb{C}(\mathcal{C}) = \mathbb{C}(x, y, z) = \mathbb{C}(\lambda, y) = \mathbb{C}(\lambda, \sqrt{f(\lambda)})$. From all of this we see that the above map makes $\mathcal{C}'$ a cover of $\mathcal{C}$ which is unramified except at the points corresponding to $f(\lambda) = 0$; in turn, it may be checked that these points are the points at infinity on $\mathcal{C}$ and $\mathcal{C}'$. Now, a celebrated theorem of Chevalley and Weil (see [17] or [77]) says (roughly speaking) that the integral points on a curve (or variety) lift to integral points on any cover of the curve which is unramified except (perhaps) at infinity. In our context this means that the integral points on $\mathcal{C}$ lift to integral points of the Thue curve $\mathcal{C}'$, which falls under the statements which we have previously proved.

The Chevalley-Weil theorem is not merely a geometrical theorem, it combines arithmetic and geometry and may be considered as an arithmetic analogue of the lifting of maps in homotopy theory. Of course in our elementary proofs we did not meet this theorem explicitly, but the underlying principle boils down to it. We had observed how a purely algebraic argument could not reduce the proof to a Thue equation; in fact in our proof we met two arithmetical points: the first occurs when we infer that the gcd of the $\phi_i(t_1, u_1)$ is bounded; the second one occurs when we infer that if a projective point $(x_1 : y_1 : z_1) = (\phi_1(t_1, u_1) : \phi_2(t_1, u_1) : \phi_3(t_1, u_1))$ is expressed in two ways with integral coordinates with bounded gcd, then the corresponding coordinates are 'essentially' equal (*i.e.*, equal up to a factor taken from a finite set).

Corollary 2.9. *If a, b, r, s are integers with $abrs(as - br) \neq 0$, the equation $Y^2 = (aX^2 + r)(bX^2 + s)$ has only finitely many integer solutions.*

Proof. This follows easily from 2.8. In fact, let (x_0, y_0) be an integral point. Note that for varying integer x_0, the integer $d := \gcd(ax_0^2 + r, bx_0^2 + s)$ can take only finitely many values, because it must divide $as - br \neq 0$. So in proving the assertion we may assume that d is fixed. Since $(ax_0^2 + r)(bx_0^2 + s) = y_0^2$ is a square we must have $ax_0^2 + r = edu^2$, $bx_0^2 + s = edv^2$, $u, v \in \mathbb{Z}$, where $e = \pm 1$. Then we have $(du)^2 = edax_0^2 + edr$, $(dv)^2 = edax_0^2 + eds$. For each e, d these simultaneous equations have only finitely many integer solutions by the theorem, concluding the argument. $\square$

Also in this case the proof amounts to take an unramified cover, this time of the curve $Y^2 = (aX^2 + r)(bX^2 + s)$ by the curve defined by the system (2.27). The corresponding arithmetical point consists in the implication that *if the product of two integers m, n with $\gcd(m, n) = d$ is a square, then each of the integers is $\pm d$ times a square.*

A completely similar argument, which we leave to the interested reader, proves:

Corollary 2.10. *Let $f(X) \in \mathbb{Z}[X]$ have at least three rational simple roots. Then the equation $Y^2 = f(X)$ has only a finite number of integer solutions.*

This result, which asserts the finiteness of integral points on affine hyperelliptic curves, may be extended on dropping the assumption that the mentioned roots are rational. It is of course a special case of Siegel's theorem, but may be reduced to diophantine approximation more easily than the general case, as was done by Siegel; see the next chapter for a proof depending on the S-unit Theorem 3.13, which in turn will be completely proved in the final chapter.

As to this S-unit Theorem, for the moment let us see one of the simplest (though highly nontrivial) examples of it, by proving

Corollary 2.11. *Let S be a finite set of prime numbers and let c be a nonzero integer. Then the equation $x - y = c$ has only finitely many solutions in integers x, y composed only of prime factors from S.*

Proof. Let $S = \{p_1, \ldots, p_s\}$. Then every integer composed only of primes from S may be clearly written in the shape $\pm p_1^{e_1} \cdots p_s^{e_s} t^3$, where t is an integer and $e_i \in \{0, 1, 2\}$: this follows on division by 3 of the exponents of the p_i appearing in the prime decomposition. Since the integer $\pm p_1^{e_1} \cdots p_s^{e_s}$ has only finitely many possibilities for fixed S, the equation in the statement gives rise to finitely many equations of the form $au^3 - bv^3 = c$ in the integer variables u, v. By Thue's theorem each of these equations has only finitely many solutions and the conclusion follows. $\square$

Once more we see implicit use of an unramified cover, this time $x \mapsto x^3$; this is unramified outside $\{0, \infty\}$ and the present integral points 'essentially' lift to integral points (they are cubes up to finitely many factors). We shall see this kind of example in more detail in the next chapter.

We conclude by mentioning that the present proofs are ineffective because they rely on ineffective proofs for Thue's Theorem. Baker's mentioned methods have produced effective proofs for such results; however the full Siegel's theorem still awaits a general effective proof.

Exercise 2.20. Prove that if $as \neq br$ and $abrs \neq 0$ the equations (2.27) define an irreducible nonsingular curve C of genus 1 in affine 3-space $\mathbb{A}^3$, with four points at infinity.

Exercise 2.21. Notation as in the previous exercise, prove that $2(A_i - A_j)$ is the divisor of some rational function on C. (Hint: Use quotients of linear forms in $1, x, y, z$.)

Exercise 2.22. Let $f \in \mathbb{C}[X, Y]$ be homogeneous without multiple factors and let $c \in \mathbb{C}^*$. Prove that the curve defined in $\mathbb{P}_2$ by $f(X, Y) = cZ^{\deg f}$ is nonsingular, and in particular irreducible.

Exercise 2.23. Prove Corollary 2.10. (Hint: if $a_i X - b_i$, $i = 1, 2, 3$ are non-proportional simple linear factors of $f(X)$ with integer coefficients, prove that for an integer solution (x, y) we have $a_i x - b_i = d_i u_i^2$, where d_1, d_2, d_3 take only finitely many integer values. Eliminating x then one reduces to 2.8.)

Exercise 2.24. Prove in some elementary self-contained way that if a polynomial $f(X) \in \mathbb{Z}[X]$ takes square values for all integer values of X then $f(X)$ is the square of a polynomial in $\mathbb{Z}[X]$. (Hint: here several methods, more or less elementary and direct, are available. For instance one may observe that on the assumptions each prime p dividing a value $f(n)$ is such that p^2 divides $f(n)$. But then p^2 divides also $f(n + p)$ and so p divides $f'(n)$... See also the exercises after Runge's theorem below.)

Exercise 2.25. Let c be a nonzero integer. Prove that the greatest prime factor of $m^3 + c$ tends to infinity as $m \to \infty$. (Hint: imitate the proof of Corollary 2.11. The conclusion is true on replacing $m^3 + c$ by any polynomial with at least two distinct roots.)

Exercise 2.26. Find in a self-contained simple way all the integer solutions of $2^m - 3^n = \pm 1$. (Hint: consider for instance the parity of m, n by looking at suitable congruences.)

Exercise 2.27. Find all the integer solutions of $11^a = 1 + 2^b 5^c$. (Hint: observe that 5^{c-1} divides a and that a is even if $b > 1$.)

Exercise 2.28. Prove in a self-contained way that the only integer solution of the equation $2^a + 3^b = 5^c$ are $a = b = c = 1$ and $a = 4, b = c = 2$. (Hint: arguing modulo 4 one finds that b must be even if $a \geq 2$. Modulo 5 one finds a to be even and modulo 3 also c must be even, leading to an easy case. Then we may assume $a = 1$. Now, if $c \geq 2$, on looking modulo 25 we must have $b \equiv 3$ (mod 5). And finally one derives a contradiction modulo 11.)

Effective decision for an infinity of integral points in genus zero

We have recalled Siegel's Theorem, which predicts only finitely many integral points on an affine irreducible curve having either genus > 0 or more than two points at infinity (as in Remark 2.13, (i), it suffices that there are more than two in a desingularization). Here we briefly sketch an algorithm to establish whether there are infinitely many integral points on a given irreducible curve C over $\mathbb{Q}$, of genus 0. We leave a few easy details to fill in by the interested reader.

Suppose that C has infinitely many integral points $P_1, P_2, \ldots$. Then (*e.g.* by Lüroth Theorem) its function field over $\mathbb{Q}$ is $\mathbb{Q}(t)$. For a given curve, to find a suitable generator t only a single nonsingular rational point is needed, which may be sought effectively by known methods. Let us explain this: to start with, one may take a canonical divisor D on C, of the form $\operatorname{div}(dx)$, where x is a nonconstant function in $\mathbb{Q}(C)$. This divisor D is rational and has degree -2; by the Riemann-Roch theorem we may use this divisor to embed the curve as a conic in $\mathbb{P}_2$.[5] But now we may use the Supplements to Chapter 1 to check effectively the existence of a rational point on such a conic. If there is one such point P_0 we may parametrize C (which means that we may find t) by projecting from P_0 to a rational line, as in the proof of Theorem 2.8.

Coming back to integral points, observe that P_i is in particular a rational point on C so $t(P_i)$ is a rational number, which we write in lowest terms as p_i/q_i. Let $x(t) \in \mathbb{Q}(t)$ be a coordinate function on C assuming integral values at all the P_i. We may write $x(p_i/q_i) = f(p_i, q_i)/g(p_i, q_i)$, where f, g are coprime homogeneous polynomials in $\mathbb{Z}[X, Y]$. The set of poles of x, which may be assumed to be precisely the set of points at infinity on C, correspond to the zeros of g in $\mathbb{P}_1$. By Exercise 2.2 the $\gcd(f(p_i, q_i), g(p_i, q_i))$ divides a fixed integer, so for an infinity of the P_i we must have $g(p_i, q_i) = c$ where c is a fixed nonzero integer, taken from a finite set which can be computed. By Thue's Theorem g must be a power of a linear or quadratic factor. (We see here without Siegel's Theorem that, in genus 0, there may be at most two points at infinity.) Now, for any given c, the theory of Chapter 1 gives an algorithm to decide whether such an equation has or not infinitely many integral solutions and to parametrize them in the former case.

Exercise 2.29. Fill in the details in the above proof.

A theorem of Runge

In 1887 Runge proved a theorem giving the finiteness of integral points on certain plane curves; an interesting feature was that when it applied it gave effective estimates for the magnitude of the solutions. We state it in the following form (see below for other equivalent formulations):

Theorem 2.12 (Runge's Theorem). *Let $\tilde{C}$ be an absolutely irreducible nonsingular projective curve defined over $\mathbb{Q}$, and let $x \in \mathbb{Q}(\tilde{C})$ have a pole Q such that $-\operatorname{ord}_Q(x)[\mathbb{Q}(Q) : \mathbb{Q}] < \deg(x)$. Then there are only finitely many rational points $P \in \tilde{C}(\mathbb{Q})$ such that $x(P) \in \mathbb{Z}$ (and they can be effectively found).*

---

[5] We take it for granted that these calculations with divisors may be performed effectively. A general algorithm has been given by Coates [27].

As will appear from the proofs below, it is a result of a less deep nature than Thue's and related ones. Also, assumptions and conclusions are not 'geometrical', *i.e.* they strongly depend on the field of definition (both of the points at infinity and of the sought solutions), a defect which does not pertain to Thue's or Siegel's theorems: we have stated them in the rational case, but they hold over any number field. Among the prototypes and most obvious cases of Runge's theorem are Thue Equations $Y^d - X^d = c$; now the finiteness follows (even for $d = 2$) from the fact that for an integral solution (x, y), $y - x$ must divide c. Other, less obvious, applications are to equations $Y^d = f(X)$ where the leading term of f is of the shape $(aX)^d$ for $a \in \mathbb{Z}$ (see the exercises below). On the other hand, even Thue's equations as simple as $X^3 - 2Y^3 = c$ completely escape from the theorem and from its proof-method.

Proof. Note that, since $x \in \mathbb{Q}(\mathcal{C})$, if R is a pole of x then all the conjugates of R over $\mathbb{Q}$ are also poles of x, with the same multiplicity. Hence the divisor of poles of x may be written in the form $(x)_\infty = -\sum_Q \mathrm{ord}_Q(x) \sum_{Q^\sigma \in \Omega_Q} Q^\sigma$, where Q runs over a system of non-conjugate poles and Q^σ runs through the set Ω_Q of conjugates of Q. The divisor $\sum_{Q^\sigma \in \Omega_Q} Q^\sigma$ is defined over $\mathbb{Q}$ and has degree $[\mathbb{Q}(Q) : \mathbb{Q}]$, whereas $\deg(x)_\infty = \deg(x)$. Hence the assumption implies that there are at least two nonconjugate poles, so the divisor of poles of x may be written as a sum $(x)_\infty = D_1 + D_2$ of two strictly positive divisors D_1, D_2 with disjoint supports and both defined over $\mathbb{Q}$. We thus see that Theorem 2.12 follows from the following (in fact equivalent) version:

Theorem 2.13 (Runge's Theorem II). *Let $\tilde{C}$ be an absolutely irreducible projective nonsingular curve defined over $\mathbb{Q}$, and let $x \in \mathbb{Q}(\tilde{C})$ be such that its pole divisor is a sum of two strictly positive divisors D_1, D_2 with disjoint supports and defined over $\mathbb{Q}$. Then there are only finitely many rational points $P \in \tilde{C}(\mathbb{Q})$ such that $x(P) \in \mathbb{Z}$ (and they can be effectively found).*

We go on by proving this second version. Let $P_i \in C(\mathbb{Q})$ run through an infinite sequence of distinct rational points of C such that $x(P_i) \in \mathbb{Z}$. Since $\tilde{C}(\mathbb{C})$ is compact, we may assume by going to a subsequence that P_i converges to some point $R \in \tilde{C}(\mathbb{C})$.

Now the proof can be very quickly completed by appealing to a (weak) form of the Riemann-Roch theorem (over $\mathbb{Q}$). In fact, by symmetry we may suppose that R does not belong to the support of D_1. Let then $y \in \mathbb{Q}(\mathcal{C})$ be a nonconstant function whose pole divisor is $\geq -ND_1$ for some large integer N; this certainly exists by Riemann-Roch. Since each pole of y is a pole of x we have that y is integral over $\mathbb{Q}[x]$ and by multiplying it by a nonzero integer we may assume that y is integral over $\mathbb{Z}[x]$. Now, on the one hand $y(P_i) \in \mathbb{Q}$ (because $P_i \in C(\mathbb{Q})$ and $y \in \mathbb{Q}(\mathcal{C})$) and on the other hand $y(P_i)$ is integral over $\mathbb{Z}$ (because $x(P_i) \in \mathbb{Z}$). Hence $y(P_i) \in \mathbb{Z}$. But since $P_i \to R$ and since $y(R)$ is finite, we have that $y(P_i)$ are bounded integers, and we obtain a contradiction because y is nonconstant and the P_i are distinct.

This argument may be easily shown to be effective, which concludes the proof. $\qquad\square$

To avoid the Riemann-Roch theorem, we can argue explicitly as follows (which is nearer to the original proofs): let $u \in \mathbb{Q}(\mathcal{C})$ be a local parameter at R and represent x as a Laurent series $x = \xi(u) \in k((u))$ where k is a field of definition for R. Note that we may assume that R is a pole of x and thus that it is an algebraic point. Let z be a primitive element for $\mathbb{Q}(\mathcal{C})/\mathbb{Q}(x)$, which we may assume to be integral over $\mathbb{Z}[x]$. We may expand also z as a Laurent series $z = \mu(u) \in k((u))$. Now we choose a large integer M and, putting $d := \deg(x) = [\mathbb{Q}(\mathcal{C}) : \mathbb{Q}(x)]$, we construct an auxiliary function of the shape

$$\phi = \phi_Q = A_0(x) + A_1(x)z + \ldots + A_{d-1}(x)z^{d-1}$$

where the A_i are polynomials over $\mathbb{Z}$ of degree $< M$; let us view their coefficients as undetermined integers $u_1, \ldots, u_{dM}$. The expansion of ϕ at R will be in $k((u))$ with pole of order $\leq Mm + O(1)$, where m is the multiplicity of R as a pole of x. We now impose that ϕ is regular and vanishes at R: this corresponds to a system of $Mm + O(1)$ linear equations in the u_j, over k, i.e. of $(Mm + O(1))[k : \mathbb{Q}]$ linear equations over $\mathbb{Q}$. These equations may be solved nontrivially over $\mathbb{Q}$ if $(Mm+O(1))[k : \mathbb{Q}] < Md$. Now, the assumption for version I (or the opening argument, if we deal with version II) yields $m[k : \mathbb{Q}] < d$ so indeed the inequality will be satisfied for large M. Evaluating ϕ at the P_i gives then a contradiction as in the above proof: the $\phi(P_i)$ are integers tending to $\phi(R) = 0$, so they must vanish for large i; so ϕ has to be identically 0, which implies $A_i = 0$ for all i, against the construction.

It will be noticed that this proof also invokes some auxiliary function, similarly to Thue's arguments. However, the matter here is easier, since we only need a nontrivial solution of a linear system, without any bound on it, i.e. we do not need Siegel's lemma. Also, here the auxiliary function is constructed once for all, whereas in Thue's proof it varies with the integral points.

Note that defining $\mathcal{C}$ as the complement in $\tilde{\mathcal{C}}$ of the set of poles of x, we obtain an affine curve whose 'divisor at infinity' in the previous sense is just the set of poles of x. We may then apply Theorem 2.12 to deduce the finiteness of $\mathcal{C}(\mathbb{Z})$. Conversely, given an affine subset $\mathcal{C}$ of $\tilde{\mathcal{C}}$ defined over $\mathbb{Q}$, the divisor at infinity is $\neq 0$ and defined over $\mathbb{Q}$; by taking a linear combination of the affine coordinate functions, we obtain a nonconstant rational function x whose pole-divisor is supported exactly at infinity. Hence we obtain a further (equivalent) version:

Theorem 2.14 (Runge's Theorem III). *Let $\mathcal{C}$ be an absolutely irreducible affine curve defined over $\mathbb{Q}$, such that the divisor at infinity is a sum of two strictly positive divisors D_1, D_2 with disjoint supports and both defined over $\mathbb{Q}$. Then $\mathcal{C}(\mathbb{Z})$ is finite (and effectively computable).*

Remark 2.15.

(i) Note that, in contrast with Siegel's Theorem, it suffices here to split the divisor at infinity into *two* components rather than *more than two*, in order to ensure finiteness. Naturally this is because we are taking into account the field of definition of the divisors; this discrepancy with Siegel's Theorem again indicates that the present theorem is not 'geometrical'.

(ii) The original formulation concerned, in particular, an irreducible plane curve $f(X, Y) = 0$ ($f \in \mathbb{Z}[X, Y]$) with the assumption that the highest homogeneous part of f is not a constant times a power of an irreducible polynomial over $\mathbb{Q}$. (See [59].) Note that this implies that this highest part splits over $\mathbb{Q}$ into a product of coprime factors, so that the set of points at infinity also splits into two (or more) coprime parts defined over $\mathbb{Q}$; this shows the link with the above statements.

Exercise 2.30. Prove in full detail the equivalence of the three above versions of Runge's Theorem.

Exercise 2.31. Deduce from any of the above versions the original statement of Runge, namely prove that: *If $f \in \mathbb{Z}[X, Y]$ is an irreducible polynomial such that its highest homogeneous part is not a constant times a power of an irreducible polynomial over $\mathbb{Q}$, then the equation $f(X, Y) = 0$ has only finitely many integral solutions.*

Exercise 2.32. Following the arguments in the above proofs of Runge's Theorem, obtain some explicit bound for the integer solutions, in terms of quantities associated to a set of defining equations for C. (This will involve the simple observation that a function y is bounded on the complement of any open subset of $\tilde{C}$ containing its poles.)

Exercise 2.33. Let $f \in \mathbb{Q}[X]$ be a polynomial of degree b, not a perfect power, and denote by l its leading coefficient. Let a be a positive integer and suppose that l is a perfect q-th power for some divisor $q > 1$ of $\gcd(a, b)$. Prove that then the equation $Y^a = f(X)$ has only finitely many integer solutions, which can be found. (Note also that it suffices to do the case $a = q | b$.)

Exercise 2.34. Let $d > 1$ be an integer and let $f \in \mathbb{Z}[X]$ be nonconstant and with no repeated roots. Prove that there are only finitely many integers x such that $f(x)$ and $f(x + 1)$ are both d-th powers in $\mathbb{Q}$. Deduce the result of Exercise 2.24. (Of course all of this follows from Siegel's Theorem, but Runge's Theorem suffices, giving moreover effectivity. Hint: apply the previous result to a suitable product $f(X)^r f(X + 1)^s$.)

Exercise 2.35. Let $f(X, Y) \in \mathbb{Q}[X, Y]$ be absolutely irreducible, of degree > 1 in Y. Prove that for large enough integer L there are only finitely many integers x such that every equation $f(x+l, Y) = 0, l = 0, \ldots, L$, has a rational solution. (Hint: consider curves $C = C_L$ which are irreducible components of the curve defined in $\mathbb{A}^{L+2}$ by the system $f(X+l, Y_l) = 0, l = 0, \ldots, L$. Observe that the ramification index in C_L of the point $X = \infty$ of $\mathbb{P}_1$ is bounded independently of L and that the same holds for the degree of any point in the fiber. Consider for this *e.g.* Puiseux series. Also, by considering ramification points inductively for $L = 1, \ldots$, deduce that the degree of the x-map from C_L is $\geq 2^L$ and so tends to infinity with L. Conclude that Theorem 2.12 may be applied for large enough L. See [17] for a similar and more refined argument.)

Exercise 2.36. Prove the following higher-dimensional version of Runge's Theorem: *Let X be an irreducible affine surface such that $\tilde{X}$ is nonsingular and the divisor at infinity $\tilde{X} \setminus X$ splits into the sum of divisors defined over $\mathbb{Q}$, with* *strictly positive self-intersection and whose supports have empty intersection. Prove that the set of integral points of $\mathcal{X}$ is not Zariski-dense in $\mathcal{X}$.* (Hint: mimic the proof of Theorem 2.13 and exploit the assumption on the divisors by using the Riemann-Roch Theorem for $\mathcal{X}$. An analogous argument holds in any dimension.)

Exercise 2.37. By applying the previous result to the symmetric square of an affine curve $\mathcal{C}$ (namely the quotient of $\mathcal{C} \times \mathcal{C}$ by the map $(u, v) \mapsto (v, u)$), prove, under suitable assumptions, a finiteness theorem for the 'quadratic integral points' on $\mathcal{C}$, namely points whose coordinates are algebraic integers in a quadratic extension of $\mathbb{Q}$. (Here some knowledge of valuations and heights should be useful. See [101] for similar results in the spirit of Siegel's Theorem.)

A Thue Equation in polynomials

Similarly to the case of Pell Equation treated at the end of the previous chapter, we may consider Thue's equations with polynomial coefficients, to be solved in polynomials. This is a special case of Thue's equations over function fields. In the next chapters we shall see some elementary solutions of this by means of the S-unit equation. At the moment let us limit ourselves to the special equations

$$X^d - \Delta(t)Y^d = c(t), \qquad \Delta, c \in \mathbb{C}[t] \setminus \{0\}, \qquad (2.28)$$

where $d \geq 3$, to be solved with $X, Y \in \mathbb{C}[t]$. By absorbing d-th powers in Y we also assume that Δ is not divisible by any non-constant d-th power. If $X = x(t)$, $Y = y(t)$ is a non-constant solution, on dividing by $c(t)$ and differentiating the equation we obtain $x^{d-1}(dx'c - xc') = y^{d-1}(\Delta'yc + d\Delta y'c - \Delta yc')$. Now, since $\gcd(x^d, y^d)$ divides c we have that y^{d-1} divides $c(dx'c - xc')$. Hence, either $dx'c = xc'$ or $(d-1)\deg y \leq 2\deg c + \deg x - 1 \leq 3\deg c + \deg y + d^{-1}\deg\Delta - 1$. In the first case x^d/c is constant and $\deg x, \deg y \leq \deg c$. In the second case $(d-2)\deg y \leq 3\deg c + \deg\Delta - 1$. In any case, since $d \geq 3$ this shows that x, y have bounded degree (in terms of $\deg c, \deg\Delta$).

Now, if Δ is a constant a^d, factoring $x^d - a^d = \prod_{\theta^d=1}(x - \theta ay) = c$, we see that the differences $x - \theta ay$ have only finitely many possibilities up to constant factors, and then the same holds for x, y. If Δ is not constant, things are fairly similar. Let us consider the curve $\mathcal{C}$ defined by $u^d = \Delta(t)$, which is irreducible in view of our assumptions. Again we may factor $x^d - \Delta y^d$ as $\prod(x - \theta uy)$. Since this product equals c, we see that the zeros of any factor $x - \theta uy$ may only lie either among the zeros of c or among the poles of t in $\mathcal{C}$. But since $\deg x, \deg y$ have bounded degree, the divisor of $x - \theta uy$ on $\mathcal{C}$ has only a finite number of possibilities, and so the $x - \theta uy$ have only finitely many possibilities up to constants. Since such constant factors are determined up to d-th roots of 1, we see that the equation has only finitely many polynomial solutions. The argument also shows how to determine these solutions when c, Δ are 'effectively' known.

Exercise 2.38. (Diophantine Approximation in function fields) Let $\Delta \in \mathbb{C}[t]$ be non constant, not a d-th power and such that $\Delta(0) \neq 0$. Also, let $\xi \in \mathbb{C}[[t]]$ be such that $\xi^d = \Delta$. Prove that for $x, y \in \mathbb{C}[t]$ we have $\mathrm{ord}_{t=0}(x(t) - \xi y(t)) \leq \deg x(t) + \deg y(t) + \deg\Delta(t) \leq 2\max(\deg x, \deg y) + \deg\Delta$. (Hint: differentiate an approximation $y - \xi x = O(t^m)$ and eliminate ξ. You will need to show that a Wronskian does not vanish. This result provides an explicit Roth's

Theorem for rational approximations - in $\mathbb{C}(t)$, with respect to the topology of $\mathbb{C}[[t]]$ - to ξ and is rather similar to the above argument for Thue's equation over $\mathbb{C}[t]$. See also Exercise 2.10 above.)

Notes to Chapter 2

Already Lagrange sought solutions to $f(X, Y) = 1$ by looking at the continued fraction for real roots ξ of $f(X, 1) = 0$, in practice seeking good rational approximations to ξ. See [37].

Thue's method of proof had a deep influence on the whole Diophantine Approximation. As mentioned above, the same principles appear (with new technical difficulties) in the later sharpenings. Even in recent times the germs of these proofs have been recognized in certain sophisticated methods of Diophantine Geometry: we are alluding here at Bombieri's proof of the Mordell conjecture [14].

We have remarked that the above proofs lead to explicit estimates for the number of solutions of Thue's equations $f(X, Y) = c$; such estimates would depend on f and on c. Some special equations (as $X^d - aY^d = 1$) were shown to have *e.g.* at most one solution, either by explicit construction of the auxiliary polynomial by Padé approximations (as done by Siegel, see [79]) or by p-adic methods (as done *e.g.* by Skolem, see [59]). Bombieri and Schmidt were the first to prove that, for instance, for $c = 1$ the number of solutions is bounded only in terms of deg f (see [70]). We shall mention in the last chapter some uniform estimates obtained by the theory of S-unit equations.

For variations on the concept of 'exponent' of approximation and for many recent advances see the survey paper by Bugeaud and Laurent in [104].

For a proof of Roth's Theorem see [25], and [17] for a modern version (see also next chapter for a relevant statement). A higher-dimensional far-reaching extension of Roth's Theorem was found and proved by W.M. Schmidt around 1970. This result and its subsequent improvements (by several authors) are commonly called 'Subspace Theorem': this bounds from below the distance of rational points in $\mathbb{P}_n$ to fixed hyperplanes defined over $\overline{\mathbb{Q}}$; the exceptional approximations are shown to belong to a certain finite union of proper linear subspaces. (Roth's theorem is the case $n = 1$.) See [17] for a proof and [101] for applications to diophantine equations.

A modern version of Siegel's Lemma is due to Bombieri and Vaaler (see [17]). This works for systems with coefficients in a number field and gives estimates for a full set of linearly independent solutions. Also, the bounds depend on certain invariants of the matrix of the system, more natural than the sup of the entries.

For Gelfond's inequality see also [47] and [17, page 26]. The argument in 2.17 goes back to Siegel. Boyd has later found a best-possible constant.

Versions of Dyson lemma have been proved by Bombieri (see [12]), by Viola (see [89]) and in higher dimensions by Esnault - Vieweg (see [43]). A somewhat different version in higher dimension is the so-called *Faltings Product Theorem*, introduced by Faltings in [46] and extremely useful in a number of different contexts, including Roth's theorem. See the paper of van der Put in [41] for an account.

Bombieri's latter method for effective diophantine approximation reverses, so to say, the roles of the target ξ and its approximations: the auxiliary polynomial now is constructed to vanish at the approximations and one draws deductions on the behaviour at ξ. See [16].

We have used the simplest definition of integral points. To make it more intrinsic (*i.e.* not dependent on the embedding) one may consider for example 'sets Σ of *quasi-integral* points for a projective variety $\tilde{\mathcal{X}}$ with respect to a divisor D *at infinity*' defined by requiring that for any rational function $f \in k(\tilde{\mathcal{X}})$ regular outside D, the set $f(\Sigma)$ consists of elements of k with bounded denominator (depending on f). See *e.g.* [77]. Another very useful (though essentially equivalent) notion, for $\tilde{\mathcal{X}}$ embedded in some $\mathbb{P}_n$, is to say that a point $P \in \tilde{\mathcal{X}}(k)$ is S-integral with respect to a divisor if its reduction modulo each $v \notin S$ does not lie in the reduction $D(v)$ of D; in turn, $D(v)$ may be defined as the the Zariski closure of the reductions of points in $D(\bar{k})$. (Note that every point in $\mathbb{P}_n(\bar{k})$ has a well-defined reduction at places above v.) All of these notions make sense also if D is not a divisor but merely a subvariety.

It has been noted by Osgood, Reyssat and especially by Vojta, that Siegel's Theorem may be considered as an arithmetic analogue of Picard's Theorem in complex function theory. In Vojta's far-reaching viewpoint a sequence of integral points on an affine variety $\mathcal{X}$ is something analogous to a holomorphic map $f : \mathbb{C} \to \mathcal{X}$. In the same context, Vojta also interpreted Roth's Theorem and Diophantine Approximations as analogues of Nevanlinna Theory. See for instance [90].

For a modern expositions of Siegel's proof of his theorem see *e.g.* [77] and [17] (the original argument by Siegel was more delicate because he did not dispose of Roth's theorem, but only of a weaker result). Another proof in non-standard language is due to Robinson and Roquette. A further proof depending on the Subspace Theorem is due to Corvaja and Zannier. See [101] and [105] for sketches of these proofs and references. Siegel's theorem may also be derived rather easily from Faltings's theorem (ex Mordell Conjecture) that curves of genus ≥ 2 have only finitely many rational points over a number field. (See [17] for Bombieri's proof of this result; in genus 1, Siegel's theorem follows on applying Faltings's result after taking a cover ramified at infinity, in order to increase the genus.)

An effective version of Siegel's Theorem is still missing, even for the general case of genus 2. For genus 0 with > 2 points at infinity and for genus 1, an effective proof follows from Baker's mentioned effective treatment of Thue's and related equations (see [7]). Effective proofs (due to Bilu and to Dvornicich - Zannier) are also known (again as applications of Baker's results) in the case of a Galois cover of the projective line. This may be reformulated by saying that *the integers x such that the equation $f(x, Y) = 0$ has only rational roots may be found (or parametrized) effectively*. See [10], also for other effective results, and [99].

Also, as sketched in the Supplements, Siegel's Theorem implicitly yields an algorithm to establish *whether a given curve has infinitely many integral points*. In fact, one may calculate effectively the genus and the number of points at infinity. If we have genus zero and at most two points at infinity, as shown in the Supplements one may reduce the question of the infinity of the integral points to the case of certain affine conics, which may be dealt with as in Chapter 1. A general criterion for the existence of infinitely many integral points, considering also their distribution in connected components of a real curve, is given in [84].

The idea of using a parametrization for the rational points on a conic in the proof of Theorem 2.8 is taken from [59].

Corollary 2.11 is a special case of a theorem of Mahler (who allowed c to vary in the same way as x, y). Siegel was among the first to consider unit equations, over algebraic number fields; he reduced the hyperelliptic equations to them. The above method of proof goes back to Siegel. Both Baker's and Bombieri's mentioned methods lead to an effective proof (*e.g.* along the above lines, or directly with Baker's lower bounds for linear form in logarithms - see [7]). Such methods also lead to explicit (and of the 'correct' order of magnitude) quantitative lower bounds for a difference $x - y$ where $x \neq y$ are integers composed only of primes from S. See also Chapter 3 for more on this.

The above effective procedure for points on curves of genus zero uses the birational reduction of a (nonsingular) curve of genus zero to a conic. This is due to Hilbert and Hurwitz.

For Runge's Theorem see *e.g.* [60] and [17] for more general refined versions. For a connection with Weil's Decomposition Theorem see [13]. See also [65] for an improvement.

For Thue's equation over function field see also [56] and [39]. For an effective Roth's Theorem over function fields see [92].

Chapter 3
Heights and diophantine equations over number fields

In the previous chapters we have worked essentially with 'classically' integral solutions, that is over $\mathbb{Z}$. However, since the times of Kummer (and even of Gauss) it has been recognized that diophantine equations are most advantageously dealt with by going out of $\mathbb{Q}$ and using tools from Algebraic Number Theory; this also led to consider solutions in integers of number fields, and even in S-integers therein, *i.e.* those which have a denominator composed only of primes in the finite set S. In turn, new concepts have been created for studying these more general solutions.

In the present chapter we shall move in this direction. We shall start by recalling some basics of the theory of valuations and heights; this last concept provides a way of measuring the 'arithmetical size', so to say, of an algebraic number. This is obtained by taking suitably into account simultaneously all the absolute values of $\overline{\mathbb{Q}}$. It evolved through various definitions, until Weil gave a 'canonical' one.[1] In developing the elements of this theory, we shall assume throughout some basic facts from elementary valuation theory and algebraic Number Theory.

Then, using this language, we shall formulate, usually with few proofs, some important results for diophantine approximations and diophantine equations over number fields, generalizing those which we have met so far. In particular, we shall recall a general version of Roth's theorem and the *S-unit theorem*, which quickly leads to a finiteness theorem for Thue's equations in S-integers over number fields. The S-unit theorem will be derived here from Roth's, but we shall completely prove a sharp quantitative form of it in the last chapter. This modern proof is substantially elementary, but will be postponed because it also uses some results from next chapter.

Finally, we shall deal with heights on finitely generated subgroups Γ of $(\overline{\mathbb{Q}}^*)^n$. We shall see that the height corresponds to a norm on an euclidean

---

[1] Actually, there have been many further sophisticated evolutions, of geometrical nature, which we shall not touch in these elementary lecture notes.

space $\mathbb{R}^r$ associated to Γ. This not only provides new insight into the concept of height, but will be relevant in the final chapter.

3.1. Fields with a product formula

3.1.1. Valuations and the product formula

Absolute values

Let k be a field. An *absolute value* (or *valuation*) on k is a map $|\cdot| : k \to \mathbb{R}^+$ with the properties

- $|a| = 0 \iff a = 0$;
- $|ab| = |a|\,|b|$ (multiplicativity);
- $|a + b| \leqslant |a| + |b|$ (triangle inequality).

An absolute value is said *non-archimedean* or *ultrametric* if it satisfies

- $|a + b| \leqslant \max\{|a|\,, |b|\}$ (ultrametric inequality).

If moreover $\kappa \subset k$ is a subfield such that $|x| = 1$ for all $x \in \kappa^*$ then $|\cdot|$ is said to be a valuation over κ or of k/κ.

We also say that the absolute value is *trivial on κ*. The absolute value trivial on the whole k is simply called *trivial*. Of course we shall be interested in nontrivial absolute values.

It is not the purpose of these lecture notes to enter into details of the classical theory of absolute values; we shall only recall some basic facts for the reader's convenience and refer *e.g.* to [61] for a more complete account and for proofs.

Any absolute value defines a distance on k by $d(x, y) := |x - y|$ and in particular induces a topology on k. Two absolute values $|\cdot|_1, |\cdot|_2$ on k are said to be *equivalent* if they induce the same topology on k. It may be proved (see [61]) that this happens if and only if there exists $l > 0$ such that $|x|_1 = |x|_2^l$ for all $x \in k$. An equivalence class of valuations is called a *place* .

For a valuation v we may form the completion of k with respect to the above defined distance. This is a field, denoted by k_v, to which the valuation extends; we have a natural embedding of k in k_v.

For a given choice of $c > 1$ we can define $v : k \to \mathbb{R} \cup \{\infty\}$ by $|x| = c^{-v(x)}$, so $v(x) = -\log|x|/\log c$. This function v satisfies properties corresponding to the ones listed above, which we omit, leaving their formulation to the interested reader. Viceversa, we can define $|\cdot|$ from the knowledge of v by the above formula; this depends on c, but different choices lead to equivalent absolute values.

We shall often deal with several absolute values at the same time and then we shall use the letter v to refer to these functions, writing *e.g.* $|x|_v$.

Product formula ([77])

Let M be a set of places of k, *i.e.* of pairwise inequivalent absolute values on k, all but a finite number being non-archimedean. We shall often write M^∞ for the set of the archimedean ones and M^0 for $M \setminus M^\infty$.

Suppose now that for every $x \in k^*$ the absolute value $|x|_v$ is 1 for all but a finite number of $v \in M$, and for every $v \in M$ let $\lambda_v > 0$ be a positive constant. We say that the λ_v define a *product formula* on k if the following holds:

$$\prod_{v \in M} |x|_v^{\lambda_v} = 1 \qquad \forall x \in k^*. \tag{3.1}$$

The cases more interesting for us occur when k is either a number field or a function field.

Example 1.   Let $k = \mathbb{Q}$. Apart from the usual absolute value, it may be shown that all the others may be obtained as follows. For a prime p and for $x \in \mathbb{Q}^*$ there is a unique integer m such that $x = p^m \frac{a}{b}$ with $a, b \in \mathbb{Z}$ and $p \nmid ab$; we define the p-adic absolute value of x as $|x|_p := p^{-m}$ and we put $v_p(x) = m$. The usual absolute value will be called the *infinite* one and we shall sometimes indicate it as $|\cdot|_\infty$.

Note that the completion $\mathbb{Q}_\infty$ is just $\mathbb{R}$; the completion at the p-adic place is called the field of p-adic numbers and denoted by $\mathbb{Q}_p$.

Now let M be the set of the usual absolute value and of all p-adic places; since each nonzero integer is divisible only by finitely many primes, the condition $|x|_v = 1$ for all but finitely many $v \in M$ is satisfied for $x \in \mathbb{Q}^*$. We assert that, taking $\lambda_v = 1$ for all $v \in M$, (3.1) holds. In fact, by multiplicativity we just need to check it for nonzero $x \in \mathbb{Z}$. If this is the case, indeed, writing the prime decomposition $x = \pm \prod_i p_i^{e_i}$ we have

$$\prod_{v \in M} |x|_v = |x| \prod_i |x|_{p_i} = \prod_i p_i^{e_i} \prod_i p_i^{-e_i} = 1.$$

Example 2.   Let $k = \kappa(t)$ and suppose for instance $\bar\kappa = \kappa$. Then it is a known easy fact that all the valuations of k/κ are associated to points P of $\mathbb{P}_1(\kappa)$: the corresponding $v = v_P$ is (up to a constant factor) $v_P(f(t)) = \mathrm{ord}_{t=P} f$. Note that the completion is now the field of Laurent series $k((t - t_0))$ if P corresponds to $t_0 \in \kappa$, whereas it is $\kappa((1/t))$ if P is the point at infinity $t = \infty$.

The product formula takes the convenient well-known additive shape

$$\sum_P v_P(f) = 0 \qquad \forall f \in \kappa(t)^*,$$

expressing the fact that a rational function has as many zeros as poles. Of course the multiplicative shape is recovered by associating to v_P an absolute value $| \cdot |_P$ as above, by the choice of a constant $c = c_P$, this time the same for all P.

If $\bar{\kappa} \neq \kappa$, the formula holds by using the valuations v associated to the irreducible monic polynomials in $\kappa[t]$, and keeping the valuation associated to ∞, namely $v_\infty(f) = -\deg f$ for a polynomial $f \in \kappa[t]$. This time however the *weight* λ_v associated to $v \neq \infty$ will be the degree $\deg(v)$ of the irreducible polynomial corresponding to v. Then the *sum* formula $\sum_v \deg(v)v(f) = 0$ continues to hold for $f \in k^*$ (actually it is a consequence of the above one).

Example 3.   More generally, let $k = \kappa(\mathcal{X})$ be the function field of a normal projective variety $\mathcal{X}$ defined over a field κ ($\mathcal{X} = \mathbb{P}_1$ in the previous example), which we shall usually assume to be algebraically closed. Let M be the set of prime Weil divisors on $\mathcal{X}$ (*i.e.* the irreducible subvarieties of codimension 1).

Let f be any non-zero rational function on $\mathcal{X}$, *i.e.* $f \in k^*$. Since $\mathcal{X}$ is normal, for $D \in M$ we may define $v_D(f)$ as the order of f at D; as above, we also put $|f|_D := c^{-v_D(f)}$. The divisor of f is defined as $\mathrm{div}(f) = \sum_{D \in M} v_D(f)D$. This makes sense, for we have $v_D(f) = 0$ for all but a finite number of prime divisors D.

We then want to choose the λ_D so that a product formula holds. In dimension 1 there is essentially no freedom: it may be shown that for $\mathbb{P}_1$ we have only the previous example and that the same holds for all curves. However, when the dimension of $\mathcal{X}$ is bigger than 1 we have some choices, as we now show.

Let C be an irreducible curve on $\mathcal{X}$, assumed to intersect every divisor (it can always be obtained by cutting $\mathcal{X}$ with general linear varieties of the correct dimension). Then we can take $\lambda_D := C.D$, the intersection multiplicity.[2] Now, $C.\mathrm{div}(f)$ is well-known to be zero. Using linearity we then have

$$0 = C.\mathrm{div}(f) = \sum_D \lambda_D v_D(f), \qquad \forall f \in K^*.$$

As in the previous example, taking $|f|_D := c^{-v_D(f)}$ we obtain a multiplicative product formula:

$$\prod_D |f|_D^{\lambda_D} = 1, \qquad \forall f \in K^*.$$

---

[2] We recall that this is defined as the cardinality of the set-theoretic intersection $C \cap D$ when the intersections are transversal. In general one reduces to this case by moving D into its linear equivalence class.

When $\mathcal{X}$ is a curve, we have $\mathcal{C} = \mathcal{X}$, the prime divisors are points, $\mathcal{C}.D$ is 1 and we get $\lambda_D = 1$ for every D. When $\mathcal{X}$ is a rational curve we find back the previous example.

When $\mathcal{X}$ is a surface we may have different product formulas corresponding to the same set of absolute values. This does not happen on $\mathbb{P}_2$ (because all divisors are linearly equivalent to a multiple of a line) but already in the case $\mathbb{P}_1 \times \mathbb{P}_1$ (isomorphic to the quadric surface defined by $\{xy = zw\} \subset \mathbb{P}_3$) we have the family $\mathbb{P}_1 \times P$ of linearly equivalent horizontal divisors and the family of linearly equivalent vertical divisors $P \times \mathbb{P}_1$. The self intersection is 0 for both families and the mutual intersection is 1, which determines the intersection product on $\mathcal{X}$. If the curve $\mathcal{C}$ corresponds to (a, b) (*i.e.* a horizontal line counted a times and a vertical one counted b times) then $\mathcal{C}$ satisfies the assumptions if $a, b > 0$. (This amounts to $\mathcal{C}$ being defined by a homogeneous polynomial of bidegree (a, b).) By varying a, b we find two nonequivalent product formulas.[3]

3.1.2. Finite extensions

Let (k, M) be a field with a product formula on a set M of places and let L/k be a finite, separable extension of degree d. We want a product formula on L. We now recall how this can be always done, by 'extending' the product formula on k.

Again, we shall usually not give proofs of these results, which represent standard material and fall out of the main scope of these lecture notes.

Let M_L be the set of all places of L extending some place $v \in M = M_k$; we write $w|v$ to denote that $w \in M_L$ extends $v \in M$. Moreover, let L_w (resp. k_v) be the completion of L (resp. k) with respect to w (resp. to v); we also denote by $d_w := [L_w : k_v]$, the so-called *local degree*. For every $v \in M$ we have $\sum_{w|v} d_w = d$. Now we define, for every $x \in L^*$,

$$\|x\|_w := \left| N_{k_v}^{L_w}(x) \right|_v^{\frac{1}{d}}.$$

It turns out that the right side defines indeed an absolute value on L_w and also that

$$\|x\|_w = |x|_w^{\frac{d_w}{d}},$$

---

[3] Similarly, for more general varieties the structure of equivalence classes of product formulas so obtained will depend on the rank of divisors modulo numerical equivalence.

where $|x|_w$ extends $|x|_v$ on k. We remark that the choice of the exponent $1/d$ produces a normalization which we shall find useful when dealing with heights.

We further note that, since $d_w \leq d$, the function $\|\cdot\|_w$ continues to verify the triangle inequality even when $|\cdot|_v$ is archimedean.

Now, it can be proven that for $x \in L^*$, $\prod_{w|v} \|x\|_w = \prod_{w|v} \left| N_{k_v}^{L_w}(x) \right|_v^{\frac{1}{d}} = \left| N_k^L(x) \right|_v^{\frac{1}{d}}$. This implies

$$\prod_{w \in M_L} \|x\|_w^{\lambda_v} = \prod_{v \in M} \prod_{w|v} \|x\|_w^{\lambda_v} = \prod_{v \in M} \left| N_k^L(x) \right|_v^{\frac{\lambda_v}{d}} = \left(\prod_{v \in M} \left| N_k^L(x) \right|_v^{\lambda_v} \right)^{\frac{1}{d}} = 1,$$

because of the product formula on k.

Remark 3.1.

(i) It may be proved that the product formula is unique on $\mathbb{Q}$ (easy exercise); in turn, this implies that the extension to a given number field, that we have just seen, is also unique (which can be proved e.g. by the so-called 'Strong Approximation Theorem', a sharpening of the Chinese Remainder Theorem). We note however that in general the extension we have described will not be unique. A construction can be obtained from Example 3 above, where we produce two product formulas on $\mathbb{P}_1 \times \mathbb{P}_1$. Now, we may view $\mathbb{P}_2$ as the symmetric square of $\mathbb{P}_1$; namely, $\mathbb{P}_1 \times \mathbb{P}_1$ is a double cover of $\mathbb{P}_2$. Then we can take two distinct product formulas on $\kappa(\mathbb{P}_1^2)$ which induce the same one on $\kappa(\mathbb{P}_2)$. See [28].

(ii) We do not have a product formula on $\overline{\mathbb{Q}}$. The reason behind this is that every place of $\mathbb{Q}$ may be extended to $\overline{\mathbb{Q}}$ in infinitely many ways, (e.g., through an increasing sequence of number fields). This implies that, if we take all the places into account the condition that $|x|_v = 1$ for all but finitely many v (for $x \neq 0$) is not satisfied.

Example 1'.   When k is a number field, the set M_k consists of the union of two sets:

(i) the set M_k^∞ of infinite (*i.e.* archimedean) places corresponding to the embeddings of k in $\overline{\mathbb{Q}}$, up to complex conjugation; there are $r_1 + r_2$ such embeddings, where r_1 (resp. $2r_2$) is the number of real (resp. non-real) embeddings, so $r_1 + 2r_2 = [k : \mathbb{Q}]$.

(ii) The set M_k^0 of finite (ultrametric) places, corresponding to the prime ideals of the ring of integers of k. If L is a finite extension of k, a finite place v of k extends to L according to the splitting of the prime ideal corresponding to v in the ring of integers of L.

Let for instance $k = \mathbb{Q}$, $L = \mathbb{Q}(i)$ and $M_\mathbb{Q}$ be the set, described in Example 1 above, of the usual absolute value and all p-adic valuations. The degree $d = [L : k]$ is 2.

The usual absolute value ∞ extends uniquely to the complex absolute value on L. The completion is now $\mathbb{C}$, so the local degree is 2.

The p-adic absolute value corresponding to a $p \equiv 3 \pmod 4$ extends uniquely to an absolute value of L; the local degree is now 2 and we have $\|x\|_p = |N_k^L(x)|_p^{1/2}$. The p-adic absolute value corresponding to a $p \equiv 1 \pmod 4$ extends in two ways to L: now $p = \pi\pi'$ decomposes into the product of two (conjugate) primes π, π' of $\mathbb{Z}[i]$, the completion L_π equals $\mathbb{Q}_p$ and $\|x\|_\pi = |x|_p^{1/2}$. Finally, if $p = 2$, we have ramification, expressed by the equality $2 = -i(1 + i)^2 = -i\eta^2$, say. The completion has degree 2 over $\mathbb{Q}_2$ and $\|x\|_2 = |x|_2^{1/2}$.

Example 2′.   We have seen that when $k = \kappa(t)$ (κ alg. closed) the set of valuations of k/κ corresponds to $\mathbb{P}_1(\kappa)$. In general, if L/k is a finite extension, L is the function field of a complete nonsingular curve $\mathcal{C}$ over κ. The inclusion $k \subset L$ corresponds to a nonconstant rational map $\phi : \mathcal{C} \to \mathbb{P}_1$. Also, the places of L/κ correspond to the points $Q \in \mathcal{C}(\kappa)$ and, as in the case of $\mathbb{P}_1$, the corresponding $v = v_Q$ is the order function at the point Q. For κ algebraically closed the situation is (similar but) simpler than for number fields: each place v of k/κ (*i.e.* point $P \in \mathbb{P}_1(\kappa)$) extends to $[L : k]$ places (*i.e.* the points in the inverse image $\phi^{-1}(P)$), except for a finite number of ramified points. Also, the local degrees d_v are exactly the ramification indices, so are 1 for all but finitely many places.

3.2. Heights

3.2.1. Weil height

Let k be a field with product formula. We shall tacitly use the symbol $\|\cdot\|_v$, to indicate the fact that the valuations v have been normalized so that the product formula holds with the weights $\lambda_v = 1$.[4]
Following Weil, we can first define a *height* on $\mathbb{P}_n(k)$. For any point $P = (x_0 : x_1 : \ldots : x_n)$, with the x_i not all zero, we define

$$H_k(P) := \prod_{v \in M_k} \sup_i \|x_i\|_v . \tag{3.2}$$

A most important fact is that indeed this definition is independent of the choice of coordinates; this comes from the product formula. In fact, let

---

[4] Note that this still does not determine the normalizations uniquely.

$\alpha \in k^*$ and $P = (\alpha x_0 : \alpha x_1 : \ldots : \alpha x_n)$; we have $\sup_i \|\alpha x_i\|_v = \|\alpha\|_v \sup_i \|x\|_v$ and

$$\prod_{v \in M_k} \sup_i \|\alpha x_i\|_v = \left(\prod_{v \in M_k} \|\alpha\|_v \right) \left(\prod_{v \in M_k} \sup_i \|x_i\|_v \right) = \prod_{v \in M_k} \sup_i \|x_i\|_v ,$$

as asserted. In particular, we may suppose that some coordinate is 1, whence $H_k(P) \geq 1$.

Now let L/k be a finite, separable extension of degree $d := [L : k]$, and take a point $P = (x_0 : x_1 : \ldots : x_n)$ in $\mathbb{P}^n(k) \subset \mathbb{P}^n(L)$. We want to compare the heights of P with respect to L and to k. In doing this we assume that the valuations of k have been extended to L as indicated above, and again denote the normalized valuations with respect to L as $\|\cdot\|_w$, where now however w indicates a place of L. So, for $w|v$ and for $x \in k$ we shall have $\|x\|_w = \|x\|_v^{d_w/d}$, where $d_w = [L_w : k_v]$ is the local degree. Then, the formulas that we have recalled earlier yield, for $x \in k$, $\prod_{w|v} \|x\|_w = \|x\|_v$, because $\sum_{w|v} d_w = d$. Now, for every valuation $v \in M_k$ we take a coordinate i_v with maximal value, i.e. $\|x_{i_v}\|_v \geq \|x_j\|_v$ for every j; note that for $w|v$ this implies $\|x_{i_v}\|_w \geq \|x_j\|_w$ for every j. We can write $H_k(P) = \prod_{v \in M_k} \|x_{i_v}\|_v$, whence,

$$H_L(P) = \prod_{v \in M_k} \prod_{w|v} \sup_j \|x_j\|_w = \prod_{v \in M_k} \prod_{w|v} \|x_{i_v}\|_w = \prod_{v \in M_k} \|x_{i_v}\|_v = H_k(P).$$

This implies that we have in fact defined an *absolute* height. In fact, we have the following

Consequence: Let k be a field with a product formula. For every finite separable extension L of k we can define a height H_L on $\mathbb{P}_n(L)$. By the above (applied with L_1 in place of k, L_2 in place of L) this is found to be compatible with any inclusion $\mathbb{P}_n(L_1) \subset \mathbb{P}_n(L_2)$, for finite extensions $L_1 \subset L_2$ of k. Therefore we can define an *absolute height* on $H = H_{\bar{k}}$ on $\mathbb{P}^n(\bar{k})$.

Note that this height will depend on k and the chosen product formula, in the sense that starting with another field k' and a product formula on k' with $\bar{k}' = \bar{k}$ we could arrive at different heights (for algebraic numbers we have however a 'canonical' choice).

When the field k and the product formula on it are given in advance we usually drop the suffix and write $H(P)$ for $H_k(P)$. We shall always do that for $k = \mathbb{Q}$, choosing the standard product formula seen in the previous subsection.

Remark 3.2.

(i) We note that several authors use a different normalization, so for number fields their $H_k(x)$ is in fact our $H_k(x)^{[k:\mathbb{Q}]}$. We do not bother, since from now on we shall only use the absolute height.

(ii) To compute the height of any point we just need to consider its minimal field of definition; nonetheless it can be sometimes advantageous to work in a bigger field; for instance when dealing with more than one point one can take a common field of definition for all points in question.

(iii) We have observed that there is no product formula on $\overline{\mathbb{Q}}$; however we have just seen that there is a height on $\overline{\mathbb{Q}}$.

It is often notationally convenient to denote $H((x_0 : \ldots : x_n))$ by $H(x_0 : \ldots : x_n)$ and to work with the logarithmic height:

$$h = h_k := \log H_k. \tag{3.3}$$

This makes sense since we have seen that $H_k(P) \geq 1$, which also implies $h_k(P) \geq 0$. This logarithmic height comes from an additive formulation of the product formula, more natural for instance in function fields. Another definition concerns the height of an element $x \in \bar{k}$:

$$H(x) := H(1 : x), \qquad h(x) := h(1 : x) = \log H(x). \tag{3.4}$$

Hence for instance $H(0) = H(1) = 1$. Note the inequalities, valid for all $x_1, \ldots, x_n \in k$,

$$0 \leq \sup_{i=1}^{n} h(x_i) \leq h(1 : x_1 : \ldots : x_n) \leq h(x_1) + \ldots + h(x_n). \tag{3.5}$$

Example 1. Let $k = \mathbb{Q}$ and take a point $P \in \mathbb{P}_n(\mathbb{Q})$. We may express $P = (x_0 : \ldots : x_n)$ where the x_i are coprime integers. Then it is an easy matter (exercise) to check that $H(P) = \max(|x_i|)$. In particular, for a rational number p/q, where p, q are coprime integers, we have $H(p/q) := H(1 : p/q) = H(q : p) = \max(|p|, |q|)$. We shall soon see some practical way of computing heights, but let us find directly the height of $\sqrt{2}$, for instance. It is an algebraic integer, so only the infinite places contribute in (3.2). Now, $\mathbb{Q}(\sqrt{2})$ has two infinite places $v_\pm$; they are real, so the local degrees are 1 and we find $|\sqrt{2}|_{v_\pm} = |\sqrt{2}|^{1/2}$ in both cases. Hence $H(\sqrt{2}) = \sqrt{2}$. Similarly for $1 + \sqrt{2}$. Now the absolute values are $|1 + \sqrt{2}|_{v_\pm} = \sqrt{|1 \pm \sqrt{2}|}$. Only one is > 1 so $H(1 + \sqrt{2}) = \sqrt{|1 + \sqrt{2}|}$. Finally, consider $3 + \sqrt{2}$; it is > 1 at both v_+, v_- so $H(3 + \sqrt{2}) = \sqrt{|3 + \sqrt{2}||3 - \sqrt{2}|} = \sqrt{7}$.

Other very interesting examples come from the roots of unity. A root of unity is an algebraic integer, and its reciprocal and all of its conjugates are roots of unity. Hence the absolute value is 1 at all places. In particular, $H(\zeta) = 1$ for all roots of unity ζ. More generally, this proves that $H(x\zeta) = H(x)$ for any $x \in \overline{\mathbb{Q}}$ and any root of unity ζ. We shall soon see that conversely the algebraic numbers ξ having $H(\xi) = 1$ are precisely 0 and the roots of unity.

As remarked earlier, in the case of the algebraic numbers we have a canonical choice $k = \mathbb{Q}$. In the next example this is not so, and the height will depend (up to a constant factor) on the choice of a 'ground field'.

Exercise 3.1. Prove that if $x_0, \ldots, x_n \in \mathbb{Z}$ are coprime integers we have $H(x_0 : \ldots : x_n) = \max(|x_i|)$.

Example 2. Now let us take $k = \kappa(t)$, a rational function field over a 'constant' field κ, supposed here to be algebraically closed. Here the additive notation will be certainly convenient, as for the product formula; since the absolute values of x are obtained by raising a constant $c > 1$ to the power $-v(x)$, we shall have to replace the 'sup' in the definition with an 'inf'. For $p, q \in \kappa[t]$ coprime polynomials, let us compute the height of $p/q \in k$, according to the natural product formula on k. We have $h(p(t)/q(t)) = h(1 : p(t)/q(t)) = h(q(t) : p(t)) = -\sum_{P \in \mathbb{P}_1(\kappa)} \min(v_P(q), v_P(p)) = \max(\deg q, \deg p)$. In fact, only the place $t = \infty$ gives a contribution, since at all other places $\min(v_P(q), v_P(p)) = 0$ because p, q have no common zero. This shows the nature of the height as a degree. Let now L be a separable extension of k of finite degree $d = [L : k]$. This corresponds to a nonsingular projective curve $\mathcal{C}$ over κ, and t becomes a nonconstant function of degree d on this curve. Let $u \in L \setminus \kappa$ be another nonconstant element. With the above normalization for the prolongation of the height to a finite extension field, the height reads as $h(u) = d^{-1} \sum_{P \in \mathcal{C}} -\min(0, v_P(u)) = d^{-1} \deg \mathrm{div}_\infty(u) = d^{-1} \deg u$. (Here v_P denotes the order function at the point $P \in \mathcal{C}$.) Hence $h(u) = \deg_L(u)/\deg_L(t)$. Note therefore that if we agree to take the standard normalization on a rational function field k, the 'absolute' height on $\bar{k}$ will depend on k (not only on $\bar{k}$).

Remark 3.3. Before this definition of Weil (motivated also by the geometric case) there were other similar notions of 'height' of an algebraic number ξ, defined via the coefficients of the minimal polynomial P, or via the conjugates ξ^σ (e.g., $\|P\|$ or $\sup_\sigma \{|\xi^\sigma|\}$). The Weil height is related to these others by quite simple inequalities; its advantage comes also from a much better behaviour with respect to mappings and operations (as in the following proposition).

Here are some very useful properties of the height. For simplicity we shall refer only to the height on $\overline{\mathbb{Q}}$, which is the one we shall mostly use in the sequel. In general there are similar properties (occasionally even stronger, as for the inequality for $h(x_1 + \ldots + x_r)$ at n. 3, which on function fields holds without the $\log r$); we leave this inspection to the interested reader.

Proposition 3.1. *For all* $x, x_1, \ldots, x_r \in \overline{\mathbb{Q}}$ *we have:*
1. $h(x) \geqslant 0$.
2. $h(1/x) = h(x)$ *and more generally* $h(x^m) = |m|\, h(x)$ *for all* $m \in \mathbb{Z}$.
3. $h(x_1 \cdots x_r) \leqslant h(x_1) + \ldots + h(x_r)$ *and* $h(x_1 + \ldots + x_r) \leqslant h(1 : x_1 : \ldots : x_r) + \log r$.
4. $h(x^\sigma) = h(x) \quad \forall \sigma \in \mathrm{Gal}(\overline{\mathbb{Q}}/\mathbb{Q})$.

Proof.
1. We have already observed this; we have $H(x) = \prod_v \sup(\|x\|_v, \|1\|_v) \geqslant \prod_v 1 = 1$ and $h(x) = \log H(x) \geqslant 0$.
2. Note that $H(1/x) := H(1 : 1/x) = H(x : 1) = H(x)$. (In essence we are using the product formula.) This proves the first part, and therefore we may suppose $m \geq 0$ in proving the rest. For this just note that $\sup(\|x\|_v^m, 1) = \sup(\|x\|_v, 1)^m$.
3. The first part follows from

$$H(x_1 \cdots x_r) = \prod_v \sup(1, \|x_1 \cdots x_r\|_v)$$

$$\leqslant \prod_v \prod_{i=1}^r \sup(1, \|x_i\|_v) = H(x_1) \cdots H(x_r).$$

For the rest, we distinguish between the cases when v is archimedean or not. In the first case,

$$\|x_1 + \ldots + x_r\|_v \leqslant \sup_{i=1}^r(\|r x_i\|_v) = \|r\|_v \sup_{i=1}^r(\|x_i\|_v).$$

In the second case,

$$\|x_1 + \ldots + x_r\|_v \leqslant \sup_{i=1}^r(\|x_i\|_v).$$

By the present normalizations, we have $\prod_{v \text{ arch.}} \|r\|_v = r$, whence

$$H(x_1 + \ldots + x_r) = \prod_v \sup(1, \|x_1 + \ldots + x_r\|_v)$$

$$\leqslant r \prod_v \sup(1, \sup_{i=1}^r \|x_i\|_v)$$

$$= r H(1 : x_1 : \ldots : x_r)$$

Taking logarithms we obtain the required properties for the $h(x_i)$.

4. Note that for every automorphism σ we have $\|x^\sigma\|_v = \|x\|_{v^\sigma}$ (essentially by definition); in other words, a conjugation acts on the (normalized) valuations as a permutation and thus the product defining $H(x)$ is invariant by $x \mapsto x^\sigma$. $\square$

Remark 3.4.   Note that nos. 2,3 imply for instance that for a fixed ξ we have $h((x - \xi)^{-1}) = h(x - \xi) = h(x) + O(1)$. See Proposition 3.2 for a generalization.

Exercise 3.2. (Projective transformations) Let $\sigma \in PGL_n(\overline{\mathbb{Q}})$ be an invertible linear projective transformation defined over $\overline{\mathbb{Q}}$. Prove that for $P \in \mathbb{P}_n(\overline{\mathbb{Q}})$ we have $h(\sigma(P)) = h(P) + O(1)$, where the implied constant depends only on σ.

Exercise 3.3. Prove that the inequalities at n. 3 cannot be sharpened on $\overline{\mathbb{Q}}$, but prove that the latter can be sharpened on fields where all valuations are ultrametric (so e.g. on function fields).[5]

Exercise 3.4. (Roots of unity) Let $x \in \overline{\mathbb{Q}}$ and let ζ be a root of unity. Prove that $h(x\zeta) = h(x)$, so in particular $h(\zeta) = 0$. (Hint: this may be checked directly using the definition of $h(x)$; otherwise, if $\zeta^n = 1$, by n. 2 we have $|n|h(x) = h(x^n) = h((x\zeta)^n) = |n|h(x\zeta)$.)

Exercise 3.5. Let $x_0, \ldots, x_r, \xi \in \overline{\mathbb{Q}}$. Extend n. 3 of the above proposition by proving that $h(x_0 + x_1\xi + \ldots + x_r\xi^r) \leq rh(\xi) + h(1 : x_0 : \ldots : x_r) + \log(r+1)$.

Exercise 3.6. (Heights of zeros of polynomials) Let

$$f(X) = X^d + a_1 X^{d-1} + \ldots + a_d \in \overline{\mathbb{Q}}[X]$$

and let $f(\xi) = 0$. Prove that $h(\xi) \leq h(1 : a_1 : \ldots : a_d) + \log d$. (Hint: use the previous exercise and $\xi^d = -a_1\xi^{d-1} - \ldots$.)

Exercise 3.7. (Heights of solutions of lacunary equations) Let $m_0 > m_1 > \ldots > m_r = 0$ be integers, $\xi, a_1, \ldots, a_r \in \overline{\mathbb{Q}}$ be such that $\xi^{m_0} + a_1\xi^{m_1} + \ldots + a_r = 0$. Also, suppose that no proper subsum $\xi^{m_0} + a_1\xi^{m_1} + \ldots + a_l\xi^{m_l}$, $0 \leq l < r$, vanishes.

(i) Prove that for $0 \leq l < r$, we have $m_l h(\xi) \leq (m_0 - m_l)h(\xi) + m_{l+1}h(\xi) + B$, where B depends only on r and $\max h(a_i)$. (Hint: start by using the equation for ξ, observing that $h(\xi^{m_0} + a_1\xi^{m_1} + \ldots + a_l\xi^{m_l}) \geq h(\xi^{m_l}) - h(\xi^{m_0-m_l} + a_1\xi^{m_1-m_l} + \ldots + a_l)$. Then use the result of the Exercise 3.5 to bound the second term. One finds that B may be bounded e.g. by $h(a_1) + \ldots + h(a_r) + 2\log r$.)

(ii) Conclude that $h(\xi) \leq Cm_0^{-1}$ where C depends only on r and $\max h(a_i)$. Also, find an explicit bound for C in terms of the said quantities. (Hint: Prove that $\mu := \max_{0 \leq l < r}(2m_l - m_0 - m_{l+1}) \geq m_0/(2^r - 1)$. For this prove inductively that $m_i \geq m_0 - (2^i - 1)\mu$. See also the Appendix to [67] for a somewhat different argument.)

---

[5] Artin and Whaples have shown that the only exception to this is the number field case.

(iii) Show that the assumption on the vanishing subsum cannot be eliminated. (Hint: if $f(\xi) = 0$ consider $t^m f(t) + f(t)$ for large m.)

Remark 3.5. Nos. 2, 3 of Proposition 3.1 express a 'good' behaviour of the height with respect to multiplication, which is the group operation of the algebraic group $\mathbb{G}_m$ (*i.e.* $\mathbb{P}_1 \setminus \{0, \infty\}$ endowed with the multiplication law). Now, similar and even sharper properties are valid for 'canonical' heights (so-called *Néron-Tate heights*) on other algebraic groups, namely on abelian varieties and in particular elliptic curves. These properties admit fundamental applications to the proof of many important arithmetical theorems. We do not enter here into any detail about this; see [83] especially for elliptic curves and [17] and [77] for the general case.

We go on by proving an important result on the behaviour of the height of an algebraic number under a rational function. This somewhat generalizes (in a weak sense) property 2 of the last proposition and the subsequent remark.

To illustrate what can happen, consider first the case of a given rational function $r(x) = p(x)/q(x) \in \mathbb{Q}(x)$ (p, q coprime polynomials) evaluated at a variable rational number m/n (m, n coprime integers). We have $r(m/n) = f(m, n)/g(m, n)$, where $f, g \in \mathbb{Z}[X, Y]$ are coprime homogeneous polynomials, associated in a simple way to p, q. Now, recall Exercise 2.2: it states that the values $f(m, n), g(m, n)$ have a bounded gcd (*i.e.*, there is 'little' cancellation!); but then the height $h(r(m/n))$ is clearly $\log \max(|f(m, n)|, |g(m, n)|) + O(1)$. Also, the maximum is $\asymp \max(|m|, |n|)^{\deg(r)}$, as is easy to see. (Again, this depends on f, g being coprime.) Hence $h(r(m/n)) = \deg(r)h(m/n) + O(1)$. The following result reproduces this simple idea in greater detail and in the general case of algebraic (rather than rational) numbers.

Proposition 3.2. *Let $r \in \overline{\mathbb{Q}}(X)$ be a rational function. Then, for $\alpha \in \overline{\mathbb{Q}}$ we have $h(r(\alpha)) = \deg(r)h(\alpha) + O(1)$ where the implied constant depends only on r (and can be explicitly estimated in terms of the height of the coefficients involved in $r(X)$).*

Note that $\deg(r) = \max(\deg(p), \deg(q))$ is the height of r in the function field $\overline{\mathbb{Q}}(X)$; the formula above is a sort of behaviour of 'composition' of heights in different fields.

Proof. As above we write $r(t) = p(t)/q(t)$ for coprime polynomials $p, q \in k[t]$ (k a number field) of maximum degree $d =: \deg(r)$.

The general case works in almost the same way as for rationals, but we have to be more careful: the degree of α can grow, so we need to use appropriate normalizations on the valuations of this variable field $K = K_\alpha := k(\alpha)$ and check that various numbers do not depend on it.

For clarity we set, for each place v of K, $\|x\|_v = |x|_v^{\lambda_v}$, where $|\cdot|_v$ extends some 'standard' valuation $|\cdot|_v$ on $\mathbb{Q}$ and $\lambda_v = \frac{[K_v : \mathbb{Q}_v]}{[K : \mathbb{Q}]}$. We have

$$H(r(\alpha)) = H(r(\alpha) : 1) = H(p(\alpha) : q(\alpha))$$
$$= \prod_{v \in M_K} \sup(\|p(\alpha)\|_v, \|q(\alpha)\|_v)$$

To ease notation, let us denote $\sigma_v = \sup(|p(\alpha)|_v, |q(\alpha)|_v)$.

Let $p(X) = p_0 X^d + p_1 X^{d-1} + \ldots + p_d, q(X) = q_0 X^d + q_1 X^{d-1} + \ldots + q_d$. Then

$$\sigma_v \leqslant \sup(1, |\alpha|_v)^d \sup\{|p_0|_v, \ldots, |p_d|_v, |q_0|_v, \ldots, |q_d|_v\} \sup(1, |d+1|_v).$$

Hence

$$H(r(\alpha)) = \prod_{v \in M_K} \sigma_v^{\lambda_v} \leqslant H^d(\alpha) H(p_0 : \ldots : p_d : q_0 : \ldots : q_d) |d+1|,$$

whence $h(r(\alpha)) \leqslant dh(\alpha) + O(1)$.

The converse inequality is more delicate, and in fact will have to take into account that $p(X), q(X)$ are coprime. Note that a big contribution to $H(\alpha)$ comes from those places v for which $\|\alpha\|_v$ is 'big'; when $\|\alpha\|_v$ is smaller, we need to verify that σ_v is not too small; this means that α is not v-adically near simultaneously to a zero of p and a zero of q. In fact this cannot happen because p, q have no common zero.

To carry out this program, assume e.g. $p_0 \neq 0$ (the case $q_0 \neq 0$ being similar) and define

$$c_v = \begin{cases} 2(d+1) \sup_i \dfrac{|p_i|_v}{|p_0|_v} & \text{for } v \in M_K^\infty \text{ (archimedean)} \\[2ex] \sup_i \dfrac{|p_i|_v}{|p_0|_v} & \text{for } v \in M_K^0 \text{ (non-archimedean)} \end{cases}$$

Since p and q are coprime, there exist polynomials $a, b \in k[X]$ of degree $\leqslant d$ such that $1 = a(X)p(X) + b(X)q(X)$. We shall denote by $a_0, \ldots, b_0, \ldots$ the coefficients of a, b. Then

$$1 \leqslant |a(\alpha)p(\alpha) + b(\alpha)q(\alpha)|_v \leqslant \sigma_v \sup(|a(\alpha)|_v, |b(\alpha)|_v) \sup(1, |2|_v)$$
$$\leqslant \sigma_v \sup(1, |\alpha|_v)^d \sup(|a_0|_v, \ldots, |a_d|_v, |b_0|_v, \ldots, |b_d|_v) \sup(1, |2|_v).$$

We estimate σ_v by partitioning the set $M = M_K$ in two subsets M', M''.[6] We define M' as the set of v such that $|\alpha|_v \leqslant c_v$ and $M'' := M \setminus M'$.

---

[6] By using the homogeneizations of $p(X), q(X)$ we could avoid this distinction and make the proof more uniform.

For $v \in M'$ we have $\sup(1, \|\alpha\|_v) \leqslant c_v^{\lambda_v}$ and $\prod_v c_v^{\lambda_v} \leqslant H(p_0 : \ldots : p_d) 2(d+1)$. So, raising to the powers λ_v and taking the product of the last displayed inequality for all $v \in M'$, we obtain

$$1 \leqslant C_1 \prod_{v \in M'} \sigma_v^{\lambda_v},$$

where $C_1 = 2H(p_0 : \ldots : p_d)^d H(a_0 : \ldots : b_0 : \ldots : b_d)(2(d+1))^d$ depends only on p, q.

For $v \in M''$, i.e. $|\alpha|_v > c_v$, we have

$$|p(\alpha)|_v = |p_0|_v |\alpha|_v^d \left| 1 + \frac{p_1}{p_0} \frac{1}{\alpha} + \frac{p_2}{p_0} \frac{1}{\alpha^2} + \ldots + \frac{p_d}{p_0} \frac{1}{\alpha^d} \right|_v.$$

For an 'infinite' place v, $|p(\alpha)|_v \geqslant \frac{|p_0|_v}{2} |\alpha|_v^d$, whereas in the finite case $|p(\alpha)|_v = |p_0|_v |\alpha|_v^d$. Then $\|\alpha\|_v^d \leqslant \|p(\alpha)\|^{\lambda_v} \|p_0^{-1}\|_v \sup(1, \|2\|_v) \leqslant \sigma_v^{\lambda_v} \|p_0^{-1}\|_v \sup(1, \|2\|_v)$.
Taking the product over $v \in M''$ we obtain

$$\prod_{v \in M''} \|\alpha\|_v^d \leqslant 2H(p_0) \prod_{v \in M''} \sigma_v^{\lambda_v}.$$

Further, we note that

$$H(\alpha) = \left(\prod_{v \in M'} \sup(1, \|\alpha\|_v) \right) \left(\prod_{v \in M''} \sup(1, \|\alpha\|_v) \right) \leqslant C_1 \prod_{v \in M''} \|\alpha\|_v.$$

Taking all of this into account we finally get

$$H(r(\alpha)) = \prod_v \sigma_v^{\lambda_v} \geq C_1^{-1} (2H(p_0))^{-1} \prod_{v \in M''} \|\alpha\|_v^d$$

$$\geq C_1^{-d-1} (2H(p_0))^{-1} H(\alpha)^d$$

concluding the proof. $\qquad\qquad\qquad\qquad\qquad\qquad\qquad\qquad\qquad\square$

Corollary 3.3. *Let $R : \mathbb{P}_1 \to \mathbb{P}_n$ be a rational function defined over $\overline{\mathbb{Q}}$. Then for $P \in \mathbb{P}_1(\overline{\mathbb{Q}})$ we have $h(R(P)) = \deg(R)h(P) + O(1)$, where the implied constant depends only on R.*

Proof. For $t \in \overline{\mathbb{Q}}$ we may write $R(t) = (P_0(t) : \ldots : P_n(t))$, where the $P_i \in \overline{\mathbb{Q}}[X]$ are polynomials with no common zero, of maximal degree $d := \deg(R)$. The required upper bound for $h(R(P))$ follows easily as in the proof of the proposition: for $P_i \neq 0$ one just uses $\|P_i(t)\|_v \leq C_{i,v} \sup(1, \|t\|_v)^{\deg P_i}$, where the constants $C_{i,v}$ are estimated depending on whether v is or is not archimedean and satisfy $\prod_v C_{i,v} \leq (\deg P_i + 1)H(P_i)$, where $H(P_i)$ denotes the height of the vector of coefficients of P_i.

The opposite inequality may be deduced directly from the proposition: By using a projective linear transformation and Exercise 3.2 one may assume that $\deg P_0 = d$ and that P_0, P_1 are coprime. Then clearly $h(R(t)) \geq h(P_0(t) : P_1(t))$. Now, the right side is $\geq dh(t) + O(1)$ in virtue of the proposition.

Naturally, the corollary may be also proved directly, similarly to the proposition. □

Remark 3.6.

(i) More general results are known, for rational functions and algebraic points on an irreducible curve C defined over $\overline{\mathbb{Q}}$. Taking two nonconstant functions $\varphi, \psi \in \overline{\mathbb{Q}}(C)$ we can ask how $h(\varphi(P))$ and $h(\psi(P))$ are related for P varying in $C(\overline{\mathbb{Q}})$. Generalizing Proposition 3.2, Siegel proved that $\frac{h(\varphi(P))}{h(\psi(P))} \sim \frac{\deg \varphi}{\deg \psi}$ as $h(\psi(P)) \to \infty$. A proof is not very complicated but is subtle and relies on a few geometric properties of heights which we do not treat here.

Proposition 3.2 represents the case when $C = \mathbb{P}_1$ has genus zero; now, taking $\varphi = r(t)$, $\psi = t$, the asymptotic formula holds in a more precise form, namely $h(\varphi(P)) = h(\psi(P))\frac{\deg \varphi}{\deg \psi} + O(1)$. After Siegel, Néron proved that for a higher genus we have $h(\varphi(P)) = h(\psi(P))\frac{\deg \varphi}{\deg \psi} + O(\sqrt{h(\psi(P))})$, and it may be shown that this result cannot be strenghtened.[7] See [17] or [77] for proofs. Bombieri interpreted these results as quantitative forms of Weil's Decomposition Theorem, and gave a new proof in terms of his theory of G-functions. See [13].

(ii) Note that the places v where $r(\alpha)$ is big are those for which α is v-adically near a pole of $r(x)$. The above result says that the magnitude of the height of $r(\alpha)$ essentially depends only on the number of poles; this yields that each pole of r will be be approached by α more or less with the same 'frequency', as the place v varies; and the same phenomenon happens for any rational function on a curve, a fact more delicate than for ordinary rational functions. This may be read as a sort of equidistribution, pointed out by Bombieri in [13].

Exercise 3.8. Let $C/\overline{\mathbb{Q}}$ be a curve and let $\varphi, \psi \in \overline{\mathbb{Q}}(C)$ be nonconstant. Prove that for $P \in C(\overline{\mathbb{Q}})$ we have $h(\varphi(P)) \leq \deg(\varphi)h(\psi(P)) + O(1)$, where the implied constant depends only on φ, ψ, not on P. (Hint: there is an equation $\varphi^d + a_1(\psi)\varphi^{d-1} + \ldots + a_d(\psi) = 0$, where $d \leq \deg \psi$ and the a_i are rational functions of degree $\leq \deg(\varphi)$. Now specialize at P and use the result of Exercise

---

[7] This is the case already for genus 1, as may be seen by composing with a translation map and using the quadratic behaviour of the canonical height; for this, see [83] for a self-contained approach.

3.6 and Corollary 3.3.) In view of the last Remark, this conclusion is rather weak, but it admits a simple explicit proof and may be useful in some cases, as in the next exercise.

Exercise 3.9. Let $C/\overline{\mathbb{Q}}$ be a curve and let $\psi_1, \ldots, \psi_n \in \overline{\mathbb{Q}}(C)$ be nonconstant. Prove that for any $B > 0$ there are points $P \in C(\overline{\mathbb{Q}})$ such that $\min_j h(\psi_j(P)) > B$. (Hint: pick any nonconstant $\varphi \in \overline{\mathbb{Q}}(C)$ and note that for all but finitely many $\xi \in \overline{\mathbb{Q}}$ we can find points $P \in C(\overline{\mathbb{Q}})$ with $\varphi(P) = \xi$. Then by the previous exercise we have $h(\xi) \ll \min \psi_i(P) + O(1)$. Now it suffices to choose ξ with large enough height.)

Note that the conclusion is not necessarily true if we use a fixed absolute value in place of the height function. (Consider for instance the case $C = \mathbb{P}_1$, $\psi_1(t) = t$, $\psi_2(t) = t^{-1}$.)

Exercise 3.10. Obtain another proof of the result in the last exercise, by picking infinitely many points P of bounded degree and using Northcott finiteness Theorem, to be stated and proved below. (Hint: note that the degree of the $\psi_i(P)$ is then bounded, hence if some $\psi_i(P)$ has bounded height it must lie in a finite set for varying P, by Northcott Theorem, a contradiction.)

We conclude this subsection by stating a very easy, but often very useful, basic principle, which generalizes the *fundamental theorem of transcendence* (in Masser's description), namely the assertion that "*an integer* > 0 *must be* ≥ 1". This generalization is sometimes called **Liouville inequality** [8] and reads:

Proposition 3.4. *Let ξ be a nonzero algebraic number. Then, for all absolute values v_0 of a number field K containing ξ we have $\|\xi\|_{K,v_0} \geq H(\xi)^{-1}$.*

Here with the notation $\|\cdot\|_{K,v_0}$ we mean that the absolute value has been normalized to K so that the height is given by (3.2).

It is easy to prove this proposition: from the product formula $\prod_v \|\xi\|_v = 1$ we deduce the inequality $\|\xi\|_{v_0} \prod_{v \neq v_0} \sup(1, \|\xi\|_v) \geq 1$, whence $\|\xi\|_{v_0} \geq (\prod_{v \neq v_0} \sup(1, \|\xi\|_v))^{-1} \geq H(\xi)^{-1}$.

Exercise 3.11. Formulate a version of Proposition 3.4 with $\alpha - \beta$ in place of ξ, where α, β are elements of number fields k, L. (This is 'Liouville inequality' as stated in [17], p. 21.)

Exercise 3.12. In the case $k = \mathbb{Q}$, $L = \mathbb{Q}(\xi)$, $\alpha = p/q$, a rational approximation to β, find back a version of Liouville Theorem 2.3 from the preceding exercise and Proposition 3.4.

---

[8] This comes from the relation with Liouville Theorem, as in the following exercises.

3.2.2. Mahler's measure

In this subsection we present a relationship between the Weil height of algebraic numbers and the so-called Mahler's measure of polynomials. This has also the advantage that it gives a convenient way to calculate the height of explicit numbers. Again, this notion was introduced after similar, less convenient, ones. We define the *Mahler's measure* $M(f)$ of a nonzero polynomial $f \in \mathbb{C}[x]$ by

$$M(f) = \exp \int_0^1 \log \left| f(e^{2\pi i \theta}) \right| d\theta.$$

We leave to the interested reader to show that the integral converges for all nonzero polynomials $f \in \mathbb{C}[x]$. Clearly $0 < M(f) \le \sup_{|z|=1} |f(z)|$. Note the multiplicative property

$$M(fg) = M(f)M(g) \qquad \text{for all} \quad f, g \in \mathbb{C}[X], \ fg \ne 0. \tag{3.6}$$

As a simple application, this allows to express $M(f)$ in terms of the complex roots of f. We have in fact the following proposition, where we use the useful convention $\log^+ t := \max(0, \log t)$ for a real number $t > 0$.

Proposition 3.5. *Let* $f(x) = a \prod_{i=1}^d (x - \xi_i)$, *where* $a \in \mathbb{C}^*$ *and* $\xi_i \in \mathbb{C}$. *We have*

$$\log M(f) = \log |a| + \sum_{i=1}^d \log^+ |\xi_i|.$$

Proof. In view of (3.6), it suffices to show the conclusion when $f(x) = x - \xi$. For this we could invoke standard result from elementary complex-function theory, but we shall instead argue directly. Suppose first $|\xi| > 1$. Then, putting for notational convenience $e(\theta) = \exp(2\pi i \theta)$, we have $|f(e(\theta))|^2 = |\xi|^2 (1 + \xi^{-1} e(\theta))(1 + \bar{\xi} e(-\theta))$. Taking logarithms and expanding $\log(1 - z) = -\sum_{r>0} z^r / r$ for $|z| < 1$ we obtain

$$2 \log M(f) = 2 \log |\xi|$$

$$- \sum_{r>0} \frac{(-1)^r}{r} \left(\xi^{-r} \int_0^1 e(r\theta) d\theta + (\bar{\xi})^{-r} \int_0^1 e(-r\theta) d\theta \right)$$

$$= 2 \log |\xi|,$$

as desired. The case $|\xi| < 1$ is similar, on writing $f(x) = x(1 - \xi x^{-1})$. Finally, the case $|\xi| = 1$ may be obtained by continuity. (See exercise below.) □

Exercise 3.13. Prove that $\lim_{r\to 1^-} \int_0^1 \log|1 - re^{2\pi i\theta}|\, d\theta = \int_0^1 \log|1 - e^{2\pi i\theta}|\, d\theta$. (Hint: estimate the integrals in a small neighborhood I of $\theta = 0$ and then use uniform convergence of the left integrand outside I.)

Remark 3.7. **(Mahler measure and Gelfond's inequality)** Among many other things, the Mahler measure can be used to give an alternative proof of Gelfond's inequality (improved with respect to Exercise 2.17).

For this, let A_f denote the maximum of the absolute values of the coefficients of a polynomial $f \in \mathbb{C}[X]$ (the 'norm' of Chapter 2). In the first place, by convexity of the exponential function and Parseval's formula we have

$$M(f) \leq \int_0^1 |f(e^{2\pi i\theta})|\, d\theta \leq \left(\int_0^1 |f(e^{2\pi i\theta})|^2 d\theta \right)^{1/2} \tag{3.7}$$

$$\leq \sqrt{(\deg f + 1)} A_f.$$

Now, by the expression of the coefficients as symmetric functions of the roots and by Proposition 3.5 we easily find

$$A_f \leq \binom{\deg f}{[\deg f/2]} M(f).$$

Further, using all of this for f, g, fg in place of f, where f, g have degrees m, n, we derive

$$A_f A_g \leq \binom{m}{[m/2]}\binom{n}{[n/2]} M(f)M(g) = \binom{m}{[m/2]}\binom{n}{[n/2]} M(fg)$$

$$\leq \binom{m}{[m/2]}\binom{n}{[n/2]} \sqrt{m + n + 1} A_{fg}.$$

Finally, using the easy estimates $\binom{A}{a}\binom{B}{b} \leq \binom{A+B}{a+b}$ and $\binom{r}{[r/2]}(r+1)^{1/2} \leq 2^r$ (see e.g. [17, page 27] or next exercise) we conclude

$$A_f A_g \leq 2^{\deg fg} A_{fg}.$$

Exercise 3.14. Prove that for positive integers A, a, B, b, r we have $\binom{A}{a}\binom{B}{b} \leq \binom{A+B}{a+b}$ and $\binom{r}{[r/2]}(r+1)^{1/2} \leq 2^r$. (Hint: the first part follows from $(1+x)^A(1+x)^B = (1+x)^{A+B}$. For the second part one can use an easy induction.)

Exercise 3.15. **(Ultrametric Gelfond's inequality)** Prove that for a finite place v, Gelfond's inequality holds in a more precise form, actually as an equality: $A_f^v A_g^v = A_{fg}^v$ for $f, g \in k_v[X]$, where A_f^v is the v-adic sup-norm of the coefficients, often called 'Gauss norm'. (Hint: this boils down to Gauss' Lemma.)

Exercise 3.16. **(Height of polynomials)** For a nonzero polynomial $f(X) = a_0 X^d + \ldots + a_d \in k[X]$ over a number field k, define the height $H(f) := \prod_v \sup(\|a_0\|, \ldots, \|a_d\|)$, with the usual normalizations, i.e. the projective height of the coefficient-vector.

(i) Prove that $c_1 H(f_1)H(f_2) \leq H(f_1 f_2) \leq c_2 H(f_1)H(f_2)$, where $c_1, c_2 > 0$ depend only on $\deg(f_1 f_2)$. (Hint: use Gelfond's inequality with respect to all places of k.)

(ii) Deduce that if $\xi_1, \ldots, \xi_d$ are the roots of f we have $\sum_i h(\xi_i) = \log H(f) + O(1)$, where the implied constant depends only on $d = \deg f$. (Hint: factor f over $\overline{\mathbb{Q}}$. This is related to Exercise 3.6; apart from the $O(1)$-term, it improves on that result.)

The Mahler measure has a close relationship with the absolute Weil height on $\overline{\mathbb{Q}}$. Combining it with Proposition 3.5 above, we obtain a very useful device for actually computing the height of an algebraic number in terms of its minimal polynomial over $\mathbb{Z}$.

Proposition 3.6. *Let $\xi \in \overline{\mathbb{Q}}$ be an algebraic number of degree d and let $f \in \mathbb{Z}[X]$ be its minimal polynomial over $\mathbb{Z}$, with integer, coprime coefficients, and positive leading coefficient a. Then $H^d(\xi) = M(f) = a \prod_{i=1}^{d} \sup(1, |\xi_i|)$, where ξ_i are the conjugates of ξ.*

Proof. We give the proof only for the easier case of algebraic integers (*i.e.* $a = 1$) and refer to [17] for the general case. We compute the height by working in $K = \mathbb{Q}(\xi)$, with the normalizations that we have seen before. If ξ is an algebraic integer, we have $\|\xi\|_v \leq 1$ for every finite place v, so these places contribute trivially to $H(\xi)$. As to the infinite places, they correspond to the embeddings σ of K in $\mathbb{C}$ up to complex conjugation. Moreover, the normalizations are such that, if v corresponds to σ, $\|\xi\|_v = |\xi^\sigma|^{d_v/d}$, where $|\xi^\sigma|$ is the complex absolute value of the conjugate ξ^σ of ξ (in a fixed embedding of $K \subset \overline{\mathbb{Q}} \subset \mathbb{C}$) and where d_v is the local degree: this is 1 if σ is real and 2 if σ is not real. So we have

$$H^d(\xi) = \prod_v \sup(1, |\xi^\sigma|^{d_v}) = \prod_\sigma \sup(1, |\xi^\sigma|),$$

where the last product is taken over all embeddings σ: in fact, in the first product we have only one representative for each pair of complex conjugate embeddings, but with the exponent 2, which is the same as taking both conjugates with the exponent 1.

Now, it suffices to apply Proposition 3.5 above. □

Remark 3.8.

(i) The 'Ultrametric Gelfond inequality' of the above exercise may be used to obtain a full proof of this last proposition, by applying it to a complete factorization of the minimal polynomial $f(X)$.

(ii) Before Weil, Mahler defined his height of an algebraic number ξ precisely as $M(f)$, as in Proposition 3.6. This proposition is useful for computing the height of specific numbers, but it may be not helpful for proving general properties of the height, as for instance properties 2,3 in Proposition 3.1.

Let us see some further examples.

- For every integer $m > 0$ we have $h(2^{\frac{1}{m}}) = \frac{1}{m}h(2) = \frac{1}{m}\log 2$. This follows immediately from n. 2 of Proposition 3.1 or (in a much less direct way) from Proposition 3.6, taking into account that the minimal polynomial of $2^{1/m}$ is $X^m - 2$.
- Let us compute the height of $1 + \sqrt{2}$, applying this last method. The minimal polynomial of $1 + \sqrt{2}$ is $X^2 - 2X - 1$. We have $a = 1$ and the only other conjugate is $1 - \sqrt{2}$, which is < 1 in absolute value. Therefore $H^2(1 + \sqrt{2}) = |1 + \sqrt{2}| = 1 + \sqrt{2}$.
- The height of $1 + i = 1 + \sqrt{-1}$. Again this is an algebraic integer, but its conjugate has the same absolute value, $i.e.$ $\sqrt{2}$. Then $H^2(1+i)=2$.

3.2.3. Further properties of the height on $\overline{\mathbb{Q}}$

The height is a very useful notion for expressing properties related to Diophantine Approximation; actually, the study of the height may be itself seen as a part of Diophantine Approximation. With this in mind, we shall now prove some quantitative and finiteness properties of the height. We start with an easy, though very important result.

Theorem 3.7 (Northcott). *For every fixed positive integers B and d, there exist only finitely many algebraic numbers $\alpha \in \overline{\mathbb{Q}}$ such that $h(\alpha) \leqslant B$ and $\deg(\alpha) \leqslant d$.*

Consequently, for any given n, there are only finitely many points $P \in \mathbb{P}_n(\overline{\mathbb{Q}})$ with bounded height and bounded degree.

Proof. For instance for $d = 1$ the first assertion reduces to the fact that there are only finitely many rational numbers with bounded height. This is immediate from the formula $h(p/q) = \log\max(|p|, |q|)$ for coprime integers p, q, proved above.

The general case can be reduced to this one. Let α have degree $\leq d$ and height $h(\alpha) \leq B$. Recall from Proposition 3.1 that the height is invariant under Galois automorphism, so if $h(\alpha) \leqslant B$ then $h(\alpha^\sigma) \leqslant B$ for all conjugates α^σ of α. Also, recall from the same proposition, n. 3, the inequalities for the height of sums and products. On combining these inequalities we obtain that the height of the elementary symmetric function of degree r of the conjugates of α is bounded by $\binom{d}{r}(rh(\alpha) + 1) \leq 2^d(dB + 1)$. Now, any symmetric function of the conjugates of α is a rational number and for given d, B there are only a finite number of them with height $\leq 2^d(dB + 1)$. Thus, also the minimal polynomial of any such α lies in a finite set, and the same holds for the root α.

Now, the assertion about projective points follows immediately from the left inequality in (3.5): $h(1 : x_1 : \ldots : x_n) \geq \sup h(x_i)$. □

Note that the proof is effective, in the sense that it produces explicit bounds and allows to calculate the algebraic numbers in question, for any given d, B. However the estimates so obtained for their number will be rather poor; we do not pursue here with refinements in this direction and refer to the relevant (vast) literature (see e.g. [17] and [77] for further results and references).

Remark 3.9.

(i) Note that a bound just on the height does not imply finiteness, of course: consider e.g. the roots of unity, which have zero height. Actually, if we do not bound the degree, the (logarithmic) heights of algebraic numbers are dense in $\mathbb{R}_{\geq 0}$: consider for instance $h(2^r) = |r| \log 2$, for $r \in \mathbb{Q}$.

(ii) Often one applies the result to deduce that there are only finitely many elements of bounded height in a given number field; however the full assertion is much stronger.

(iii) Northcott Theorem does not generally hold over function fields. Let for instance k/κ be the function field of a curve $\mathcal{C}$ over κ; we have seen that here the height is proportional to the degree. Hence if κ is infinite there are infinitely many elements of bounded height (think e.g. of the 'family' of rational functions in $\kappa(t)$ of given degree). The analogy is strict for function fields over finite fields. However, it remains true in general that, as in the case $k = \kappa(t)$, the elements of bounded height fall into finitely many 'families' (a concept which can be easily imagined by the reader, but that we do not formally define here).

Exercise 3.17. List and count all the algebraic numbers ξ of degree ≤ 2 such that $H(\xi) < \sqrt{2}$. Can you explain *a priori* the fact that their number is $\equiv 1$ (mod 4)?

Exercise 3.18. Let Ω be a finite subset of $\overline{\mathbb{Q}}$. Prove that for every $\epsilon > 0$ there exists a closed subset $\tilde{\Omega}$ of $\mathbb{C} \setminus \Omega$, with the following property: for every $B > 0$ there is a finite set F such that if $\alpha \in \overline{\mathbb{Q}} \setminus F$ and $h(\alpha) \leq B$, then α has at least $(1 - \epsilon) \deg(\alpha)$ conjugates in $\tilde{\Omega}$. (Hint: if 'many' conjugates of α go near to a number ξ in Ω, then the height of α becomes large, because $h(\alpha) \geq h((\alpha - \xi)^{-1}) + O(1)$.)

Exercise 3.19. Formulate and prove a version of Runge's Theorem (see the Supplements to the previous chapter) for a number field k. (Hint: if the divisor at infinity of an affine irreducible curve over k splits into $\geq r_1 + r_2 + 1$ strictly positive divisors defined over k and with pairwise disjoint supports, then the set of $\mathcal{O}_k$-integral points is finite.)

Exercise 3.20. Let $f_n(X) = X^n + X + 1$ and let $\xi_n \in \overline{\mathbb{Q}}$ be one of its roots. Prove that $H(\xi_n) \to 1$ as $n \to \infty$ (easy case of Exercise 3.7). Then deduce that for $n \not\equiv 2 \pmod 3$ the minimal degree of an irreducible factor (over $\mathbb{Q}$) of f_n tends to ∞ with n.[9] What happens for $n \equiv 2 \pmod 3$?

---

[9] Selmer proved that actually f_n is irreducible for such n. Here is a sketch of an argument due to Ljunggren. For $F(x) = \sum a_m x^m \in \mathbb{C}[x, 1/x]$, set $N(F) = |a_0|^2$. Suppose now $f = f_n =$

Especially at the light of Northcott Theorem, it is interesting to study the numbers with small height. As we have remarked, the smallest possible (logarithmic) height is 0; this lower bound is attained by 0 and the roots of unity: this has been observed in Example 1 at page 85, by noting that all absolute values of a root of unity are 1. Alternatively, note that for a root of unity ζ of order $n > 0$ we have $nh(\zeta) = h(\zeta^n) = h(1) = 0$, so $h(\zeta) = 0$.

It is an important result of Kronecker (anticipated in the said example) that the converse holds:

Theorem 3.8 (Kronecker). *Let $\xi \in \overline{\mathbb{Q}}^*$ be a nonzero algebraic number with $h(\xi) = 0$. Then ξ is a root of unity.*

Proof. Let $h(\xi) = 0$. By n. 2 of Proposition 3.1, for every positive n we have $h(\xi^n) = 0$. Also, the powers ξ^n lie in $\mathbb{Q}(\xi)$ and in particular have bounded degree. By Northcott's Theorem the set of such powers is finite. Thus there exist two integers $n > m \geq 0$ with $\xi^n = \xi^m$, so that ξ, being nonzero, must be a root of 1. $\square$

Remark 3.10.

(i) Kronecker's original statement was not formulated in terms of heights and read: *if all the conjugates of an algebraic integer $\xi \neq 0$ have absolute value ≤ 1, then ξ is a root of unity.* This is plainly equivalent with the present statement: to say that ξ is an algebraic integer means that $\|\xi\|_v \leq 1$ for all finite places v; so Kronecker's assumptions amount to $\|\xi\|_v \leq 1$ for all places, which in turn corresponds to $h(\xi) = 0$.

(ii) Also in connection with (i), we observe that Proposition 3.6 yields that a number $\xi \in \overline{\mathbb{Q}}$ with 'small' $h(\xi)$ must have 'many' conjugates near the unit circle. In fact, if ξ_i are the conjugates, the said proposition implies $d^{-1} \sum \log^+ |\xi_i| \leq h(\xi)$; applying this with $1/\xi$ in place of ξ and recalling that $h(\xi) = h(1/\xi)$ we get $d^{-1} \sum |\log |\xi_i|| \leq 2h(\xi)$, and our claim follows. A much more precise result by Bilu, which we shall discuss in some detail in the next chapter, asserts, roughly speaking, that *the conjugates of numbers of small height tend to be equidistributed around the unit circle.*

(iii) As a consequence of Northcott Theorem, we also conclude that the degree of roots of unity tends to infinity; this is of course a trivial consequence of

---

$g(x)h(x)$, with $g, h \in \mathbb{Z}[x]$ of degrees > 0. Put $G(x) := g(x)h(1/x) = c_1 x^{m_1} + c_2 x^{m_2} + \ldots + c_l x^{m_l}$, with nonzero $c_i \in \mathbb{Z}$ and distinct m_i. We have $F(x) := f(x)f(1/x) = G(x)G(1/x)$ whence $3 = N(F) = \sum c_i^2$. But the c_i are integers, whence $l = 3$ and $c_i = \pm 1$ for $i = 1, 2, 3$. Then $G(x) = ax^r + bx^s + cx^t$ with integers $r > s > t$ and $a, b, c = \pm 1$. Now, $F(x) = G(x)G(1/x)$ yields $x^n + x^{n-1} + x + 3 + x^{-1} + x^{1-n} + x^{-n} = acx^{r-t} + \ldots$. One now compares coefficients and exponents, obtaining $a = b = c = \pm 1$ and $r - t = n$, $\{r - s, s - t\} = \{r - s, n - (r - s)\} = \{1, n - 1\}$. So either $r = s + 1$ or $r = n - 1 + s$. In the first case $G(x) = \pm x^r f(1/x)$, whence $g(x) = \pm x^r g(1/x)$ and similarly in the second case. This yields that g or h has a root ρ with ρ^{-1} also a root. The same must then hold for f, leading to cubic roots of unity, possible only for $n \equiv 2 \pmod 3$.

Gauss' well-known result that $[\mathbb{Q}(\zeta) : \mathbb{Q}] = \phi(n)$ for a ζ of order n, but it is worth observing that it also follows from consideration of heights.

(iv) The roots of unity are the torsion points on the algebraic group $\mathbb{G}_m$. With this in mind, we note that the above results admit analogues in the case of abelian varieties: the torsion points therein can be characterized as the points of zero 'canonical' height.

(v) The roots of unity have trivial valuation at all places of $\overline{\mathbb{Q}}$. Over function fields, *e.g.* of a complete curve $\mathcal{C}$ over an algebraically closed 'constant' field κ, precisely the elements of κ^* have this property: together with 0, they are the elements of zero (logarithmic) height. In that case such set makes up a field, contrary to the case of algebraic numbers.

Exercise 3.21. Prove directly that if all the conjugates of a nonzero algebraic integer have absolute value ≤ 1, then they all have absolute value $= 1$. (Hint: consider the norm.)

Exercise 3.22. Prove the following quantitative version of Kronecker's Theorem: *Let ξ be an algebraic integer of degree d such that $|\xi^\sigma| \leq 1 + 3^{-d-1}$ for all conjugates ξ^σ of ξ. Then ξ is a root of unity.* (Hint: write $\xi = re^{2\pi i\theta}$, with real $r \geq 0, \theta$. Now use Dirichlet Lemma 1.1 to find a positive integer $q \leq Q = Q(d)$ such that $q\theta$ is nearly integer; then ξ^q will be near to 1. Finally take the norm of $\xi^q - 1$.)

Exercise 3.23. (**A 'real' version of Kronecker's theorem**) Let ξ be a totally real algebraic integer such that all of its conjugates lie in the interval $[-2, 2]$. Prove that $\xi = \zeta + \zeta^{-1}$ for a suitable root of unity ζ. (Hint: consider a root ζ of the equation $x^2 - \xi x + 1 = 0$; it is an algebraic integer and all of its conjugates have absolute value ≤ 1. Now apply 3.8 to ζ.)

Exercise 3.24. Prove that for every proper closed subinterval I of $[-2,2]$ there are only finitely many totally real algebraic integers all of whose conjugates lie in I. (Hint: apply the result of the previous exercise.)

Exercise 3.25. Sharpen the previous conclusion as follows:

(i) Let $\rho_1, \ldots, \rho_d$ be real numbers in $[-2, 2]$. Prove that the Vandermonde determinant $V(\rho_1, \ldots, \rho_d) := \det(\rho_i^s)_{s=0,\ldots,d-1} = \pm \prod_{i<j}(\rho_i - \rho_j)$ satisfies
$$|V(\rho_1, \ldots, \rho_d)| \leq 2^d d^{(d-1)/2}.$$
(Hint: set $\rho_i = \zeta_i + \zeta_i^{-1}$, where $\zeta_i = e^{\sqrt{-1}\theta_i}$ lies in the unit circle. Then, if T_s is the Chebishev polynomial, prove $V(\rho_1, \ldots, \rho_d) = \pm \det(T_s(\rho_i)) = \det(\zeta_i^s + \zeta_i^{-s}) = 2^d \det(\cos s\theta_i)$, and the result follows by Hadamard's inequality.)

(ii) Observe that the inequality continues to hold if all the ρ_i lie in an interval of length ≤ 4. (Hint: note that $V(\rho_1, \ldots, \rho_d)$ depends only on the differences $\rho_i - \rho_j$.)

(iii) Let $\delta > 0$. Prove that up to addition of an integer, there exist only finitely many totally real algebraic integers ξ all of whose conjugates lie in an interval of length $\leq 4 - \delta$. (Hint: let $\xi_1, \ldots, \xi_d$ be the conjugates of ξ, and set $\rho_i = 4\xi_i/(4-\delta)$. Use the previous parts to show that $|V(\xi_1, \ldots, \xi_d)|^2 \leq (1 - \frac{\delta}{4})^{d(d-1)} 4^d d^{d-1} \to 0$ for $d \to \infty$. But $|V(\xi_1, \ldots, \xi_d)|^2$ is an integer > 0, so d is bounded. Now observe that the fractional part of ξ has bounded height.)

Exercise 3.26. (Preperiodic points) Let $r \in \overline{\mathbb{Q}}(X)$ be a rational function of degree > 1. Prove that the *preperiodic* points for r (*i.e.* those algebraic numbers ξ such that two iterates of r coincide on ξ) have degree tending to infinity. In particular, deduce again that the roots of unity have degree tending to infinity. (Hint: use Proposition 3.2 to prove that they have bounded height. Then note that the roots of unity are preperiodic for the map $x \mapsto x^2$.)

Having classified all elements of $\overline{\mathbb{Q}}$ with height 0, we turn to the following question: *assuming* $h(\xi)$ *is not zero, how 'small' can it be with respect to the degree* $d = [\deg(\xi) := \mathbb{Q}(\xi) : \mathbb{Q}]$?

Note that from Northcott Theorem it follows that the height of any $\xi \in \overline{\mathbb{Q}}$, if not 0, is bounded below by a strictly positive number depending only on $\deg(\xi)$. Just for the sake of example, in the following exercises we give a possible estimate in this direction.

Exercise 3.27. Following the proof of Northcott Theorem, produce an explicit upper bound for the number of algebraic numbers of degree $\leq d$ and height $\leq B$.

Exercise 3.28. Using the last exercise, prove an explicit strictly positive lower bound for $h(\xi)$, where ξ is an algebraic number of degree $\leq d$, not zero or a root of unity. Note that the bound will depend only on $\deg(\xi)$. Compare this with Exercise 3.22. (Hint: the numbers ξ^j, with integer $j \leq B/h(\xi)$, are distinct, have height $\leq B$ and degree $\leq d$. Now apply the above bound...)

The lower bound coming from the method suggested in the exercises is extremely poor, and indeed very far from the truth. What about some upper bounds? The numbers $2^{1/d}$ have degree d and height $h(2^{1/d}) = (\log 2)d^{-1}$. A similar behaviour occurs *e.g.* for solutions of equations $x^d + ax^m + b = 0$ for 'small' m and $a, b \in \overline{\mathbb{Q}}$ of small height. It seems not easy to find numbers with even smaller (but nonzero) height, compared to the degree; indeed, a celebrated conjecture reads as follows:

Conjecture 3.9 (D.H. Lehmer). *There exists an absolute constant* $c > 0$ *such that if* $\xi \in \overline{\mathbb{Q}}^*$ *has degree* d *and is not a root of unity we have* $h(\xi) \geq \frac{c}{d}$.

Remark 3.11.
  (i) Actually Lehmer formulated an equivalent conjecture for the Mahler Measure. The best known result in the direction of the conjecture is due to Dobrowolski, who obtained (under the same assumptions) the lower bound $h(\xi) \geq \frac{c}{d}(\frac{\log\log(3d)}{\log d})^3$. Another very interesting result is due to Breusch and to Smyth, who proved the lower bound of the conjecture (with a certain explicit constant, best-possible in Smyth's result), but assuming ξ to be *non-reciprocal*, namely not conjugate to ξ^{-1}. Moreover, for special subfields of $\overline{\mathbb{Q}}$, like the field generated by all roots of unity, even sharper (*e.g.* absolute) lower bounds hold. See [17], Chapter 4, the survey [85] and the notes below for informations on the relevant literature.

(ii) Similar lower bounds may be proved (or conjectured) for algebraic points in higher dimensional spaces. Actually, as we shall see in the next chapter, if we confine the points in a given algebraic variety, an absolute lower bound > 0 holds for the height (that is, dependent only on the variety, not on the degree of the point), provided we stay out of certain exceptional subvarieties.

Exercise 3.29. Prove that if $\xi \in \overline{\mathbb{Q}}^*$ has degree d and $h(\xi) < \frac{\log 2}{d}$, then ξ and ξ^{-1} are algebraic integers. (This shows that in testing Lehmer's conjecture one may consider only units. Hint: use Proposition 3.6.)

Exercise 3.30. Using Example 2 at page 86 prove a kind of 'Lehmer conjecture for function fields of curves'.

3.3. Some diophantine analysis over number fields

In this section, at the light of the notion of height, we shall reformulate some theorems in diophantine approximation in greater generality, for approximations by algebraic, rather than rational, numbers.

3.3.1. A generalized Roth Theorem

Recall that Roth Theorem 2.5 states that for every given $\xi \in \overline{\mathbb{Q}} \smallsetminus \mathbb{Q}$ and every fixed $\epsilon > 0$, the inequality $|q\xi - p| > q^{-1-\epsilon}$ holds for all but finitely many integer pairs (p, q).

This result is expressed in terms of *one* absolute value and one 'target' ξ, to be approached by the rational fraction p/q. Mahler and his student Ridout found a v-adic generalization, with several absolute values v and corresponding algebraic targets depending on v. As a simple example, let us state a special case of a result by Ridout:

Theorem 3.10. *(Ridout). *If* $\xi \in \overline{\mathbb{Q}} \setminus \mathbb{Q}, \epsilon > 0$ *are as above and all the prime factors of q (or of p) lie in a prescribed finite set S, then* $|q\xi - p| > q^{-\epsilon}$, *with only finitely many exceptions.*

So, the restriction on q to have all prime factors in S improves by 1 the exponent of approximation in Roth Theorem.

Exercise 3.31. Consider the decimal expansion of $\sqrt{2}$ and let a_n be the maximum number of consecutive 0 digits among the first n digits. Show that $\lim_{n \to \infty} \frac{a_n}{n} = 0$. (Hint: apply Ridout Theorem with $S = \{2, 5\}$, approximating $\sqrt{2}$ by a truncation of the decimal expansion.)

A view on Ridout's result is as follows. Let l be a prime dividing q to a high order. Then p/q is near ∞ in the l-adic valuation. If p/q is near ξ in the usual absolute value, we have a *simultaneous* approximation by p/q, to ∞ and to ξ, with respect to different absolute values. In Ridout

Theorem this approximation is 'distributed', so to say, among the usual absolute value and the ones corresponding to the primes l in S, whereas in Roth Theorem all the approximation is with respect to the usual place. This explains the apparent 'gain' in the exponent: rather, in the above statement of Theorem 3.10 the approximation to ∞ with respect to the places in S is hidden in the special shape of q.

After some intermediate steps by Mahler, Lang found a general formulation of an approximation theorem, for arbitrary number field and finite sets of places. In this version the strength of the approximation was measured by using the Weil height of the approximants. Here it is:

Theorem 3.11. *(Generalized Roth-Lang). *Let k be a number field and let S be a finite subset of M_k. Fix an $\epsilon > 0$ and, for every $v \in S$, let $\xi_v \in k_v \cap \bar{k}$. Then the following inequality holds for all but a finite number of $\beta \in k$:*

$$\prod_{v \in S} \min(1, \|\xi_v - \beta\|_v) \geqslant \frac{1}{H(\beta)^{2+\varepsilon}}.$$

The absolute values here are normalized with respect to k. This holds even for $\xi_v = \infty$, on using the convention $\|\beta - \infty\| := 1/\|\beta\|$. It is a result on simultaneous approximations, with respect to different valuations. Note that the theorem becomes stronger if we enlarge S, because every factor is ≤ 1. The above Roth and Ridout Theorems are immediate consequences of this (*e.g.* take $k = \mathbb{Q}$ and $S = \{\infty\}$ for Roth Theorem). The proof-technique is very similar (though more complicated) with respect to Roth's theorem, and in turn to Thue's theorem that we have proved above: one constructs an auxiliary polynomial, now with zeros of high order at all the ξ_v, and then compares upper and lower bounds by evaluating at the approximation β, and making use of the product formula. (Of course the full Roth's nonvanishing lemma, which appears only in a very simple way in Thue's proof, is needed here.) See [17] for a complete proof.

Exercise 3.32. Using the above result, show that if ξ is an irrational algebraic number, if $\epsilon > 0$ and if all prime factors of both integers p and q are in a prescribed finite set S, then $|q\xi - p| > q^{1-\epsilon}$, with only finitely many exceptions.

Exercise 3.33. Prove in an elementary direct way that not all the integers $[10^n \sqrt{2}]$ can be composed only of primes from a fixed finite set. (Hint: consider for instance $[10^{n+s} \sqrt{2}] - 10^s [10^n \sqrt{2}]$.)

Exercise 3.34. Now show that the greatest prime factor of $[10^n \sqrt{2}]$ tends to infinity with n. Prove the same conclusion *e.g.* for $[10^n \sqrt{2}] + 3^n + n$. (Hint: use Ex. 3.32.)

Exercise 3.35.
  (i) Prove that for every positive $\delta < 1$ there are only finitely many integers $n > 0$ such that the fractional part of $(3/2)^n$ is $< \delta^n$. (Hint: consider the approximation of 1 by the fraction $[(3/2)^n]2^n/3^n$.)
  (ii) More generally, show that for large enough n and all integers $p, q > 0$ we have $q^2|(3/2)^n - (p/q)| > \delta^n$. Deduce Pourchet's result: *The length of Euclid algorithm for $3^n : 2^n$ tends to infinity as $n \to \infty$.* (Hint: Use the properties of continued fractions to show that the partial quotients for $3^n/2^n$ are small compared to 2^n.)

Exercise 3.36. Let $a \in \mathbb{Z}$. Prove that $5^n + 2^n + a$ can be a square only for finitely many $n \in \mathbb{N}$. (Hint: Use the generalized Roth Theorem to bound the distance of $\sqrt{5^n}$ from an integer. That $\sqrt{5} > 2$ is crucial with this approach. This kind of restriction can be removed with Schmidt Subspace Theorem; see [29] and [101, Chapter 4].)

Exercise 3.37. Use the original Roth Theorem to prove that $\sum_{j=0}^{\infty} 2^{-3^j}$ is transcendental. Use the generalized version to prove that $\sum_{j=0}^{\infty} 2^{-2^j}$ and $\sum_{j=0}^{\infty} 2^{-F_j}$ are transcendental, where F_j is the Fibonacci number. (We have noted that Liouville Theorem yields that $\sum_{j=0}^{\infty} 2^{-j!}$ is transcendental.)

3.3.2. S-integers, S-units

Also at the light of Theorem 3.11, it makes sense to focus our attention to finite sets $S \subset M_k$ of places of a number field k, and to the behaviour of numbers in k at places in S. Unless otherwise stated, from now on we shall make the assumption that S *contains all the archimedean places.* An element $x \in k$ is called an S-*integer* if $\|x\|_v \leq 1$ for all $v \notin S$.[10] The set of S-integers is denoted $\mathcal{O}_{k,S}$, or simply $\mathcal{O}_S$, when no confusion can arise.

Note that this set of S-integers in k is a ring. In fact, let $x, y \in \mathcal{O}_S$; then, for every place $v \notin S$ we have $\|xy\|_v = \|x\|_v \|y\|_v \leq 1$; also, since S contains all archimedean valuations, we have $\|x \pm y\|_v \leq \max(\|x\|_v, \|y\|_v) \leq 1$. When S consists just of the archimedean places, $\mathcal{O}_{k,S}$ equals the ring of integers $\mathcal{O}_k$ of k. This ring $\mathcal{O}_S$ may be also described as the localization of the usual ring of integers $\mathcal{O}_k$ with respect to the multiplicative subset of elements whose only ideal factors correspond to places in S. As a localization of an integrally closed ring, $\mathcal{O}_S$ is itself integrally closed, which can also be checked directly: if $x \in k$ satisfies a monic equation over $\mathcal{O}_S$ and if v is a finite place outside S, then the equation shows that $|x|_v \leq 1$; hence $x \in \mathcal{O}_S$. Further, note that for $S \subset S'$ we have $\mathcal{O}_S \subset \mathcal{O}_{S'}$.

---

[10] Note that this does not depend on the chosen normalization for the places.

The multiplicative group of invertible elements of $\mathcal{O}_S$ is denoted $\mathcal{O}_S^* = \{x \in k^* \mid x, x^{-1} \in \mathcal{O}_S\}$, and it is called the *group of S-units*. They can be characterized as those $x \in k^*$ such that $\|x\|_v = 1$ for all $v \notin S$. Note in particular that for $x \in \mathcal{O}_S^*$ the product formula holds in the form $\prod_{v \in S} \|x\|_v = 1$. Note that the roots of unity in k lie in $\mathcal{O}_{k,S}^*$ for all S.

We may view the S-integers as numbers in k whose denominator (as a fractional ideal) is divisible only by primes in S, and similarly for S-units. We remark that in the study of diophantine equations it is often convenient to allow such denominators, for instance in order to change freely the 'models' and to get more uniform statements.

For $S = M^\infty$ (the set of all infinite places of k), the ring $\mathcal{O}_S$ is just the set of algebraic integers in k, whereas $\mathcal{O}_S^*$ is the set of ordinary units; in this case we have already recalled Dirichlet Theorem on the generation of $\mathcal{O}_S^*$ (see the notes to Chapter 1). A more general result, which admits a similar proof, states that *the group $\mathcal{O}_S^*$ is finitely generated and isomorphic to $U \oplus \mathbb{Z}^{s-1}$, where U is the finite group of roots unity in k and $s := \#S$.* For proofs, see e.g. [54] or [61]; we also note that the proof that $\mathcal{O}_S^*$ is finitely generated (of rank $\leq s - 1$) is rather easier than the determination of the actual rank.

It is an easy but important observation that given any nonzero element $a \in k^*$ we may find a finite S such that $a \in \mathcal{O}_S^*$: it just suffices to include in S all the finitely many places $v \in M_k$ such that $\|a\|_v \neq 1$. By applying this to each element in a finite set of generators of a finitely generated subgroup Γ of k^*, we see that any such Γ is a subgroup of a suitable $\mathcal{O}_S^*$, with $S = S(\Gamma)$. Conversely, we have noted that $\mathcal{O}_S^*$ itself is finitely generated so the study of any finitely generated Γ often reduces to the study of the groups $\mathcal{O}_S^*$.

Remark 3.12. The S-integers and S-units can be defined also in fields other than number fields, e.g. function fields $k = \kappa(\mathcal{C})$ of curves $\mathcal{C}/\kappa$. In this last case the S-integers are those rational functions all of whose poles lie in S, while the S-units are those functions having both poles and zeroes in S (so $\mathcal{O}_S^*$ contains κ^*). In (partial) analogy with the number field case, the factor group $\mathcal{O}_S^*/\kappa^*$ is finitely generated, of rank $\leq \#S - 1$; however, if the curve has positive genus, the rank can be smaller (it will be 'usually' 0 in characteristic 0); it can be interpreted as the rank of the group generated by the classes of differences $P - Q$ of points $P, Q \in S$ in the Jacobian variety of the curve. (See the examples below and the supplement to Chapter 1 on the polynomial Pell Equation, for an interesting instance with $\#S = 2$.)

Examples. Let $k = \mathbb{Q}$. For $S = \{\infty\}$, we have $\mathcal{O}_S = \mathbb{Z}$ and $\mathcal{O}_S^* = \{\pm 1\}$. In general, for $S = \{\infty, p_1, \ldots, p_n\}$, the S-integers are those rationals whose denominator is composed only of primes among $p_1, \ldots, p_n$, i.e. $\mathcal{O}_S = \mathbb{Z}[p_1^{-1}, \ldots, p_n^{-1}] = \mathbb{Z}[(p_1 \cdots p_n)^{-1}]$. Similarly, the S-units are $\mathcal{O}_S^* = \{\pm p_1^{a_1} \cdots p_n^{a_n} \mid a_1, \ldots, a_n \in \mathbb{Z}\}$.

Let now $k = \mathbb{Q}(\sqrt{d})$, with $d > 1$ a squarefree integer, and let $S = M_k^\infty = \{\infty_+, \infty_-\}$; then the S-integers are the algebraic integers of k, given by $\mathcal{O}_S = \mathbb{Z}[\frac{1+\sqrt{d}}{2}]$ when $d \equiv 1 \pmod 4$ and $\mathcal{O}_S = \mathbb{Z}[\sqrt{d}]$ otherwise. In any case, $\mathcal{O}_S^* \cong \{\pm 1\} \times \mathbb{Z}$ by the theory of Pell Equation explained in Chapter 1.

Finally, let k/κ be the function field of a complete nonsingular curve C/κ, where the 'constant' field κ is algebraically closed. The constants κ^* are among the S-units; they are the elements without any zero or pole.[11] On the other hand, let $\phi \in \kappa(C)$ be an S-unit. If $S = \{P_1, \ldots, P_s\}$ for distinct points $P_i \in C(\kappa)$, its divisor is of the shape $\operatorname{div}(\phi) = m_1 P_1 + \ldots + m_s P_s$, with $m_1, \ldots, m_s \in \mathbb{Z}$, where $m_1 + \ldots + m_s = 0$. If $[P_i]$ denotes the image of $P_i - P_1$, say, in the Jacobian variety $J = J_C = \operatorname{Pic}_0(C)$ of C, we have $\sum m_i[P_i] = 0$. Conversely, such a relation leads to a function ϕ as above, determined uniquely up to an element of κ^*. This proves that the rank of $\mathcal{O}_S^*/\kappa^*$ can be at most $s - 1$.

If $C = \mathbb{P}^1$ (in which case J is trivial) then $\mathcal{O}_S^* \cong \kappa^* \oplus \mathbb{Z}^{s-1}$. Now the analogue of the theorem of Dirichlet holds. For C of positive genus, the rank can very well become smaller and in characteristic 0 actually will be 'usually' 0; for this it suffices that the $[P_i]$ ($i \geq 2$) are independent (over $\mathbb{Z}$) in J. Since J has the structure of a complex torus we may expect that this will happen for a 'generic' choice.[12] The analogy with the number field case remains strict (*i.e.*, $\mathcal{O}_S^*/\kappa^*$ has rank $s-1$) if κ is the algebraic closure of $\mathbb{F}_p$; we can see this for instance by noting that every point in $J(\kappa)$ is defined over a finite field and is therefore a torsion point. (Here the Riemann-Roch Theorem would suffice, without invoking the Jacobian; see also [64].)

Exercise 3.38. Let A be an abelian variety defined over $\overline{\mathbb{Q}}$. Prove that $A(\overline{\mathbb{Q}})$ has infinite rank. (Hint: for instance look at Galois group, or at heights, or...)

Heights of S-integers

In analogy with the fact that the height of an integer $a \in \mathbb{Z}$ is just $H(a) = |a|$, the height of S-integers can be computed by taking into account only the places in S. Namely, we have the simple useful formula

$$H(\alpha) = \prod_{v \in S} \sup(1, \|\alpha\|_v), \qquad \text{for all } \alpha \in \mathcal{O}_S. \tag{3.8}$$

---

[11] They are the elements of zero height. In a number field they would be 0 and the roots of 1, by Kronecker's theorem. Here, together with 0 they make up the field κ.

[12] Actually this last remark applies to complex points. However, even for points P_i over $\overline{\mathbb{Q}}$, it may be seen that a 'random' choice, in various meanings, leads to the independence of $[P_i]$ over $\mathbb{Z}$.

In view of the product formula, we have also the inequality

$$\prod_{v \in S} \|\alpha\|_v \geq 1, \qquad \text{for all } \alpha \in \mathcal{O}_S \setminus \{0\}, \qquad (3.9)$$

with equality if and only if $\alpha \in \mathcal{O}_S^*$. (Like Liouville Inequality, *i.e.* Proposition 3.4, this also extends the principle that 'a rational integer > 0 must be ≥ 1'.)

3.3.3. Some diophantine applications

We shall now present some applications of the Generalized Roth Theorem to diophantine equations, generalizing what we have seen in previous chapters. We start with Thue Equations.

Theorem 3.12 (Mahler). *Let $f \in k[X, Y]$ be homogeneous, of degree at least 3, without multiple factors. Then, for any $c \in k^*$, there are only finitely many solutions to $f(x, y) = c$, with $x, y \in \mathcal{O}_S$.*

Proof. Assuming Theorem 3.11. After Theorem 3.11, the idea is very simple, analogous to Proposition 2.2 to reduce Thue's equation over $\mathbb{Z}$ to Thue's theorem in diophantine approximation.

By enlarging k and after a linear change of coordinates we may factor f as $a \prod_{i=1}^d (X - \xi_i Y)$, where $a \in k^*$ and where the ξ_i are distinct elements of k. Supposing $y \neq 0$ and setting $\beta := x/y$, the equation gives $\prod(\beta - \xi_i) = c/ay^d$.

Suppose by contradiction that there are infinitely many solutions in $\mathcal{O}_S^2$. For a solution (x, y) and for a place $v \in S$, let $\xi_v = \xi_v(x, y)$ be one of the ξ_i such that $\|\beta - \xi_v\|_v$ is minimum. Since S is finite, by going to a still infinite subset of solutions we may suppose that, for all $v \in S$, ξ_v does not depend on x, y, *i.e.* is the same for all solutions.

Let $v \in S$ be such that $\|y\|_v > 1$. Since the ξ_i are pairwise distinct we easily see (as in Proposition 2.2) that $\min(1, \|\beta - \xi_v\|_v) \leq B_1 \|y\|_v^{-d}$, for a number B_1 depending (like the subsequent $B_2, \dots$) only on f and c. In any case the minimum is ≤ 1, so, for all $v \in S$,

$$\min(1, \|\beta - \xi_v\|_v) \leq B_2 \sup(1, \|y\|_v)^{-d}.$$

Taking the product over $v \in S$ and applying (3.8) to the S-integer y we get

$$\prod_{v \in S} \min(1, \|\beta - \xi_v\|_v) \leq B_3 H(y)^{-d}.$$

We are now almost in position to apply the Generalized Roth Theorem, but first we have to compare $H(y)$ with $H(\beta)$. This is not difficult: the equation $f(x, y) = c$ easily yields $\|x\|_v \leq B_4 \|y\|_v + B_5$ for every $v \in S$. Hence $H(\beta) = H(x : y) \leq B_6 H(y)$, since we can compute the height by using only the places in S, as in (3.8). In conclusion, the last displayed inequality gives $\prod_{v \in S} \min(1, \|\beta - \xi_v\|_v) \leq B_7 H(\beta)^{-d}$. By Theorem 3.11 and the fact that $d \geq 3$ we deduce that $H(\beta)$ is bounded, whence β has only finitely many possibilities, by Northcott Theorem. Since $c = f(x, y) = y^d f(\beta, 1)$, also x, y have after all only finitely many possibilities, proving the theorem. $\square$

Remark 3.13.

(i) For this proof, we have used the Generalized Roth Theorem, which is not proved in this book. However we shall soon deduce the result from the S-unit theorem, which will be completely proved in the final chapter.

(ii) We have used Northcott Theorem, so this proof does not fully carry over to function fields. But the result still holds there: on the one hand it may be proved that there are only finitely many solutions of bounded height: see the last of the Supplements to Chapter 2 for an example, whose method can be generalized. On the other hand we have just mentioned another proof depending on the S-unit theorem; this last theorem holds also over function fields, actually with an elementary proof (given in the Supplements below).

Exercise 3.39. Following the supplement to Chapter 2 on a polynomial Thue equation, prove that a Thue equation over the function field $\kappa(t)$ has only finitely many solutions of bounded height (*i.e.* bounded degree). (Hint: factor the equation over $\overline{\kappa(t)}$ as $\prod_{i=1}^{d}(X - \xi_i Y) = c$ and consider the divisors of the factors.)

Another important diophantine theorem is the so-called *S-unit Theorem*:

Theorem 3.13. *There are only finitely many solutions $x, y \in \mathcal{O}_S^*$ to $x + y = 1$.*

To better appreciate the implications of this result, recall that $\mathcal{O}_S^*$ is a finitely generated group; then, expressing the unknowns x, y in terms of products of powers of generators $\gamma_1, \ldots, \gamma_r$, we may read the equation $x + y = 1$ as the *exponential diophantine equation* $\gamma_1^{a_1} \cdots \gamma_r^{a_r} + \gamma_1^{b_1} \cdots \gamma_r^{b_r} = 1$ in the integer unknowns $a_i, b_j \in \mathbb{Z}$; if for instance the γ_i are multiplicatively independent[13], the result implies the finiteness of the set of solutions $(a_1, \ldots, a_r, b_1, \ldots, b_r)$. In the general case $k = \mathbb{Q}$, the theorem is due to Mahler and may be stated as follows: *If $p_1, \ldots, p_t, \ldots, p_u, \ldots p_r$ are distinct primes, the equation $p_1^{a_1} \cdots p_t^{a_t} +$*

---

[13] *i.e.* they do not satisfy any nontrivial relation $\prod \gamma_i^{m_i} = 1$, with integers m_i not all zero

$p_{t+1}^{a_{t+1}} \cdots p_u^{a_u} = p_{u+1}^{a_{u+1}} \cdots p_r^{a_r}$ has only finitely many solutions in the exponents $a_i \in \mathbb{Z}$. (Observe also that a special case of this appears as Corollary 2.11 in the Supplements to Chapter 2, with a complete and easy deduction from Thue's Theorem.) As in the exercises in the Supplements to Chapter 2, some simple instances of this equation can be solved in *ad hoc* ways. However, even for $r = 3$ the assertion seems not to admit a simple self-contained treatment. (See the next exercise for further special cases.)

Finally, note that any equation $ax + by = 1$, for fixed $a, b \in k^*$ and unknowns $x, y \in \mathcal{O}_S^*$ can be seen as an S-unit equation $x' + y' = 1$, if we enlarge S enough so that $a, b \in \mathcal{O}_S^*$. So any such equation has only finitely many solutions as well.

Exercise 3.40. Prove in a direct elementary way that, for a given integer $b > 1$ and given primes $p_1, \ldots, p_h$, the equation $b^l = p_1^{a_1} \cdots p_h^{a_h} + 1$ has only finitely many solutions in integers $l, a_1, \ldots, a_h$. (Hint: prove that l must be divisible by 'large' powers of the p_i.)

In the rest of the present chapter we shall explore the mutual relationships among the above theorems: we shall see arguments to deduce the S-unit Theorem 3.13 from the Generalized Roth Theorem 3.11 and we shall see mutual implications of the former with Mahler's Theorem 3.12.

Although we shall not prove Theorem 3.11 in these lecture notes, in the last chapter we shall prove, by a different method, a complete (and very sharp) quantitative version of Theorem 3.13; because of the mentioned implication, to appear soon in this chapter, this will also prove Theorem 3.12, providing so a partial remedy to our omission of the proof of Theorem 3.11.

Proof of Theorem 3.13, *assuming Theorem* 3.11. We apply Roth's Theorem 3.11 as follows. If $(x, y = 1 - x)$ is a solution, namely $x, 1 - x$ are both S-units, we set $\beta = x$. Also, we partition the places v in S into three disjoint sets S_1, S_2, S_3 according as $\|x\|_v \leq 1/2, 1/2 < \|x\|_v \leq 2$ and $\|x\|_v > 2$ respectively. This partition of S may depend on the solution, but since S is finite we may suppose for our purposes that the partition is the same for all solutions. In these three cases we set resp. $\alpha_v = 0, 1, \infty$. We have $\min(1, \|\beta - \alpha_v\|_v) \leq \|x\|_v$ for $v \in S_1, \leq \|y\|_v$ for $v \in S_2$ and $\leq 1/\|x\|_v$ for $v \in S_3$.

On the other hand we have:

$$H(x) = H(x^{-1}) = \prod_{v \in S} \max(1, \|x\|_v^{-1}) \text{ (since } x \in \mathcal{O}_S^*)$$

(i)
$$\leq \left(\prod_{v \in S_1} \|x\|_v^{-1} \right) \cdot 2^{\#S};$$

$$\text{(ii)} \quad H(y) = H(y^{-1}) = \prod_{v \in S} \max(1, \|y\|_v^{-1}) \quad (\text{since } y \in \mathcal{O}_S^*)$$

$$\leq \left(\prod_{v \in S_2} \|y\|_v^{-1} \right) \cdot 3^{\#S};$$

$$\text{(iii)} \quad H(x) = \prod_{v \in S} \max(1, \|x\|_v) \quad (\text{since } x \in \mathcal{O}_S^*)$$

$$\leq \prod_{v \in S_3} \|x\|_v \cdot 2^{\#S}.$$

Taking into account these cases we find, using also $H(1-x) \geq H(x)/2$,

$$\prod_{v \in S} \min(1, \|\beta - \alpha_v\|_v) \leq 12^{\#S} H(x)^{-2} H(1-x)^{-1} \leq 2^{4\#S+1} H(x)^{-3}.$$

Hence for large enough $H(x)$ we get a contradiction with the inequality of Theorem 3.11 (applied with any fixed $\epsilon < 1$).[14] This shows that $H(x)$ is bounded, so x belongs to a finite set, which concludes the argument. $\square$

Remark 3.14.
(i) *S*-**integral points for** $\mathbb{P}^1 \setminus \{0, 1, \infty\}$. The solutions of the *S*-unit equation $x + y = 1$ may be seen as '*S*-integral points' for the affine curve $C := \mathbb{P}^1 \setminus \{0, 1, \infty\}$. In fact, the ring $k[C]$ of regular functions on C (over k) is $k[x, x^{-1}, (x-1)^{-1}] = k[x, (x(x-1))^{-1}]$, so C may be embedded as the affine plane curve defined by $X(X-1)Z = 1$. Now, if x and z are *S*-integers with $x(x-1)z = 1$, then both $x, x-1$ must actually be *S*-units, so we have the solution $x + (1-x) = 1$ of the *S*-unit equation (and conversely). Note that this curve C has genus 0 and *three* points at infinity, so we are in a case of Siegel's Theorem (see the Supplements to Chapter 2), and actually this finiteness result amounts to the full genus 0 case of this theorem.

(ii) **A holomorphic analogue.** Picard's Little Theorem states that there are no non-constant holomorphic functions $f: \mathbb{C} \to \mathbb{P}^1 \setminus \{0, 1, \infty\}$ (*i.e.* entire non-constant complex functions that miss two or more finite values). Especially in view of part (i) of this Remark, this plainly resembles the *S*-unit Theorem. Note also that the units in the ring of entire functions are those with no zeros; so, if f misses the values 0, 1, both f and $g := 1 - f$ are units and satisfy $f + g = 1$. As recalled in the notes to Chapter 2, this analogy was pointed out by Osgood, Reyssat and especially by Vojta, in much greater generality. In Vojta's view (see [90]), an infinite sequence of *S*-integral points on an affine variety $\mathcal{X}$ 'corresponds' to a holomorphic nonconstant map $f : \mathbb{C} \to \mathcal{X}$, which in turn is just a point on $\mathcal{X}$ with 'entire' coordinates. A simple significant example when such points exist is

---

[14] The exponent "-3" attributed to $H(x)$ corresponds to the "three" points at infinity, *i.e.* $0, 1, \infty$, for the present curve $\mathbb{P}_1 \setminus \{0, 1, \infty\}$: see next Remark, (i).

$\mathbb{P}_1 \setminus \{0, \infty\}$, which may be embedded in $\mathbb{A}^2$ as the hyperbola $XY = 1$; there are infinitely many S-integral points (for suitable k, S) and indeed there is a holomorphic point: $z \mapsto (e^z, e^{-z})$. This is also related to Pell Equation: its solutions give integral points on a hyperbola with two points at infinity, and they may be parametrized by exponential maps.

(iii) **Effectiveness.** The Generalized Roth Theorem is ineffective at present. However the S-unit theorem (and so also Mahler's Theorem 3.12) may be made effective, with a different proof based on Baker's theory of linear forms in logarithms. (See [7].)

(iv) **S-unit points on curves.** By means of Theorem 3.11 one may study more generally S-unit points on (plane) curves, *i.e.* the general equation $f(x, y) = 0$, f a given irreducible polynomial, $x, y \in \mathcal{O}_S^*$. As proved first by Lang [52] there is finiteness except for the cases $f(X, Y) = aX^m + bY^m$ or $f(X, Y) = aX^m Y^m + b$. (These shapes for $f(X, Y)$ correspond to *translates of algebraic subgroups of* $\mathbb{G}_m^2$, which we shall study in detail in the next chapter.) A proof of this result involves Puiseux series. One may obtain a 'good approximation' by expanding y as a function of x, at a place v where x is 'large'; see the exercise below. Such technique leads even to effective results if one uses Baker's theory in place of Roth-type theorems. In several variables one has a general qualitative result by M. Laurent, mentioned below, but this is not effective. (For all of this see [17, Theorem 5.4.5. and Section 7.4].)

(v) **A generalization to several variables.** Theorem 3.13 may be generalized to higher dimensions, by considering (for a given $n \geq 2$) the equation $x_1 + \ldots + x_n = 1$, to be solved with $x_i \in \mathcal{O}_S^*$. A finiteness result does not hold anymore: take for example $n = 3$ and the solutions $x + (-x) + 1 = 1$, with $x \in \mathcal{O}_S^*$. However Evertse and independently van der Poorten & Schlickewei proved that *there are only finitely many 'non-degenerate' solutions*, where we call degenerate a solution $(x_1, \ldots, x_n)$ such that there is a nonempty subsum $\sum_{i \in A} x_i = 0$. This easily yields a complete description of the structure of solutions. The proofs depend on the deep Schmidt Subspace Theorem (a far-reaching generalization to higher dimensions of Roth's theorem), as formulated by Schlickewei for several places. (See [17].) This general S-unit theorem may be applied to study S-unit points on any algebraic variety: the idea is that the equations defining the variety may be seen as linear forms in the monomials, yielding linear equations in S-units, to which the result applies. A final formulation is due to Laurent; we do not state it explicitly and instead refer to [17], §7.4 (and also [101]). As in previous results by Liardet, Laurent's conclusion actually applies to the *division group* of the S-units, namely to solutions in algebraic numbers x, y for which $x^l, y^l \in \mathcal{O}_S^*$ for some $l = l(x, y) > 0$. The deduction of this more general result may be obtained from the former by considerations of Galois Theory (rather than arithmetic); the case $n = 2$ appears in the exercises below. Note that a special case of this concerns the equation $x_1 + \ldots + x_n = 1$ in roots of unity x_i; as to this last issue, we shall recover a complete description from a general theorem by Shou-Wu Zhang, of somewhat different nature, in the next chapter. We finally remark that for $n \geq 3$ there are no known methods to determine all solutions of the S-unit equation in the general case.

Exercise 3.41. (Lang) Let $f(X, Y) \in \overline{\mathbb{Q}}[X, Y]$ be absolutely irreducible. Prove that if there are infinitely many S-units x, y with $f(x, y) = 0$ then either $f(X, Y) = aX^m + bY^m$ or $f(X, Y) = aX^mY^m + b$. (Hint: if $v \in M_k$ is such that $\|x\|_v$ is large, y is given by some Puiseux series in x, converging v-adically; from this deduce that, for some rational β and algebraic c, both in a finite set, $\|yx^\beta - c\|_v$ is very small. Since x, y are S-units, an application of Theorem 3.11 suffices to imply that $yx^\beta = c$, as required.)

Exercise 3.42. Deduce from Theorem 3.13 that *there are only finitely many algebraic numbers x, y such that $x + y = 1$ and $x^l, y^l \in \mathcal{O}_S^*$ for some $l > 0$.* (Hint: Let $x + y = 1$ be a solution, where $x^l, y^l \in \mathcal{O}_S^*$, for some $l > 0$. Conjugating over $\mathbb{Q}(x^l, y^l)$, obtain equations $\zeta x + \eta y = 1$, where ζ, η are l-th roots of unity. By considering the intersection of two circles, deduce that ζ, η have at most two possibilities, whence l may be chosen independently of x, y. Since $\mathcal{O}_S^*$ is finitely generated, this gives rise to solutions over a fixed number field, and then Theorem 3.13 applies to this field, concluding the argument.)

Exercise 3.43. Describe the solutions to $\zeta_1 + \zeta_2 + \zeta_3 = 1$, with ζ_i roots of 1. (Hint: here several methods are available; see for instance [39] and the next chapter.)

Exercise 3.44. The above proof of the S-unit theorem shows that the conclusion of Theorem 3.11 with any fixed $\epsilon < 1$ suffices to imply Theorem 3.13. Retrospectively, since Theorem 3.11 allows us to take any $\epsilon > 0$, we can improve on the S-unit theorem: in fact, prove that if we restrict x, y to have *a big enough percentage of their height coming from places in S*, there are only finitely many solutions of $x + y = 1$. (Such x, y are sometimes called *almost S-units*. Of course this is vague, and to find a rigorous statement is part of the exercise.)

Exercise 3.45. Let $p_1, \ldots, p_r, \ell$ be distinct prime numbers and let $\epsilon > 0$. Prove that the congruence $p_1^{a_1} \cdots p_r^{a_r} \equiv 1 \pmod{\ell^m}$, $m > \epsilon \max(|a_i|)$, has only finitely many solutions in integers a_i. (Hint: this is related to the previous exercise: on writing $p_1^{a_1} \cdots p_r^{a_r} + \ell^m q = 1$ we obtain a kind of *almost S-unit* equation. A proof may be given directly with Generalized Roth. For $r = 1$ there is an elementary proof, obtained by observing that $\ell^m \ll a_1$. Baker's theory would provide effectiveness and also an improved bound with respect to the approach coming from Roth; for instance, Yu has proved $m \ll \log \max |a_i|$ for fixed ℓ, p_i - see the notes - whereas Roth gives only $m = o(\max |a_i|)$.)

Exercise 3.46. Let $f \in \mathbb{Z}[X]$ have at least two distinct roots. Prove that (Pólya-Siegel) *the greatest prime factor of $f(n)$, for $n \in \mathbb{Z}$, tends to infinity with n.* (Hint: If $f(n)$ is an S-unit, then $n - \xi$ is an S-unit in $\mathbb{Q}(\xi)$, for ξ a root of f. Use this with two roots and apply Theorem 3.13. For three distinct roots one may also use Mahler's theorem. See also [77, page 105].)

Proof of Theorem 3.12, assuming Theorem 3.13. The conclusion becomes stronger if we enlarge k or S, so we shall do that freely along the argument. We may assume that f factors completely over k, so the equation becomes $\prod_{i=1}^d (\alpha_i x + \beta_i y) = c, \alpha_i, \beta_i, c \in k$. With a linear change of coordinates we can assume $\alpha_i \beta_i \neq 0$ for every i and by enlarging S

we can also assume $c, \alpha_i, \beta_i \in \mathcal{O}_S^*$. For every i we have $\alpha_i x + \beta_i y \in \mathcal{O}_S$, but their product over $i = 1, \ldots, d$ is $c \in \mathcal{O}_S^*$, so $\mu_i := \alpha_i x + \beta_i y \in \mathcal{O}_S^*$.

Recall that $d \geq 3$; hence by eliminating x, y from the equations $\mu_i = \alpha_i x + \beta_i y, i = 1, 2, 3$, we obtain

$$\mu_1 \Delta_{23} - \mu_2 \Delta_{13} + \mu_3 \Delta_{12} = 0,$$

where the coefficients $\Delta_{ij} := \alpha_i \beta_j - \alpha_j \beta_i$ are nonzero for distinct $i, j \in \{1, 2, 3\}$, because the factors $\alpha_i X + \beta_i Y$ are pairwise non-proportional. These coefficients do not depend on the solution (x, y), so by enlarging S further, we can assume that they are S-units. Then we obtain the equation

$$\frac{\mu_1 \Delta_{23}}{\mu_2 \Delta_{13}} + \frac{\mu_3 \Delta_{12}}{\mu_2 \Delta_{13}} = 1,$$

where both $\mu_1 \Delta_{23} / \mu_2 \Delta_{13}$ and $\mu_3 \Delta_{12} / \mu_2 \Delta_{13}$ are S-units.

Since the Δ_{ij} are fixed, by Theorem 3.13 the ratio μ_1 / μ_2 has only finitely many possibilities. But this ratio equals $(\alpha_1 x + \beta_1 y) / (\alpha_2 x + \beta_2 y)$, so $(x : y) \in \mathbb{P}_1(k)$ has also only finitely many possibilities. Since however $f(x, y)$ is fixed $(= c)$ and nonzero, and since f is homogeneous, it follows that x, y have only finitely many possibilities, which concludes the argument. $\square$

Proof of Theorem 3.13, *assuming Theorem* 3.12. Recall the result (essentially by Dirichlet) that $\mathcal{O}_S^*$ is finitely generated (the actual rank is immaterial for this qualitative argument). Hence, for each integer $d \geq 1$, $\mathcal{O}_S^* / \mathcal{O}_S^{*d}$ is a finitely generated abelian group of exponent d and thus is a finite group. Then there exists a finite set $\Phi \subset \mathcal{O}_S^*$ such that every $z \in \mathcal{O}_S^*$ can be written as $z = \varphi w^d$, with $\varphi \in \Phi$ and $w \in \mathcal{O}_S^*$. Using this with x, y in place of z, we can write the S-unit equation $x + y = 1$, $x, y \in \mathcal{O}_S^*$, as $\varphi_1 w_1^d + \varphi_2 w_2^d = 1$, where φ_1, φ_2 lie in the finite set Φ and where $w_1, w_2 \in \mathcal{O}_S^*$. If there were infinitely many solutions (x, y), then for an infinity of them φ_1, φ_2 would be fixed. Then this would give infinitely many solutions to the Thue-Mahler equation $\varphi_1 X^d + \varphi_2 Y^d = 1$, $X, Y \in \mathcal{O}_S$, contradicting Theorem 3.12 if we choose $d \geq 3$. $\square$

Remark 3.15.   Note that in this argument we have taken into account the Thue Equations merely of the special shape $\varphi_1 X^d + \varphi_2 Y^d = 1$. So, combining this deduction with the previous arguments used to deduce Theorem 3.12 from Theorem 3.13, we find that, somewhat surprisingly, the finiteness of the set of solutions (over $\mathcal{O}_S$) of the special Thue Equations directly implies the same finiteness for the general ones. This may be useful since the special Thue Equation is related to diophantine approximations to radicals, which may be easier to study.

Another consequence of Theorem 3.13 is a theorem due to Siegel (a 'hyperelliptic' case of his general theorem on integral points on curves), of which Corollary 2.10 (of the Supplements to Chapter 2) is a special case.

Theorem 3.14 (Siegel). *Let $f \in k[X]$ be a polynomial with at least three simple roots. Then there are only finitely many solutions to $y^2 = f(x)$, with $x, y \in \mathcal{O}_S$.*

Proof. Let (x, y) run through an infinite sequence of solutions in $\mathcal{O}_{k,S}^2$. As before, on enlarging k we can factor f and assume $y^2 = c^2(x - \alpha_1) \cdots (x - \alpha_d)$, where $c \in k^*$ and the roots $\alpha_i \in k$, $i = 1, 2, 3$, are distinct and simple. By enlarging S we can also assume $c \in \mathcal{O}_S^*$ and $\alpha_i \in \mathcal{O}_S$. By enlarging S further, we can assume that $\mathcal{O}_S$ is a principal ideal domain: in fact, it is well known that the ring of integers of k has only finitely many ideal classes, and it suffices that S contains a set of primes generating this finite class group. (See also the exercise below.)

Let us now consider the common divisors in $\mathcal{O}_S$ of any two factors $(x - \alpha_i)$ and $(x - \alpha_j)$, $i \in \{1, 2, 3\}$, $j \neq i$. Any prime dividing both factors divides also $(\alpha_i - \alpha_j)$. Hence it divides $\delta := \prod_{i=1}^3 \prod_{j \neq i} (\alpha_i - \alpha_j) \neq 0$.

Hence, since the product of the $x - \alpha_i$ is a square in $\mathcal{O}_S$ and since α_i occurs only once for $i = 1, 2, 3$, we can write $x - \alpha_i = \eta_i \mu_i^2 \varepsilon_i$, $i = 1, 2, 3$, where $\varepsilon_i \in \mathcal{O}_S^*$, where η_i is a product of distinct primes dividing δ, thus lying in a finite set, and where $\mu_i \in \mathcal{O}_S$. Since the group $\mathcal{O}_S^*$ is finitely generated, we can further write $\varepsilon_i = \varepsilon_i' \varepsilon_i''^2$, with $\varepsilon_i', \varepsilon_i'' \in \mathcal{O}_S^*$ and ε_i' in a finite set. Putting $v_i := \mu_i \varepsilon_i'' \in \mathcal{O}_S$ and $\delta_i := \eta_i \varepsilon_i'$ we have $x - \alpha_i = \delta_i v_i^2$, with δ_i in a prescribed finite set.

Now, this yields, for $i, j = 1, 2, 3$, $\alpha_i - \alpha_j = \delta_j v_j^2 - \delta_i v_i^2$. Enlarging further the number field k and the finite set S, the δ_i can be assumed to be squares, $\delta_i = \gamma_i^2$, and the differences $\alpha_i - \alpha_j$ to be S-units. Then from $\alpha_i - \alpha_j = (\gamma_j v_j + \gamma_i v_i)(\gamma_j v_j - \gamma_i v_i)$ we deduce that $\gamma_j v_j \pm \gamma_i v_i \in \mathcal{O}_S^*$.

Finally, we obtain an equation $u + v = w$ in S-units ($u := \gamma_1 v_1 + \gamma_2 v_2$, $v := \gamma_3 v_3 - \gamma_2 v_2$, $w := \gamma_1 v_1 + \gamma_3 v_3$). Dividing by w we deduce that u/w lies in a finite set. Hence, setting $\beta_i := \gamma_i v_i$, $i = 1, 2, 3$, for infinitely many solutions we would have a relation $\beta_1 = c_2 \beta_2 + c_3 \beta_3$ with fixed c_2, c_3. However $\beta_1^2 = \beta_2^2 + (\alpha_2 - \alpha_1)$ and $\beta_3^2 = \beta_2^2 - (\alpha_2 - \alpha_3)$, which yields $\xi_1 \beta_1^2 + \xi_2 \beta_2^2 + \xi_3 \beta_3^2 = 0$ for certain nonzero fixed ξ_i. Substituting for β_1, it easily follows that β_2/β_3 takes only finitely many values, whence the same must be true of x, a contradiction which proves the assertion. $\square$

Exercise 3.47. Using the finiteness of the set of ideal classes in $\mathcal{O}_k$, prove that $\mathcal{O}_S$ is a principal domain for large enough S. (Hint: let S be large enough so that prime ideals in S generate the ideal class group of $\mathcal{O}_k$. Let I_S be an ideal of $\mathcal{O}_S$, so $I = I_S \cap \mathcal{O}_k$ is an ideal of $\mathcal{O}_k$. We have $P_1 \cdots P_m I = \alpha \mathcal{O}_k$ for suitable $P_i \in S$ and an $\alpha \in \mathcal{O}_k$. Since P_i^h is principal for some $h > 0$, we have $P_i^h = \pi_i \mathcal{O}_k$, for a $\pi_i \in \mathcal{O}_k \cap \mathcal{O}_S^*$. Then, if $b \in I$, $\pi_1 \cdots \pi_m b \in \alpha \mathcal{O}_k$. This easily proves that $\alpha \mathcal{O}_S = I_S$.)

Remark 3.16. Naturally, a similar and simpler argument applies to superelliptic equations $Y^m = f(X)$, with suitable assumptions on m, f.

As for previous theorems, this result becomes effective on using an effective proof of Theorem 3.13 (which, as we have recalled, may be obtained by Baker's method).

For $\deg f = 3$ we obtain the important result that any affine subset of a curve of genus 1, defined over a number field, has only finitely many S-integral points (over k): in fact, such an affine subset may be embedded in $\mathbb{A}^2$ in Weierstrass form, *i.e.* by an equation $y^2 = f(x)$ with f a cubic polynomial without repeated roots. In turn, this result may be used to prove that up to isomorphism over $\mathcal{O}_S$ there are only finitely many elliptic curves over k with good reduction outside S. (See [83].)

3.4. Heights on finitely generated subgroups of $\mathbb{G}_m^n$

In this section we shall study the height on the algebraic group $\mathbb{G}_m^n$, noting that it defines a semi-distance there; especially, we shall focus on finitely generated subgroups. The results below may be viewed as a kind of setting up a 'Diophantine Approximation theory on finitely generated subgroups'. This will admit an important application to the S-unit theorem, in the final chapter, but the conclusions and methods have independent interest, and also provide an instructive insight into the notion of height. We recall that the *multiplicative algebraic group* $\mathbb{G}_m$, as an algebraic variety is just $\mathbb{P}_1 \smallsetminus \{0, \infty\} \cong {}^1 \smallsetminus \{0\}$; the denomination comes from the fact that $\mathbb{G}_m$ is equipped with the multiplication law $(x_1, x_2) \mapsto x_1 x_2$, expressed as a polynomial function of the coordinates. As an affine variety, $\mathbb{G}_m$ may be embedded into $\mathbb{A}^2$ as the hyperbola $XY = 1$. By $\mathbb{G}_m(R)$ we mean the abelian group of points of $\mathbb{G}_m$ over the ring R, namely just R^*; thus $\mathbb{G}_m^n(\overline{\mathbb{Q}})$ denotes just the set of n-dimensional vectors with nonzero algebraic coordinates, equipped with coordinatewise multiplication. So, for a point $P = (\alpha_1, \ldots, \alpha_n) \in \mathbb{G}_m^n(\overline{\mathbb{Q}})$ we have $P^m = (\alpha_1^m, \ldots, \alpha_n^m)$; P is a *torsion* point if P^m is the identity (*i.e.* $(1, \ldots, 1) \in \mathbb{G}_m^n$) for some nonzero $m \in \mathbb{Z}$; this is the case if and only if all coordinates α_i are roots of unity. Such points are clearly Zariski-dense in $\mathbb{G}_m^n$.

By 'abuse of language' we shall occasionally identify $\mathbb{G}_m$ with the set $\mathbb{G}_m(\overline{\mathbb{Q}})$, which will be our ambient space very often, and similarly for $\mathbb{G}_m^n$.

There are various (more or less natural) ways of defining a height on $\mathbb{G}_m^n(\overline{\mathbb{Q}})$; for our purposes they are all equivalent, and we shall choose the following one:

$$\widehat{h}(\alpha_1, \ldots, \alpha_n) := h(\alpha_1) + \ldots + h(\alpha_n). \tag{3.10}$$

Remark 3.17. This height is inherited (in a certain precise sense on which we do not pause in these notes) from the natural immersion of $\mathbb{G}_m^n$ in $\mathbb{P}_1^n$. Alternatively, one could use the natural embedding $\mathbb{G}_m^n \hookrightarrow \mathbb{P}_n$, which would produce the height $h(\alpha_1, \ldots, \alpha_n) = h(1 : \alpha_1 : \ldots : \alpha_n)$.

Combining Proposition 3.1, Northcott's and Kronecker's theorems 3.7, 3.8, we immediately find that this height has the properties listed in the following

Proposition 3.15. *For all* $P = (\alpha_1, \ldots, \alpha_n), Q, P_1, \ldots, P_s \in \mathbb{G}_m^n(\overline{\mathbb{Q}})$ *we have*

1. $\widehat{h}(P) \geqslant 0$, *with equality holding if and only if P is torsion, i.e. all α_i are roots of* 1.
2. $\widehat{h}(P^{-1}) = \widehat{h}(P)$ *and more generally* $\widehat{h}(P^m) = |m|\widehat{h}(P)$ *for all* $m \in \mathbb{Z}$.
3. $\widehat{h}(PQ^{-1}) \leq \widehat{h}(P) + \widehat{h}(Q)$ *and* $\widehat{h}(P_1 \cdots P_s) \leqslant \widehat{h}(P_1) + \ldots + \widehat{h}(P_s)$.
4. *There are only finitely many points of bounded height and bounded degree.*

For algebraic points $P, Q \in \mathbb{G}_m^n$ let us put $d(P, Q) := \widehat{h}(PQ^{-1}) \geq 0$. Then, by n. 1 we have that $d(P, Q) = 0$ if and only if $P = QZ$ for a torsion point Z; also, n. 3 implies that $d(P, Q)$ satisfies the triangle inequality. In other words, $d(P, Q)$ is a semi-distance on $\mathbb{G}_m^n(\overline{\mathbb{Q}})$, and it becomes actually a distance on the quotient group $\mathbb{G}_m^n(\overline{\mathbb{Q}})/T$, where T is the torsion subgroup.

In the sequel we shall apply this especially to finitely generated subgroups Γ of $\mathbb{G}_m^n(\overline{\mathbb{Q}})$, of finite rank r.[15] In this case the alluded distance will be read as a norm on a finite dimensional euclidean space.

An important example is $\Gamma = (\mathcal{O}_{k,S}^*)^n$ for any number field k and finite set $S \subset M_k$. Actually, any finitely generated $\Gamma \subset \mathbb{G}_m^n(\overline{\mathbb{Q}})$ is contained in $(\mathcal{O}_{k,S}^*)^n$, for suitable k, S. (Here is a simple example of this situation, in $\mathbb{G}_m^2$: $\Gamma := \{(\pm 2^a, 3^b 7^c) \mid a, b, c \in \mathbb{Z}\}$; this is indeed a finitely generated

---

[15] Recall that the rank of an abelian group, here written in multiplicative notation, is the maximal number of multiplicatively independent elements in it. For instance the group of all roots of unity has rank 0 (although it is not finitely generated). Of course there is no general relation between the rank of Γ and the dimension n of the ambient space $\mathbb{G}_m^n$.

subgroup, of rank 3, generated by $(2, 1)$, $(1, 3)$, $(1, 7)$ and $(-1, 1)$. It is contained in $(\mathcal{O}^*_{\mathbb{Q},\{2,3,7\}})^2$ which has rank 6.)

It turns out that in several issues such finitely generated groups play a companion role with *algebraic subgroups* of $\mathbb{G}_m^n$ (recall *e.g.* Remark 3.14), that we shall study in the next chapter (a simple instance illustrating their general shape is the curve defined by $x^4 y^6 = 1$ in $\mathbb{G}_m^2$). However these notions must absolutely not be confused; for instance, $\mathbb{G}_m$ is an algebraic subgroup of itself (and also of each $\mathbb{G}_m^n$) but $\mathbb{G}_m(\mathbb{Q})$ has not even finite rank.

Exercise 3.48. Prove that $\mathbb{G}_m(k) = k^*$ is finitely generated if and only if k is a finite field. Prove that it has finite rank if and only if k is algebraic over a finite field, in which case the rank is 0. (Hint: first observe that finite rank implies that $\operatorname{char}(k)$ is > 0 and that k is algebraic over the prime field.) Prove that the conclusion holds also for algebraic groups isomorphic to $\mathbb{G}_m$ over $\bar{k}$; for instance, consider the group of $(x, y) \in k^2$ such that $x^2 - dy^2 = 1$. (Here the group law is $(x, y) \cdot (x', y') := (xx' + dyy', xy' + x'y)$.)

Let now Γ be a given finitely generated subgroup of $\mathbb{G}_m^n(\overline{\mathbb{Q}})$, of rank r. We define $\Gamma_{\mathrm{tors}} := \{g \in \Gamma : \exists m > 0, \ g^m = 1\}$ to be the set of torsion elements of Γ. We shall use the height $\widehat{h}$ to define a norm on $\mathbb{R}^r$, in which $\Gamma / \Gamma_{\mathrm{tors}}$ will be embedded as a lattice. We shall do this in three steps, starting with a norm on $\mathbb{Z}^r$, and subsequently extending it to $\mathbb{Q}^r$ and $\mathbb{R}^r$.

A norm on $\mathbb{Z}^r$

By the structure theorem for finitely generated abelian groups, the group Γ is isomorphic to the direct sum of $\mathbb{Z}^r$ with a finite group; hence $\Gamma / \Gamma_{\mathrm{tors}}$ is isomorphic to $\mathbb{Z}^r$. Fixing an isomorphism $\varphi \colon \mathbb{Z}^r \to \Gamma / \Gamma_{\mathrm{tors}}$, we can define the *norm* of an element $\underline{u}$ of $\mathbb{Z}^r$ as the height of any representative $P \in \Gamma$ of $\varphi(\underline{u}) \in \Gamma / \Gamma_{\mathrm{tors}}$, *i.e.* $\|\underline{u}\| := \widehat{h}(P)$. Note that this does not depend on the representative P, so the map is well defined: indeed, if $P' = PZ$ where Z is a torsion points, then the coordinates of P' equal those of P up to factors which are roots of unity, and the assertion follows on recalling that $h(x\zeta) = h(x)$ for any $x \in \overline{\mathbb{Q}}$ and for any root of unity ζ.

Remark 3.18. Explicitly, if $\gamma_1, \ldots, \gamma_r \in \Gamma$ generate Γ modulo torsion, and represent the isomorphism φ in the sense that $\varphi(\underline{u})$ is the class of $\gamma_i^{u_1} \cdots \gamma_r^{u_r}$ modulo torsion, for $\underline{u} = (u_1, \ldots, u_r) \in \mathbb{Z}^r$, the norm is obtained on setting $\|(u_1, \ldots, u_r)\| := \widehat{h}(\gamma_i^{u_1} \cdots \gamma_r^{u_r})$. Note that this depends on the chosen isomorphism φ, namely on the choice of the independent generators γ_i. For instance, by composing with an automorphism of $\mathbb{Z}^r$ (*i.e.* an element of $GL_r(\mathbb{Z})$), we shall generally obtain a distinct, though *equivalent*, norm.[16]

---

[16] Two norms are said to be equivalent if their ratio is bounded above and below by positive constants,

Exercise 3.49. Prove that the group of automorphisms $\sigma \in GL_r(\mathbb{Z})$ which leave the norm invariant is finite and contains $-I$. (Hint: Use Proposition 3.15, ns. 4, 2.)

Exercise 3.50. Let $\sigma \in GL_r(\mathbb{Z})$; find explicit constants $c_1 = c_1(\sigma), c_2 = c_2(\sigma) > 0$ (in terms of the entries of σ) so that $c_1 \left\| \underline{u} \right\| \leq \left\| \sigma(\underline{u}) \right\| \leq c_2 \left\| \underline{u} \right\|$ for all $\underline{u} \in \mathbb{Z}^r$.

Note that what we have defined is indeed a 'norm' in the usual sense, as follows from Proposition 3.15; namely:

(i) We have $\left\| \underline{u} \right\| = 0$ if and only if $\varphi(\underline{u})$ is represented by an element P of Γ with $\widehat{h}(P) = 0$, which holds (n. 1 of Proposition) if and only if P is torsion, namely $\underline{u} = 0$.

(ii) For any $m \in \mathbb{Z}$ we have $\left\| m\underline{u} \right\| = |m| \left\| \underline{u} \right\|$, as follows from n. 2; to express this property we say that the norm is *homogeneous*.

(iii) Finally, the triangle inequality follows from n. 3 and the fact that φ is a homomorphism.

Extending the norm to $\mathbb{Q}^r$

Let now $\underline{v}$ be an element of $\mathbb{Q}^r$ and let $m \in \mathbb{Z}$ be a nonzero integer such that $\underline{u} := m\underline{v}$ belongs to $\mathbb{Z}^r$; we define the norm of $\underline{v}$ as $\left\| \underline{v} \right\| := \left\| \underline{u} \right\| / |m|$. This is again well defined in view of the homogeneity of the norm: if m_1 and m_2 are distinct nonzero integers such that both $\underline{u}_1 = m_1\underline{v}$ and $\underline{u}_2 = m_2\underline{v}_2$ belong to $\mathbb{Z}^r$, we have $m_2\underline{u}_1 = m_1\underline{u}_2$, so that $|m_2| \left\| \underline{u}_1 \right\| = |m_1| \left\| \underline{u}_2 \right\|$ and $\left\| \underline{u}_1 \right\| / |m_1| = \left\| \underline{u}_2 \right\| / |m_2|$, as required.

The function so defined on $\mathbb{Q}^r$ plainly continues to satisfy the three mentioned properties (i), (ii), (iii) of a norm; also, the homogeneity property (ii) now holds for any $m \in \mathbb{Q}$.

We also note that, defining *the division group* Γ' of Γ as $\Gamma' := \{P \in G_m^n(\overline{\mathbb{Q}}) \mid \exists m > 0, \ P^m \in \Gamma\}$, the isomorphism φ may be extended to $\mathbb{Q}^r$ so that it yields an isomorphism $\varphi: \mathbb{Q}^r \to \Gamma'/\Gamma'_{\text{tors}}$. In particular, the points in $\mathbb{Q}^r$ still correspond under φ to algebraic points of G_m^n. (Namely, $\varphi(\mathbb{Q}^r) \subset G_m^n(\overline{\mathbb{Q}})/torsion$.)

Extending the norm to $\mathbb{R}^r$

We shall now extend our norm to $\mathbb{R}^r$. We shall easily see that the norm can be (uniquely) extended by continuity to a non-negative *function* on $\mathbb{R}^r$; however (as was pointed out by Cassels) it is not so automatic that this yields indeed a *norm*: the homogeneity property and the triangle inequality trivially continue to hold, and certainly this function cannot assume

---

out of the origin.

negative values; but, *a priori*, the strict positivity outside the origin could fail, in the course of the limiting process. Here is a simple example which shows that indeed not every norm on $\mathbb{Q}^r$ induces a norm on $\mathbb{R}^r$: the map $(a, b) \mapsto \left| a - b\sqrt{2} \right|$ is a norm on $\mathbb{Q}^2$ but not on $\mathbb{R}^2$. Nevertheless, we shall see that this degeneracy cannot happen for our actual norm.

To carry out all of this, we start to define by continuity the said extension to $\mathbb{R}^r$. Observe that the norm is a continuous real map on $\mathbb{Q}^r$ (with the topology induced from $\mathbb{R}^r$). More precisely, if $|\cdot|$ is the sup-norm on $\mathbb{R}^r$, because of homogeneity and the triangle inequality, if $B_1, \ldots, B_r$ is a fixed basis of $\mathbb{Q}^r$ over $\mathbb{Q}$, we have $\left\| \sum x_i B_i \right\| \leq C \sup |x_i|$ (for $x_i \in \mathbb{Q}$) where $C := r \sup \| B_i \|$, whence $\| P \| \leq C |P|$. Also, the triangle inequality implies $\left| \, \| P \| - \| Q \| \, \right| \leq \| P - Q \|$, whence

$$\left| \, \| P \| - \| Q \| \, \right| \leq C |P - Q|. \tag{3.11}$$

Now consider, for any $P \in \mathbb{R}^r$, a sequence of points $\{ P_i \} \subset \mathbb{Q}^r$ with limit P. By inequality 3.11, the sequence of norms $\{ \| P_i \| \}$ is a Cauchy sequence in $\mathbb{R}$; therefore the limit $\lim_i \| P_i \|$ exists and is ≥ 0. Also, it is independent of the choice of the sequence $\{ P_i \}$: if $P_i' \to P$, where P_i' are in $\mathbb{Q}^r$, then $P_i' - P_i \to 0$, and 3.11 again proves that $\| P_i' \| - \| P_i \| \to 0$. Hence we can define $\| P \|$ to be equal to this limit. Note that this definition is compatible with the previous one if $P \in \mathbb{Q}^r$, because then we may take $P_i = P$ for all i.

We now show the crucial fact that the function $\| \cdot \| : \mathbb{R}^r \to \mathbb{R}_{\geq 0}$ so defined is a norm. As pointed out above, it suffices to prove that it vanishes only at the origin, the other properties (homogeneity, now for every $m \in \mathbb{R}$, and triangle inequality) being very easily verified from the above definitions. An elegant proof (due to Cassels) based on Minkowski convex body Theorem appears in [17], p. 137. Here we give the following self-contained argument, similar at bottom but simpler.

If $\| \cdot \|$ is not strictly positive on $\mathbb{R}^r \setminus \{0\}$, let $\underline{\rho} = (\rho_1, \ldots, \rho_n) \in \mathbb{R}^r$ be a nonzero vector with $\left\| \underline{\rho} \right\| = 0$. For an integer $q > 0$, set $\underline{z}_q := ([q\rho_1], \ldots, [q\rho_r]) \in \mathbb{Z}^r$, where $[y]$ denotes the integral part of the real number y. Then $\underline{z}_q - q\underline{\rho}$ has bounded coordinates, whence, by homogeneity and the triangle inequality on $\mathbb{R}^r$, $\left\| \underline{z}_q \right\| \leq \left\| q\underline{\rho} \right\| + O(1) = |q| \left\| \underline{\rho} \right\| + O(1) = O(1)$, as $q \to \infty$.

In other words, the vectors $\underline{z}_q \in \mathbb{Z}^r$ have bounded norm. They correspond (up to torsion) resp. to elements $g_q \in \Gamma$ with bounded height. However, Γ is finitely generated so it is contained in k^{*n}, for a suitable number field k, and so by n. 4 of Proposition 3.15 (in practice by Northcott Theorem) there can be only finitely many g_q. Hence there can be only finitely many z_q, which is not the case since the sup-norm of z_q tends to infinity (because $\rho \neq 0$). This contradiction proves what we need.

To conclude we remark that we have inequalites

$$\alpha_1 \|P\| \leq |P| \leq \alpha_2 \|P\|, \qquad \forall P \in \mathbb{R}^r, \tag{3.12}$$

where $\alpha_1 > \alpha_2 > 0$ are positive numbers and $|\cdot|$ is the sup-norm. Such properties are in fact well-known to hold for every pair of norms. In particular, it follows that $\|\cdot\|$ induces the usual topology on $\mathbb{R}^r$, so in particular the unit ball $\{P \in \mathbb{R}^r : \|P\| \leq 1\}$ is compact.

For a proof, the left inequality follows as above from $\left\|\sum x_i B_i\right\| \leq \sup(|x_i|)(\sum \|B_i\|)$. As to the right inequality, since $\|\cdot\|$ is continuous it attains a minimum μ on the unit 'sphere' $\{P : |P| = 1\}$, which is compact. Since the norm vanishes only at the origin we have $\mu > 0$. We have $\|P\| \geq \mu$ for $|P| = 1$ whence by homogeneity $\|P\| \geq \mu|P|$ for all $P \in \mathbb{R}^r$, proving what we need with $\alpha_2 = \mu^{-1}$.

For quantitative applications it is important to note that the numbers α_1, α_2 depend on the group Γ and actually also on the choice of generators for it; in fact we have already noted that the norm itself depends on such choices.

Remark 3.19. **(Function fields)** For function fields k/κ ($k = k(\mathcal{X})$, $\mathcal{X}$ an algebraic variety over an algebraically closed 'constant' field κ) we do not generally dispose of Northcott Theorem. However (an analogue of) the above construction continues to be possible: We start with a finitely generated group Γ and use the height on $\Gamma/(\Gamma \cap \kappa^*)$ to define a norm on $\mathbb{Z}^r$; this norm still extends to a positive definite function on $\mathbb{R}^r$. This may be now deduced essentially from the fact that all absolute values are discrete. See Exercise 3.56 for this deduction. In this context, see also the paper [19] for a study of heights on finitely generated subgroups in fields with product formulas, in particular in function fields.

Here is a very simple example of the 'shape' of this kind of norm (for $n = 1$, $r = 2$). Consider the subgroup $\Gamma = \{\pm 2^a 3^b \mid (a, b) \in \mathbb{Z}^2\} \subset \mathbb{Q}^*$; it has rank 2 and torsion $\Gamma_{\text{tors}} = \{\pm 1\}$. The homomorphism $\psi : (\pm 2^a 3^b) \mapsto (a, b)$ induces the norm

$$\|(a, b)\| = h(2^a 3^b) = \begin{cases} |a| \log 2 + |b| \log 3 & \text{for } ab \geq 0; \\ \max(|a| \log 2, |b| \log 3) & \text{for } ab < 0. \end{cases}$$

Exercise 3.51. Draw the 'unit sphere' with respect to the extension of this norm to $\mathbb{R}^2$ (*i.e.* the set $\{(x, y) \in \mathbb{R}^2 \mid \|(x, y)\| \leq 1\}$).

Exercise 3.52. Prove that the euclidean volume of the unit sphere (relative to a norm as above, associated to a group Γ) is independent of the chosen generators for Γ (whereas the norm depends on that choice).

Exercise 3.53. Estimate the number of elements in Γ of height $\leq t$, for $t \to \infty$.

Exercise 3.54. Let $\gamma_1, \ldots, \gamma_r$ be multiplicatively independent nonzero elements of a number field k.

For $v \in M_k$, set $\underline{l}_v := (\log \|\gamma_1\|_v, \ldots, \log \|\gamma_r\|_v) \in \mathbb{R}^r$ (with the usual normalizations). Prove that $\underline{l}_v = 0$ for all but finitely many v and that the $\underline{l}_v$ generate $\mathbb{R}^r$ over $\mathbb{R}$. (Hint: prove that any vector orthogonal - in the usual sense - to all the $\underline{l}_v$ has zero norm, with respect to a suitable group $\Gamma \subset \overline{\mathbb{Q}}^*$ as above.)

Exercise 3.55. Prove that if $\gamma_1, \ldots, \gamma_r$ are multiplicatively independent S-units in a number field k, then $\#S \geq r + 1$. (Hint: use the result in the previous exercise and the product formula. Of course this is one half of the result by Dirichlet mentioned several times.)

Exercise 3.56. Extend the construction of this section (for $n = 1$) and the result of Exercise 3.54 to any function field k/κ, $k = k(\mathcal{X})$, $\mathcal{X}$ an algebraic variety over an algebraically closed 'constant' field κ, working with a finitely generated subgroup Γ of k^*, such that $\Gamma/(\Gamma \cap \kappa^*)$ has rank r. (Hint: first, the elements γ_i will be chosen to be independent modulo κ^*. Now observe that any valuation appearing in the product formula is discrete. Then it follows immediately that the rank over $\mathbb{R}$ of the matrix of the $\underline{l}_v$ is the same as over $\mathbb{Q}$. This last rank is r by the assumption of independence. Note that we cannot use the number field argument because we do not have Northcott Theorem here.)

Exercise 3.57. Let Γ be a finitely generated subgroup of $\mathbb{G}_m^n(k)$, where k/κ is a function field, as in the previous exercise. Prove that, up to κ^*, Γ contains only finitely many elements of bounded height. (Hint: note that by including Γ in a product, it suffices to deal with the case $n = 1$. For this, observe that the divisors of the elements of Γ lie in a finitely generated group. Alternatively, use the result of the previous exercise: if $\gamma_1^{a_1} \cdots \gamma_r^{a_r}$ has bounded height, then $(a_1, \ldots, a_r)$ has bounded scalar product with all the vectors l_v; these scalar products are bounded integers, so have finitely many possibilities, hence...This result is a kind of 'Northcott Theorem for Γ'.)

Supplements to Chapter 3

The S-unit equation over function fields

We shall now prove a strong version of the S-unit theorem for the case of function fields of curves. The existence of non-trivial derivations now allows an elementary quick proof, in place of the elaborated proofs for number fields.

Let $\mathcal{C}$ be a nonsingular complete curve of genus g, defined over an algebraically closed field κ of characteristic 0, with function field $k := \kappa(\mathcal{C})$. We shall study the S-unit equation $x + y = 1$, with $x, y \in \mathcal{O}_S^* \setminus \kappa^*$. Here S is a finite set of points of $\mathcal{C}$ and $\mathcal{O}_S^*$ is (as explained in 3.3.2 above) the set of rational functions in $\kappa(\mathcal{C})$ with all poles and zeroes in S. We are going to present a simple proof of a theorem of Mason, which bounds the heights of x, y only in terms of $\#S$ and g. Below we normalize the valuations so that their value group is $\mathbb{Z}$. So, for $x \in \kappa(\mathcal{C})$ the height $h(x) = h(1 : x)$ coincides with $\deg(x) = [k(\mathcal{C}) : k(x)]$ and generally $h(x_0 : \ldots : x_n) = -\sum_{P \in \mathcal{C}} \inf_i v_P(x_i)$, where v_P is the order function at the point P. (If we want to refer to the curve $\mathcal{C}$ explicitly we can use the notation $h_{\mathcal{C}}$.)

Theorem 3.16. *For any* $x, y \in \mathcal{O}_S^* \setminus \kappa^*$ *with* $x + y = 1$ *we have* $\deg(x) = h(1 : x : y) \leq 2g - 2 + \#S$.

Before the proof, let us pause for a few comments. The result expresses the fact that the equation $x + y = 1$ somewhat prevents x and y to have zeros/poles of 'too large multiplicities' on average: in fact, when 'many' of the multiplicities are large the height (in practice, the degree) is much larger than the number of distinct zeros/poles, contradicting the inequality. (The case of polynomials is particularly illustrative, as in Corollary 3.17 below.)

The inequality is entirely explicit and uniform; it is a strong analogue of what follows in the numerical case from Baker's method, mentioned several times; in applying such method, we have in practice the strong restriction to work with fixed S, otherwise the results tend to become weak.

Actually, Theorem 3.16 is a sharp function field analogue of a statement for number fields which is still unknown but would have far-reaching consequences; we are alluding to the celebrated 'abc-conjecture' of Masser and Oesterlé. In the simplest form this states: *For each* $\epsilon > 0$ *there is a number* $c(\epsilon)$ *with the following property. Let* a, b, c *be coprime integers such that* $a + b = c$. *Then*
$$\max(|a|, |b|, |c|) \leq c(\epsilon) \left(\prod_{p|abc} p \right)^{1+\epsilon}.$$

There are also versions for number fields. Several (known and unknown) spectacular statements on diophantine equations, and also in other contexts, would follow more or less easily from this. (See [17] for a discussion.) To appreciate the analogy of this conjectural statement with Theorem 3.16[17], we may use the following dictionary:

1. A point $P \in \mathcal{C}(\kappa)$ (or valuation v) corresponds to a prime p (counted with a *weight* $\log p$).
2. A rational function corresponds to a rational number. Specifically, x, y correspond to $a/c, b/c$. To have a zero (resp. pole) at P corresponds to the numerator (resp. denominator) being divisible by p.
3. The set S may be taken as the set of zeros and poles of x, y and corresponds to the set of primes p dividing abc. The number $\#S$ corresponds to $\log(\prod_{p|abc} p) = \sum_{p|abc} \log p$.
4. The height $h(1 : x : y)$ corresponds to $h(1 : a/c : b/c) = h(a : b : c) = \log(\max(|a|, |b|, |c|))$, the last equality holding because a, b, c are coprime.

Particularly near to the above original abc-conjecture over $\mathbb{Q}$ is the special but significant case of Theorem 3.16 when $\mathcal{C} = \mathbb{P}_1$, so $\kappa(\mathcal{C}) = \kappa(t)$, the field of rational functions in one variable. Let us formulate explicitly the theorem for this situation as follows:

Corollary 3.17 (*abc for polynomials*). *Let* $a(t), b(t), c(t) \in \kappa[t]$ *be coprime polynomials, not all constant and such that* $a + b = c$. *Also, let* S *be the set of zeros of abc in* κ. *Then we have the inequality* $\max(\deg a, \deg b, \deg c) \leq \#S - 1$.

---

[17] Theorem 3.16 is often called 'abc-theorem for function fields'.

Exercise 3.58. Deduce the corollary from Theorem 3.16. (Hint: set $x = a/c$. Note that the set S of the corollary is not quite the same as in the theorem.)

Proof of Theorem 3.16. To prove Theorem 3.16 we consider the differentials of x and $y = 1 - x$; we have $dx = -dy$.

Let S_1 be the set of poles of x, S_2 be the set of zeros of x, S_3 be the set of zeros of y. If x is regular at a point P, then so is dx and conversely; thus S_1 is the set of poles of dx. The three sets S_i are pairwise disjoint and contained in S. For every $v \notin S_1 \cup S_2 \cup S_3$, we have $v(dx) \geqslant 0$. We have $v(dx) = v(x) - 1$ for $v \in S_1 \cup S_2$, and $v(dx) = v(dy) = v(y) - 1$ for $v \in S_3$.

Let us then compute in two ways the degree of the divisor of the differential dx; on the one hand it is $2g - 2$, because x is nonconstant; on the other hand, taking into account the above remarks we get

$$2g - 2 = \deg(dx) = \sum v(dx) \geqslant \sum_{S_1} v(dx) + \sum_{S_2} v(dx) + \sum_{S_3} v(dx)$$

$$= \sum_{S_1} \left(v(x) - 1 \right) + \sum_{S_2} \left(v(x) - 1 \right) + \sum_{S_3} \left(v(y) - 1 \right)$$

$$= -(\#S_1 + \#S_2 + \#S_3) + \left(\sum_{S_1} v(x) + \sum_{S_2} v(x) \right) + \sum_{S_3} v(y)$$

$$\geqslant -\#S + \deg(y),$$

where the last inequality holds because (a) the S_i are pairwise disjoint and contained in S, (b) $\sum_{S_1} v(x) + \sum_{S_2} v(x) = 0$ (in view of the product formula) and (c) $\sum_{S_3} v(y)$ is the total number of zeros of y, which equals its degree.

Hence $h(y) = \deg(y) \leq \#S + 2g - 2$. Finally, since $x + y = 1$ we have that $v(x) = v(y)$ whenever v is a pole of x or y; hence $h(1 : x : y) = \sum_v -\min(0, v(x), v(y)) = h(y) = \deg(y)$, which concludes the proof. $\square$

A different proof and a generalization

Keeping the above notation, we define $\chi = \chi(\mathcal{C}, S) := 2g - 2 + \#S$, so that Theorem 3.16 reads $h(1 : x : y) \leq \chi$. This χ is the Euler characteristic of the real surface corresponding to $\mathcal{C}$, deprived of the points in S. It is a quantity which behaves well under rational maps. For instance we have the following

Theorem 3.18. *Let* $f : \mathcal{C} \to \mathcal{X}$ *be a nonconstant rational map of degree d between complete nonsingular curves over κ, let R be a finite subset of $\mathcal{X}$ and $S := f^{-1}(R)$. Then $\chi(\mathcal{C}, S) \geq d\chi(\mathcal{X}, R)$, with equality if and only if every point of $\mathcal{X}$ outside R is unbranched under f.*

Proof. The Hurwitz genus formula yields for the genera, $2g(\mathcal{C}) - 2 = d(2g(\mathcal{X}) - 2) + \sum_{P \in \mathcal{X}} \sum_{f(Q)=P} (e_Q - 1)$, where $e_Q = \mathrm{ord}_Q(f - f(Q))$ is the ramification index. Now we have $\sum_{f(Q)=P} e_Q = \deg f = d$ for every $P \in \mathcal{X}$. Therefore, if $d_P = \#f^{-1}(P)$ is the cardinal of the fiber above P, we have $2g(\mathcal{C}) - 2 = d(2g(\mathcal{X}) - 2) + \sum_{P \in \mathcal{X}} (d - d_P)$. Hence $\chi(\mathcal{C}, S) := 2g(\mathcal{C}) - 2 + \#f^{-1}(R) = 2g(\mathcal{C}) - 2 + \sum_{P \in R} d_P = d(2g(\mathcal{X}) - 2) + \sum_{P \in \mathcal{X}} (d - d_P) + \sum_{P \in R} d_P \geq$

$d(2g(\mathcal{X}) - 2 + \#R)$, the last inequality following because $d - d_P$ is always ≥ 0. If there is equality then $d - d_P = 0$ for all P outside R, and conversely, proving the contention. $\qquad\square$

This good behaviour of χ under a nonconstant rational map $f : C \to \mathcal{X}$ is shared by the height: in fact, if we have functions $x_1, \ldots, x_n$ on $\mathcal{X}$ and we view them as functions on C in the embedding $\kappa(\mathcal{X}) \subset \kappa(C)$ (namely we identify x_i with $x_i^* := x_i \circ f$) we have $h_C(x_1 : \ldots : x_n) = \deg(f) \cdot h_{\mathcal{X}}(x_1 : \ldots : x_n)$.

All of this may be used in our context. For instance, let us see how Theorem 3.18 yields an alternative proof for Theorem 3.16. Let us first verify this last result for the almost tautological case when $\mathcal{X}$ is the line defined in $\mathbb{P}_2$ by $X + Y = Z$, letting $x := X/Z$, $y := Y/Z$ be the affine coordinate functions on $\mathcal{X}$. This verification is indeed very easy: $h = h(1 : x : y)$ is just the number of poles of x or y, counted with the maximal multiplicity; but x and y have each the single pole $(1 : -1 : 0)$, which is simple for both. Hence $h(1 : x : y) = 1$. On the other hand x (resp. y) has the single simple zero $(0 : 1 : 1)$ (resp. $(1 : 0 : 1)$). Hence $\#S \geq 3$ and of course $g = 0$, so $h = 1 \leq -2 + \#S = \chi$, as stated.

Let us now see how this inequality for this special curve $\mathcal{X}$ yields the same inequality for a curve C with nonconstant functions $x, y \in \kappa(C)$ satisfying $x + y = 1$ (i.e., the general Theorem 3.16). In the sequel we define as above S to be the set of zeros/poles of x and y. We have a map $f : C \to \mathcal{X}$ given outside the poles of x by $P \mapsto (x(P) : y(P) : 1)$. This f extends to a regular map on the whole C by sending the poles of x (which are the same as of y) to the 'infinite' point $(1 : -1 : 0)$ on $\mathcal{X}$. Hence, defining $R = \{(0 : 1 : 1), (1 : 0 : 1), (1 : -1 : 0)\}$ we see that $S = f^{-1}(R)$. Moreover, the height $h(1 : x : y) = h_C(1 : x : y)$ on C is just d times the height of the same functions viewed on $\mathcal{X}$, where $d = \deg f$, hence $h(1 : x : y) = d$. Finally, by Theorem 3.18, $\chi(C, S) \geq d\chi(\mathcal{X}, R) = d = h(1 : x : y)$, as required.

More generally, we may start with a given curve $\mathcal{X}$ and R-unit functions $x_1, \ldots, x_n$ on it, for which an inequality $h_{\mathcal{X}}(1 : x_1 : \ldots : x_n) \leq c\chi(\mathcal{X}, R)$ holds, for a certain number $c > 0$. Then, the same inequality will hold for a cover $f : C \to \mathcal{X}$ of degree d, the functions $x_i \circ f$ on C, and $S := f^{-1}(R)$. Let us see an explicit example, by proving the following generalization of Theorem 3.16:

Theorem 3.19. *Let $F \in \kappa[X, Y, Z]$ be a homogeneous polynomial of degree D, defining a nonsingular curve $\mathcal{X}$ in $\mathbb{P}_2$. Suppose that $XYZ = 0$ meets $\mathcal{X}$ in $3D$ distinct points and let C be a nonsingular complete curve with nonconstant functions x, y satisfying $F(x, y, 1) = 0$. Then $h_C(1 : x^D : y^D) \leq \chi(C, S)$, where $S \subset C$ is the set of zeros and poles of x, y.*

Proof. Let us first verify the conclusion for the curve $C = \mathcal{X}$ and $x := X/Z$, $y := Y/Z$. For the genus $g = g(\mathcal{X})$ we have the familiar formula $2g - 2 = D(D - 3)$. The assumptions easily yield $\#R = 3D$ for the set R of zeros/poles of x, y and $h(1 : x : y) = D$. Hence $\chi(\mathcal{X}, R) = D^2$, proving the conclusion for $\mathcal{X}$.

In the general case, the functions x, y define a map $f : C \to \mathcal{X}$, because $F(x, y, 1) = 0$. This is everywhere defined and surjective, because it is a nonconstant map between complete nonsingular curves. Hence the set S of zeros/poles of x, y is precisely $f^{-1}(R)$, where R is the set of zeros/poles of the functions $X/Z, Y/Z$ on $\mathcal{X}$. The functions x, y are resp. the pull-back of $X/Z, Y/Z$, hence $h_{\mathcal{C}}(1 : x : y) = \deg f \cdot h_{\mathcal{X}}(1 : X/Z : Y/Z) = \deg f \cdot D \leq \deg f \cdot \frac{1}{D}\chi(\mathcal{X}, R) \leq \frac{1}{D}\chi(\mathcal{C}, S)$, where we have applied the special case $\mathcal{C} = \mathcal{X}$ and Theorem 3.18 for the last two steps. Using $h(1 : x^D : y^D) = Dh(1 : x : y)$, the result follows. $\qquad\square$

In the case of this theorem, the constant c alluded to above is D^{-1}, which is best-possible (as shown by the very proof, for the case $\mathcal{C} = \mathcal{X}$). We find back Theorem 3.16 on taking $F = X + Y - Z$. Note also that the assumptions of the theorem are certainly verified for a 'general' F; anyway, on dropping some of them one may still obtain similar results, with different constants c: the 'right' constant is gotten on checking the case where $\mathcal{C}$ is defined by $F = 0$, hence $c = h_{\mathcal{X}}(1 : x : y)/\chi(\mathcal{X}, R)$. In fact, in practice, the above proof exploits the fact that, by Theorem 3.18, the function $\chi(\mathcal{X}, R)/h_{\mathcal{X}}(1 : x : y)$ is increasing through pull-back by a rational map. For instance, for $F = X^2 + XY + XZ + Z^2$, we get a constant $c = 1$. For $F = XY + Z^2$ we have $\chi(\mathcal{X}, R) = 0$, which leads to $c = \infty$, a useless result! (And it couldn't be otherwise, in view of the equation $xy + 1 = 0$ with S-unit solutions of unbounded degree.)

Exercise 3.59. Prove that for any integer $d > 0$ there is an example of equality in Theorem 3.16, with $\deg x = \deg y = d$. (Hint: consider e.g. $x = t^d$, $y = 1 - t^d$.)

Exercise 3.60. Prove that the inequality of Theorem 3.16 is an equality if and only if the map $x : \mathcal{C} \to \mathbb{P}_1$ is unbranched outside $\{0, 1, \infty\}$. (Hint: this follows easily by inspection of the proof given above: if there is equality, the differential dx cannot have zeros outside S, and conversely. It follows also from the alternative treatment involving Theorem 3.18.)

Exercise 3.61. Prove that even if $\mathrm{char}(k) = p > 0$ Theorem 3.16 holds provided $dx \neq 0$.

Exercise 3.62. Prove the following version of Corollary 3.17 for a field κ of characteristic $p > 0$: *Let $a, b, c \in \kappa[t]$ be coprime polynomials, not all constant, such that $a + b = c$. Assume that at least one of them is not a p-th power in $k[t]$. Then* $\max(\deg a, \deg b, \deg c) \leq \#S - 1$. (Hint: just imitate the above proof of Theorem 3.16, using $d(a/c) \neq 0$.)

Exercise 3.63. Dropping the coprimality assumption in the last exercise, prove that *if a has no roots of multiplicity $\geq p$, we have* $\max(\deg a, \deg b, \deg c) \leq \deg c + N(ab) - 1$, where $N(f)$ is the number of distinct zeros of f.

Exercise 3.64. (Elliptic congruences) Let $f \in \mathbb{F}_p[X]$ have degree ≤ 3. Prove that for $p > 17$ the equation $y^2 = f(x)$ has at least one solution in $\mathbb{F}_p^2$ with $y \neq 0$. (Hint: let $g(x) \in \mathbb{F}_p[X]$ have exactly the zeros of f in $\mathbb{F}_p$, with multiplicitly 1. If there are no solutions, $g(x)(f(x)^{\frac{p-1}{2}} + 1)$ vanishes for all $x \in \mathbb{F}_p$. Hence $g(X)(f(X)^{\frac{p-1}{2}} + 1) = h(X)(X^p - X)$ for some $h \in \mathbb{F}_p[X]$ of degree $\leq (p - 3 + 2\deg(g))/2$. Write this as $g(X)f(X)^{\frac{p-1}{2}} - h(X)X^p = g(X) - Xh(X)$. Applying the result of the previous exercise one finds $3(p -$

1)/2 + deg(g) ≤ deg h + 1 + 7 + deg h ≤ p + 2 deg(g) + 5. This yields $p \leq 17$.
See [100] for a development of the method.)

Exercise 3.65. Let $f(t) \in \kappa[t]$ be a nonconstant polynomial, not a p-th power
(where char(κ) $= p$). Prove that the number of zeros of $f(t)(1 - f(t))$ is
$\geq \deg f + 1$. Also, prove that this cannot be improved and that we have infinitely
many examples of equality in which $f(x)$ has all roots of multiplicity ≥ 2,
whereas $f(x) + 1$ has precisely two simple roots. (Hint for the last question:
look at a polynomial Pell Equation.)

Exercise 3.66. For char(κ) $= 0$, prove 'Fermat Theorem for polynomials', *i.e.*
that *for $d \geq 3$, there do not exist coprime nonconstant polynomials $f, g, h \in$
$\kappa[t]$ such that $f^d(t) + g^d(t) = h^d(t)$.* (Hint: use Corollary 3.17; an alternative
approach, near to the ideas of Fermat, Kummer and others, is to use a descent
argument, on factoring $f^d - h^d = \prod_{\theta^d=1}(f - \theta h)$; if this is a d-th power, every
factor must be a d-th power,...)

Exercise 3.67. Apply Corollary 3.17 to treat the Thue Equation for polynomials
(2.28).

Exercise 3.68. Apply Theorem 3.16 to treat the hyperelliptic equation in S-
integers over a function field of a curve. (Hint: Mimic the proof of Theorem
3.14, using Theorem 3.16 in place of Theorem 3.13. Note that now we do not
generally have unique factorization in $\mathcal{O}_S$, but it suffices to remark that e.g. the
square roots of elements with divisor divisible by 2 generate a field of bounded
degree over the function field; this amounts to the fact that torsion of given order
on the Jacobian is finite.)

Exercise 3.69. Let $f, g \in \kappa[t]$ be polynomials of degrees resp. $3n, 2n$, where
char(κ) $= 0$. Prove that either $f^2 = g^3$ or $\deg(f^2 - g^3) \geq n + 1$. (Hint: use
Corollary 3.17. Alternatively, deduce that if α_i, β_j are the roots of f, g, we have
$2 \sum \alpha_i^r = 3 \sum \beta_j^r$ for $r = 0, \ldots, 6n - \deg(f^2 - g^3) - 1$, whence...)

Exercise 3.70. Suppose again char(κ) $= 0$ and let Γ be a finitely generated
subgroup of $\kappa(\mathcal{C})^*$, of rank r modulo κ^*. Estimate the number of solutions of
the S-unit equation $x + y = 1$ with $x, y \in \Gamma \backslash \kappa^*$. (Hint: apply first Theorem 3.16
to obtain a bound for the degrees of x, y. Then, let $\gamma_1, \ldots, \gamma_r$ be independent
generators for Γ/κ^* and consider the vectors $(v_P(\gamma_1), \ldots, v_P(\gamma_r)) \in \mathbb{Z}^r$, for
$P \in \mathcal{C}$. Note that, by independence, these vectors generate $\mathbb{Q}^r$ over $\mathbb{Q}$. If
$h(\gamma_1^{a_1} \cdots \gamma_r^{a_r}) \leq B$ we have in particular $|\sum a_i v_P(\gamma_i)| \leq B$ for all points P;
hence $(a_1, \ldots, a_r)$ lies in the inverse image under an invertible linear map of
the 'cube of side B' of integers, whence...This rough methods leads to a bound
depending on #S and g; in fact, the dependence on g can also be eliminated and
actually a bound dependent only on r turns out to be true. Compare also with
Exercise 3.57 and see the notes below for further informations.)

Exercise 3.71. Extend the method of proof of Theorem 3.19 to more general
cases, also when the curve $\mathcal{X}$ is defined by several polynomials in a higher dimensional space.

Remark 3.20. As noted in the above exercises, easy examples show that the inequality of Theorem 3.16 cannot be generally sharpened. Actually, the cases of
equality (in positive characteristic) can be classified according to Exercise 3.60,

which reduces the question to finding the covers of $\mathbb{P}_1$ unbranched outside three given points. The branching conditions correspond to permutations with given cycle decomposition, generating the monodromy group. Given such combinatorial data, the existence of related covers follows from the so-called Riemann Existence Theorem; this allows to realize real surfaces with certain topological data as Riemann surfaces. It is a rather deep result; it admits essentially two proofs, each using substantial analytical work. (See for instance [76] and [91].) A general algebraic proof would be highly desirable (e.g. for dealing with the case of positive characteristic), but is still missing. It turns out that every such cover may be realized over $\overline{\mathbb{Q}}$ and conversely a striking theorem of Belyi states that every algebraic curve over $\overline{\mathbb{Q}}$ may be realized as such a cover. All of this also raises the arithmetical question of the minimal number field of definition; Grothendieck pointed out that in this way certain graphs associated to the covers correspond to number fields, which gave rise to the theory of *dessins d'enfant*. (See [73].)

Detecting multiplicative dependence in $\overline{\mathbb{Q}}$

We briefly discuss the following 'algorithmic' problem: *For given algebraic numbers* $\alpha_1, \ldots, \alpha_n \neq 0$, *establish whether they are multiplicatively independent or not and more generally describe their possible multiplicative relations.* (We do not pause here on the term 'given' and leave to the interested readers the task of formulating some precise definitions.) For instance this may be useful for effective analysis of finitely generated subgroups of $\mathbb{G}_m^N$, studied earlier. In the sequel we let $k := \mathbb{Q}(\alpha_1, \ldots, \alpha_n)$.

A simple case occurs *e.g.* when $k = \mathbb{Q}$; then we can compute prime decompositions and write $\alpha_i = \pm \prod_{p \in S} p^{a_p(i)}$ for a finite set S of primes and integers $a_p(i)$. Plainly, the α_i are multiplicatively dependent if and only if the vectors $v_p := (a_p(1), \ldots, a_p(n)) \in \mathbb{Z}^n$, $p \in S$, are linearly dependent. So we reduce to a linear algebra problem which may be checked algorithmically.

Exercise 3.72. Prove that $\alpha_1, \ldots, \alpha_n \in \mathbb{Q}^*$ are multiplicatively dependent if and only if there exist integers $b_1, \ldots, b_n$, not all zero, such that $|b_i| \leq 2(n \max h(\alpha_i)/\log 2)^{n-1}$ and $\prod_{i=1}^n \alpha_i^{b_i} = 1$. (Hint: use the above remarks and solve the relevant linear system by Siegel's lemma. One can also sharpen the inequality by re-doing the proof of that lemma for this special case.)

Exercise 3.73. Let $p_1, \ldots, p_{n-1}, p$ be pairwise distinct primes, $p > n$. Setting $\alpha_i := p_i^{p+i}$ for $i \leq n - 1$ and $\alpha_n = p_1 \cdots p_{n-1}$, observe that they are multiplicatively dependent and prove that the maximal exponent in any relation is $\gg \prod_{i=1}^{n-1} (h(\alpha_i)/\log p_i)$. Observe that for fixed n, fixing $p_1, \ldots, p_{n-1}$ to be the first $n-1$ primes, and taking a 'large' p, this lower bound is $\gg_n (\max h(\alpha_i))^{n-1}$; compare with previous exercise.

Another 'easy' case occurs for $n = 1$ (and any k): now we want to determine whether a given algebraic number α is multiplicatively dependent, *i.e.* a root of 1. For example, consider $\alpha = \frac{3+4i}{5}$: is it a root of 1? The answer is no, which can be proved by several arguments:

(i) Its minimal polynomial over $\mathbb{Z}$ is $5\alpha^2 - 6\alpha + 5 = 0$, so α is not an algebraic integer; actually,

(ii) though its complex absolute value is 1, the value at the place $2 - i$ of $\mathbb{Z}[i]$ is > 1.

(iii) Assuming that α be a root of 1, its order would be ≤ 12, because the degree $\phi(n)$ of a primitive n-th root of unity is > 2 for n not dividing 12 (and we can check that $\alpha^m \neq 1$ for $m \leq 12$).

(iv) We can use congruences: it is a known (and easy) fact that the reduction of a root of unity modulo a prime ideal above p has the same prime to p-part of the order. Then, using two primes p, p' such that α is integral at them, determines the order. (In the present case, take $e.g.$ $p = 2, p' = 3$; α reduces to 1 modulo 2 and to $\pm\sqrt{-1}$ modulo 3. Thus the order would divide 4.)

(v) We can follow the argument using Northcott Theorem: the algebraic numbers of degree 2 and zero height are roots of an equation $x^2 + ax + b = 0$ where a, b are rationals with $h(a) \leq \log 2$ and $h(b) \leq 0$, so $a = \pm 2, \pm 1/2$, $b = \pm 1$, which is not the case for the equation for α. And anyway there are at most 8 possibilities, so if a β of degree 2 is a root of unity, the powers β^m lie in this set, so β has order ≤ 8.

(vi) Assuming α to be a root of 1, say $\alpha = e^{i\theta}$, the sequence $\lambda_m = \cos(2^m\theta)$ would be ultimately periodic, whereas it has increasing height, as shown by inspection of the recursion $\lambda_0 = 3/5, \lambda_{m+1} = f(\lambda_m)$, for $m \geq 0$, where $f(X) = 2X^2 - 1$.

For general k, n, the independence problem can be solved on considering factorizations, as in the case of $\mathbb{Q}$; one needs to take into account prime ideals and units, which is possible but involves considerable computational effort. Instead, we shall present another method, that has the advantage to apply also to the case of elliptic curves or abelian varieties, to check linear dependence of given algebraic points: note that in these cases the factorization method does not admit any obvious analogue.

The principle is that if there exists a nontrivial relation $\alpha_1^{m_1} \cdots \alpha_n^{m_n} = 1$, then there is one with bounded exponents. For this, we shall need the following generalization of Dirichlet Lemma to simultaneous approximations (a proof of which is also sketched as a hint to Exercise (I.7)):

Lemma 3.20. *Let $\xi_1, \ldots, \xi_r$ be real numbers. For every integer $Q > 0$ there exists an integer q, with $0 < q \leq Q^r$, and integers $p_1, \ldots, p_r$ such that $|q\xi_i - p_i| < Q^{-1}$ for every $i = 1, \ldots, r$.*

Proof. For every integer $t \in [0, Q^r]$, consider the point $P_t = (\{t\xi_1\}, \ldots, \{t\xi_r\}) \in [0, 1[^r$, where $\{y\}$ denotes the fractional part of y. We divide the cube $[0, 1[^r$ in Q^r semi-open subcubes of side Q^{-1}. Since we have a sequence $\{P_t\}$ of $Q^r + 1$ points, two of them must lie in the same subcube. As in the proof of Dirichlet Lemma, taking their difference we obtain the sought conclusion. □

It will be convenient to deal with more general (nontrivial) multiplicative relations, that is, of the shape $\alpha_1^{m_1} \cdots \alpha_n^{m_n} = \zeta$, where the m_i are integers not all 0 and where ζ is a root of unity. Note that such a ζ lies in k, so has a finite number of possibilities which can be found as above ($e.g.$ with (iii) and (v)).

Given such a nontrivial relation, let us assume $0 < m := m_n = \max |m_i|$. In particular, the number $\zeta^{-1}\alpha_1^{m_1} \cdots \alpha_{n-1}^{m_{n-1}}$ is an m-th power in k and we shall exploit this fact. We apply the lemma to approximate a small multiple of the m_i with multiples of m. In the lemma we set $r = n - 1$, $\xi_i := \frac{m_i}{m}$ for $i = 1, \ldots, n - 1$ and we let $Q > 0$ be an integer, to be chosen later. We obtain integers $q, p_1, \ldots, p_r$, with $0 < q \leqslant Q^r$ and $|\delta_i| < mQ^{-1}$, where we have put $\delta_i := qm_i - p_i m$ for $i = 1, \ldots, r$; we also put $p_n := q$ and $\delta_n = qm_n - p_n m = 0$.

We have $qm_j = mp_j + \delta_j$, so taking the q-th power of the above relation, we obtain $\zeta^q = \prod_{j=1}^n \alpha_j^{qm_j} = (\prod_{j=1}^n \alpha_j^{p_j})^m (\prod_{j=1}^n \alpha_j^{\delta_j})$. Let $\beta := \prod_{j=1}^{n-1} \alpha_j^{\delta_j}$ and $\gamma = \prod_{j=1}^n \alpha_j^{p_j}$. Then $mh(\gamma) = h(\beta) \leqslant \sum_{j=1}^{n-1} |\delta_j| h(\alpha_j) \leqslant mQ^{-1} \sum_{j=1}^{n-1} h(\alpha_j)$. Hence $h(\gamma) \leq Q^{-1} \sum_{j=1}^{n-1} h(\alpha_j)$.

Applying Northcott's theorem (Theorem 3.7) to the number field k, we obtain that there exists a number $c = c(k) > 0$, computable in terms only of $[k : \mathbb{Q}]$, such that either $h(\gamma) \geq c$ or γ is a root of 1. Then, if we choose $Q = [c^{-1} \sum_{j=1}^{n-1} h(\alpha_j)] + 1 \geq 1$, we deduce that γ is a root of 1 of k. Then $\prod_{i=1}^n \alpha_i^{p_i} = \gamma$ is a multiplicative relation of the same type, which is nontrivial (recall $p_n := q > 0$) and with exponents $|p_i| \leq q|m_i/m| + 1 < Q^{n-1} + 1$. Thus the exponents are bounded only in terms of k and the α_i, and in the worst case we can check all the possibilities to find whether there exist some nontrivial relation at all. We have proved:

Proposition 3.21. *Let $\alpha_1, \ldots, \alpha_n$ be nonzero algebraic numbers generating the number field k. If there exist integers $m_1, \ldots, m_n$, not all zero, such that $\prod \alpha_i^{m_i}$ is a root of 1, then the $|m_i|$ can be chosen $\leq (c^{-1} \sum_{i=1}^n h(\alpha_i) + 1)^{n-1}$, where c is the minimal non-zero height of an element in k.*

Note that the estimate so obtained is comparable with the one for the case $k = \mathbb{Q}$, coming from a previous exercise. Also, Exercise 3.73 above proves that this cannot be improved by much.

Exercise 3.74. Let $r_1(t), \ldots, r_n(t) \in \mathbb{Q}(t)^*$ be multiplicatively independent rational functions.

(i) Prove that for infinitely many integers t_0 the values $r_i(t_0)$ are defined, nonzero and multiplicatively independent. (Hint: by factoring the r_i, it suffices to assume that the r_i are irreducible polynomials, possibly constant, over $\mathbb{Z}$. Then prove and use the fact that the values of a nonconstant polynomial at integers cannot be composed of a finite number of primes.)

(ii) By working in a number field containing zeros/poles of the r_i, prove that the independence in fact holds for all but finitely many $t_0 \in \mathbb{Z}$ provided the r_i are multiplicatively independent modulo constants.
(Hint: the $\gcd(p(t_0), q(t_0))$ for p, q coprime polynomials, is bounded.) By using Theorem 3.13 remove this last assumption if the set of zeros/poles of the r_i in $\mathbb{P}_1$ contains at least three points.

(iii) Let k be a number field. Prove that the number of $t_0 \in k$ with $h(t_0) \leq T$ and $r_i(t_0)$ are multiplicatively dependent is $\ll T^{n^2}$. Use this estimate to recover (i) above. (Hint: apply last proposition; then observe that the number of rationals t_0 of height $\leq T$ is $\gg e^{2T}$. This method is due to Masser.)
For more general results see the next supplement, the paper [18] and also previous papers of Masser quoted therein.

(iv) Let $\alpha_1, \ldots, \alpha_n \in k^*$ be multiplicatively independent, where k is a number field. Prove that for large enough prime p, the classes of the α_i in $(k^*/k^*)^p$ remain multiplicatively independent. (Hint: Let $\alpha_1^{a_1} \cdots \alpha_n^{a_n} \beta^p = 1$, for a $\beta \in k$, where not all the a_i are divisible by p. By absorbing p-th powers in β^p we may assume that $0 \le a_i < p$, and that the a_i are not all 0. By taking heights we obtain $h(\beta) \le n \max h(\alpha_i)$. Apply now the proposition with $n + 1$ in place of n. This delivers a nontrivial relation $\alpha_1^{m_1} \cdots \alpha_n^{m_n} \beta^m = 1$ with $|m|$ bounded independently of p. Eliminating β we obtain, by independence of the α_i, $p m_i = m a_i$, hence p must divide m. This forces $m = 0$ and we have a contradiction.)

Specializations preserving multiplicative independence

This supplement concerns the following question: Let $\varphi_1, \ldots, \varphi_n$ be rational functions on a given algebraic curve C/κ, supposed to be multiplicatively independent (namely, they do not satisfy any relation $\prod \varphi_i^{m_i} = 1$ with integers m_i not all zero). *How can we describe the set of points $P \in C(\kappa)$ such that the values $\varphi_1(P), \ldots, \varphi_n(P)$ remain multiplicatively independent?* For instance, do there always exist such points? What can we say about their field of definition? And so on.

In the case when $C = \mathbb{P}_1$, *i.e.* the φ_i are usual rational functions, some answers and some references appeared in Exercise 3.74. Here we present a short proof of a result appearing as Theorem 1 in [18], and implying some conclusions in the said exercise. Actually, this works only under a slightly stronger independence assumption:

Theorem 3.22. *Let C be an algebraic curve defined over $\overline{\mathbb{Q}}$ and let $\varphi_1, \ldots, \varphi_n \in \overline{\mathbb{Q}}(C)$ be multiplicatively independent modulo $\overline{\mathbb{Q}}^*$. Then the set of points $P \in C(\overline{\mathbb{Q}})$ such that $\varphi_1(P), \ldots, \varphi_n(P)$ are multiplicatively dependent has bounded height.*

The proof will use the properties of the height developed above. Before it, we briefly comment on such result:

(i) By 'height' here we mean the Weil height, taken in some projective space in which C is embedded. It turns out that a bound for such height amounts to a bound for the height $\varphi_i(P)$ of the value at P of any fixed nonconstant function φ_i. To see this equivalence it suffices *e.g.* to invoke the result of Exercise 3.8, which compares the heights $\phi(P), \psi(P)$ for different nonconstant rational functions ϕ, ψ.

(ii) Note that for the conclusion it does not suffice that the φ_i are multiplicatively independent: in fact, let for instance $\varphi_1(t) = 2, \varphi_2(t) = t$, where t is a coordinate on $\mathbb{P}_1$. Then, at the points $t = 2^m$ of unbounded height the values are multiplicatively dependent.

(iii) The result may be interpreted in the following way, using the language of algebraic subgroups of $\mathbb{G}_m^n$ which we shall study in the next chapter: identify the curve C with its image under the map $P \mapsto (\varphi_1(P), \ldots, \varphi_n(P))$ from $C \setminus S$ to $\mathbb{G}_m^n$, where S is the finite set of zeros/poles of the φ_i. The φ_i become restrictions of the coordinate functions x_i on $\mathbb{G}_m^n$ and a multiplicative relation $x_1^{m_1} \cdots x_n^{m_n} = 1$ defines an algebraic subgroup of codimension

1 (see next chapter). The 'bad' points P where the $\varphi_i(P)$ acquire multiplicative dependence then correspond to the intersections of C with the union of the proper algebraic subgroups. The result says that the union of these intersections is a set of bounded height.

(iv) It is easy to see (see next exercise) that the set of 'bad' points $P \in C(\overline{\mathbb{Q}})$ is infinite if not all the functions are constant. However, via Northcott Theorem the present result implies in particular that the set of bad points P defined over a given number field, or even of bounded degree, is finite. In [18] it is shown that even dropping these restrictions we recover finiteness if we impose *two* independent multiplicative relations on the $\varphi_i(P)$.

Exercise 3.75.

(i) Let $C/\mathbb{Q}$ be a curve and let $\varphi_1, \ldots, \varphi_n \in \overline{\mathbb{Q}}(C)$, not all constant. Prove that there are infinitely many $P \in C(\overline{\mathbb{Q}})$ such that the $\varphi_i(P)$ are multiplicatively dependent. (Hint: for instance, if φ_1 is nonconstant, there are infinitely many P such that $\varphi_1(P)$ is a root of unity.)

(ii) Sharpen the previous result by proving that if the φ_i generate modulo $\overline{\mathbb{Q}}^*$ a multiplicative group of rank ≥ 2, there are infinitely many points $P \in C(\overline{\mathbb{Q}})$ such that there exists a relation $\prod_i \varphi_i(P)^{m_i} = 1$ with *coprime* exponents m_i. (Hint: consider for instance the zeros of the functions $\varphi_1 \varphi_2^m$, $m \in \mathbb{Z}$, assuming φ_1, φ_2 are multiplicatively independent modulo constants.)

Proof of Theorem 3.22. Let $P \in C(\overline{\mathbb{Q}})$ be a 'bad' point, *i.e.* such that there exist integers $m_1, \ldots, m_n$ not all zero and with $\varphi_1(P)^{m_1} \cdots \varphi_n(P)^{m_n} = 1$.[18] The idea is to 'mimic' this multiplicative relation with some other one with bounded exponents. For this, we apply Lemma 3.20 with $r = n$. By symmetry we may assume $m_n = \max |m_i| \geq 1$. Then we choose once and for all a large integer $Q > 1$ (independent of the m_i) and we put $\xi_i = m_i/m_n$. The lemma delivers an integer q with $0 < q \leq Q^n$ and integers $p_1, \ldots, p_n$ with $|\delta_i| \leq m_n Q^{-1}$ for $i = 1, \ldots, n$, where we have set $\delta_i := qm_i - p_i m_n$.

Define the function $\psi := \varphi_1^{p_1} \cdots \varphi_n^{p_n}$. Note that $|p_i| \leq q|m_i/m_n| + 1 \leq Q^n + 1$, hence ψ, although depending on the m_i, lies in a certain finite set dependent only on Q. Note that ψ cannot be constant: for otherwise all the p_i would vanish, in view of our independence assumption, whence $|qm_n| \leq Q^{-1} m_n$, which would force $q = 0$, which is not true.

Let us estimate the height of $\psi(P)$. We have

$$\psi^{m_n} = \prod_{i=1}^{n} \varphi_i^{p_i m_n} = \prod_{i=1}^{n} \varphi_i^{qm_i - \delta_i},$$

whence $\psi(P)^{m_n} = \prod_{i=1}^{n} \varphi_i(P)^{-\delta_i}$, so using the above estimates we obtain

$$h(\psi(P)) \leq m_n^{-1} \sum_{i=1}^{n} |\delta_i| h(\varphi_i(P)) \leq Q^{-1} \sum_{i=1}^{n} h(\varphi_i(P)).$$

---

[18] We tacitly disregard the finitely many P lying among zeros/poles of the φ_i.

Now, since ψ is not constant, ψ, φ_i are algebraically dependent for any i, and we may write $f_i(\varphi_i, \psi) = 0$ for minimal polynomials $f_i \in \overline{\mathbb{Q}}[X, Y]$. Dropping the index i, we write the relation in the form $a_0(\psi)\varphi^d + a_1(\psi)\varphi^{d-1} + \ldots + a_d(\psi) = 0$, where d is bounded only in terms of Q (actually by $\deg(\psi)$) and where the a_j are coprime polynomials with $\deg a_j \leq \deg(\varphi)$. Specializing at P and applying the result of Exercise 3.6 we obtain

$$h(\varphi(P)) \leq h(a_0(\psi(P)) : \ldots : a_d(\psi(P))) + \log d.$$

On the other hand, by Corollary 3.3 we have

$$h(a_0(\psi(P)) : \ldots : a_d(\psi(P))) \leq (\max \deg a_j)h(\psi(P)) + O(1),$$

which by the above estimates is $\leq \deg(\varphi)Q^{-1}\sum_i h(\varphi_i(P)) + O(1)$. (For this inequality we could also use directly Exercise 3.8.) Taking $\varphi(P)$ to be of maximal height among the $\varphi_i(P)$ we find

$$\max_i h(\varphi_i(P)) \leq (\max \deg(\varphi_i))n Q^{-1} \max_i h(\varphi_i(P)) + O(1).$$

The constant implied in the $O(1)$-term may depend on Q. However, on taking e.g. $Q = [2n(\max \deg(\varphi_i) + 1)]$ we obtain a bound for $h(\varphi_i(P))$ independent of P, proving the theorem.   □

Notes to Chapter 3

Our approach for defining product formulas and heights is mainly taken from [77]. For a general theory of fields with a product formula see [4]. See also [28] for some examples of product formulas on field extensions of infinite degree. For the general theory of heights we have also borrowed much from [17], where many other fine properties are presented in full detail.

The height on $\mathbb{P}_n$ gives rise by restriction to a height for algebraic points on any subvariety $\mathcal{X}$ of $\mathbb{P}_n$. If $\mathcal{X}$ instead is given abstractly, we can associate a height to any projective embedding of $\mathcal{X}$; in turn, such an embedding corresponds to a very ample divisor. All of this originates a whole geometric theory of heights, associated to divisors, for which we refer to [17, 50] and [77].

A generalization of Proposition 3.2 appears in Siegel's paper [81]. As pointed out by Bombieri, this is related to Weil's *Théorème de décomposition*, which roughly speaking states that the values of an algebraic function at algebraic points split into (ideal) factors, according to the splitting of the divisor of the function. Bombieri [13] quantified the magnitude of the factors, recognizing this as a theorem on the behaviour of heights under rational maps; he gave two completely different treatments, with best-possible error terms. For a different viewpoint related to a previous approach by Sprindzuk, see [11].

The denomination *Mahler's measure* was introduced by Waldschmidt. The present treatment and its application to Gelfond's inequality follows [17]. In that book one can also find the case of several variables and related references. (The Mahler measure of a polynomial in several variables can be used to define a 'height' of the hypersurface it defines.) See also [85] for a recent survey and history of the whole topic.

A quantitative form of Northcott Theorem is due to Schanuel, who estimated asymptotically the number of points in $\mathbb{P}_n(k)$ of height $\le T$, for a given number field k and $T \to \infty$. (See [77, page 17].) For uniform estimates see [55].

In the paper [22] the Northcott property for a subset $\Sigma \subset \overline{\mathbb{Q}}$ is introduced: *for any B there are only finitely many $\xi \in \Sigma$ with $h(\xi) \le B$*. It is proved that *For any number field k and integer d, the composite of abelian extensions of k of degree $\le d$ has the Northcott property*. So, for instance $\mathbb{Q}(\sqrt{2}, \sqrt{3}, \ldots)$ has this property. It is not known whether 'abelian' can be omitted.

As alluded in the text, on abelian varieties over $\overline{\mathbb{Q}}$ (for instance on elliptic curves) Néron and Tate have shown that one may define 'canonical' heights (associated to ample symmetric divisors). The torsion points turn out to be those of canonical height 0, extending Kronecker's Theorem. (See [17] or [77].)

Much work has been done on the Lehmer conjecture. Apart from the mentioned results by Breusch, Smyth and Dobrowolski (see the presentation in the book [67]), there are also analogue conjectures and results in higher dimensions, for algebraic points on abelian varieties, and for 'heights of algebraic varieties', obtained by Amoroso, David and others (see *e.g.* [1] and the Appendix below by Amoroso). For numbers of 'small' height, a result of Bilu asserts that the Galois conjugates 'tend' to be equidistributed 'around' the unit circle (see the next chapter for more on this). Further, the height is known to be bounded below by an absolute constant (apart from roots of unity) in totally real extensions (see [66] and also [22] for p-adic analogues), in cyclotomic extensions (see [2]) and in abelian extensions of a given number field; actually, it has been proved that a lower bound near to Dobrowolski's mentioned one, holds with d equal to the degree of ξ over any abelian extension of a given number field. (See [3].)

The result in Exercise 3.25 is due to Schur, whereas Robinson has proved a kind of converse. (See [61, Chapter 2].)

Given a rational map $f : \mathbb{P}_1 \to \mathbb{P}_1$ of degree d, one can define a corresponding height h_f on $\mathbb{P}_1(\overline{\mathbb{Q}})$ by $h_f(P) := \lim_{n \to \infty} h(f^{(n)}(P))/d^n$, where $f^{(n)}$ is the n-th iterate. This procedure mimics an idea of Tate, to obtain a 'canonical' height on elliptic curves (and general abelian varieties); see [17]. One can prove that the *preperiodic points* for f are those with $h_f(P) = 0$. (See [51] for generalizations to several variables.)

The result in Exercise 3.35(i), concerning the fractional part of $(3/2)^n$, is related to *Waring's problem*. (See [17, page 153].) The result in part (ii) has been generalized in [31] using the Subspace Theorem. The applications to transcendence of the subsequent exercises are rather standard. For more recent such applications of results in Diophantine Approximation, especially the Subspace Theorem, see [30, 32, 87] and [101].

The equation $x + y = 1$ in units of a number field was already studied by Siegel, who noted its relations to Thue's and other diophantine equations. For an effective tratment of the S-unit equation, see [7], and see also [94] for p-adic versions and applications to the congruence $u \equiv 1$ (mod $\mathcal{P}^m$) in S-units u.

Recently a completely new, quite sophisticated, proof of the rational case of Theorem 3.13 has been found by M. Kim. See Faltings' paper in [104] for a version of this proof.

In analogy with finitely generated subgroups of $\mathbb{G}_m^n$, the height induces a norm also on finitely generated groups of rational points on an elliptic curve (or abelian variety). In these cases one uses a 'canonical' height, (already mentioned above), which gives rise to an even richer structure, since it induces a positive definite quadratic form on the relevant euclidean space (not merely a norm). See [17, 77] and [83].

Theorem 3.16 is due to Mason and independently to Stothers (see [56, 86]). The cases of equality (as in Exercise 3.60) were investigated by Stothers and in [96]; the result of Exercise 3.69 is due to Davenport, who used the second of the sketched methods. The existence of cases of equality for this and similar situations was studied in [98].

See also [102] for the estimate 9^r for the number of nonconstant solutions of $x+y = 1$ in a finitely generated subgroup of rank r, over function fields. The method used therein is rather different from the one sketched in Exercise 3.70.

Theorem 3.22 appears as in [18, Theorem 1]. See this paper also for references concerning previous results by Masser. An explicit version has recently been worked out by Habegger in his Ph.D. thesis (see [48]). This result admits analogues in the case of intersections of a curve, inside an abelian variety, with the abelian subvarieties. A correct analogue for subvarities of higher dimensions proved to be more difficult and was eventually solved by Habegger; see for instance his paper *On the bounded height conjecture*, IMRN 5 (2009), 860–886. See also the author's book "Some Problems of Unlikely Intersections in Arithmetic and Geometry", Annals of Mathematics Studies, Princeton Univ. Press, 2011, for further informations and references on these problems.

Chapter 4
Heights on subvarieties of $\mathbb{G}_m^n$

In this chapter we shall deal with algebraic subvarieties $\mathcal{X}$ of $\mathbb{G}_m^n$ and heights of algebraic points therein, without any restriction on their degree. Starting with an old problem of Lang on torsion points, we shall consider the distribution of points in $\mathcal{X}(\overline{\mathbb{Q}})$ with 'small height': although algebraic points in $\mathbb{G}_m^n$ can have arbitrarily small height, it turns out that the restriction of lying in $\mathcal{X}$ forces the height to be bounded below by a number $c(\mathcal{X}) > 0$, provided however we stay out of a certain exceptional Zariski-closed set in $\mathcal{X}$. This is the content of a theorem by Shou-Wu Zhang, actually analogue to a former conjecture by Bogomolov in the context of abelian varieties; it may be read as predicting a 'discrete' distribution of algebraic points on varieties. The alluded 'exceptional' subvarieties of $\mathcal{X}$ are finite unions of translates (by torsion points) of connected algebraic subgroups of $\mathbb{G}_m^n$, themselves isomorphic (in the algebraic group sense) to powers of $\mathbb{G}_m$. We shall present an elementary proof of this theorem, studying along the way the simple theory of algebraic subgroups of $\mathbb{G}_m^n$. We shall also sketch a different approach to Zhang's theorem, through equidistribution of the Galois conjugates of points of small height. Finally, Zhang's theorem will be applied to gain uniformity in the estimation of the number of solutions of small height of the S-unit equation.

4.1. A problem of Lang

We have seen that the set of heights of algebraic numbers is dense in the positive reals (*e.g.* $h(2^r) = |r| \log 2$ for $r \in \mathbb{Q}$); in particular, the height can be arbitrarily small (and nonzero) if we do not impose bounds on the degree, and of course the same is true for heights of algebraic points in higher dimensional spaces. On the other hand it is not so clear *a priori* what happens if the coordinates of such points are not allowed to vary independently but are subject to some restrictions, *e.g.* of algebraic nature. This amounts to consider the height of points lying on a given algebraic variety; for instance, in the case of an irreducible curve $\mathcal{C}$, say defined by $f(x, y) = 0$, we have recalled in Remark 3.6 the theorem of Siegel that for algebraic points $P \in \mathcal{C}, h(x(P))/h(y(P))$ is asymptotic to $\deg(x)/\deg(y)$ as $h(x(P)) \to \infty$. However this is not useful for points of small height, and it is their distribution which we shall study in this chapter. Zero coordinates do not affect the height, and we shall work in $\mathbb{G}_m^n(\overline{\mathbb{Q}})$, equipped with its group law and the height $\widehat{h}$ introduced in 3.10. Recall from the previous chapter that the height defines a (semi)distance on the set of algebraic points (actually a distance if we consider them modulo torsion). Here we shall be concerned with the phenomena of 'discreteness' of the resulting space. For instance we shall see that points of small height on a given variety are in a sense very 'rare'.

Let us see some simple examples, starting with dimension 2 and with points $P = (x, y) \in \mathbb{G}_m^2(\overline{\mathbb{Q}})$, subject to some nontrivial algebraic relation and having the smallest possible height $\widehat{h}(P) = h(x) + h(y) = 0$; this amounts to x, y being both roots of unity, namely P being a torsion point of $\mathbb{G}_m^2$. For instance, if the alluded algebraic relation between x, y is of 'multiplicative nature', like $x^m = y^n$, we obtain (for any fixed m, n) as many torsion points as we wish: if $n \neq 0$ we may take x an arbitrary root of unity and solve for y. On the other hand, trying with more general equations we immediately encounter difficulties in producing 'several' such points; for instance, a line $ax + by = 1$ $(ab \neq 0)$ has at most two torsion points: actually, at most two points with $|x| = |y| = 1$, corresponding to the intersection of the circles in $\mathbb{C}$ with radii $|a|, |b|$ and centers $0, 1$.

For general curves in $\mathbb{G}_m^2$, this matter actually formed the object of a problem posed by Lang around 1960. Solutions were soon provided, *e.g.* by Ihara, Serre and Tate. We resume a final conclusion in the following statement:

Proposition 4.1. *Let $f \in \overline{\mathbb{Q}}[X, Y]$ be irreducible. Then the equation $f(x, y) = 0$ has infinitely many solutions in roots of unity x, y if and only if $f(X, Y)$ is either of the shape $aX^m Y^n - b$ or $aX^m - bY^n$, with a/b a root of 1.*

Note that in case of infinitely many solutions we recover the 'multiplicative type' that we have met earlier. We shall soon rephrase this result in geometrical terms, better suited for the broad generalizations that we shall meet later in this chapter (*e.g.* Corollary 4.13). But at the moment we pause to present an elementary proof of this statement.[1]

---

[1] The argument which follows is similar to Liardet's, as presented in [53].

Proof. Assume $f(\zeta, \eta) = 0$, with ζ and η roots of 1, say of minimal common order N. We can write $\zeta = \xi^r$ and $\eta = \xi^s$, for a primitive N-th root of unity ξ, where $0 \leqslant r, s < N$, $\gcd(r, s, N) = 1$.

Consider the homomorphism $\lambda : \mathbb{Z}^2 \to \mathbb{C}^*$ defined by $\lambda(t, u) := \zeta^t \eta^u = \xi^{rt+su}$; its kernel is the lattice $\Lambda = \{(t, u) \in \mathbb{Z}^2 \mid rt + su \equiv 0 \pmod{N}\}$; the image of λ has (at most) N elements, so $[\mathbb{Z}^2 : \Lambda] \leq N$.

Let us seek a 'small' element in Λ: consider the integer points $(a, b) \in \mathbb{N}^2$ with $0 \leq a, b \leq [\sqrt{N}]$; their number is $[\sqrt{N} + 1]^2 > N$, so two of them must be congruent modulo Λ. Their difference produces $(t, u) \in \Lambda \setminus \{0\}$ with $|t|, |u| \leqslant \sqrt{N}$. Let now $g(X, Y) = X^t Y^u - 1$. The numerator $\tilde{g}$ of g is a polynomial of degree $\leq |t| + |u| \leq 2\sqrt{N}$.

The point $P = (\zeta, \eta)$ is a zero of both f and $\tilde{g}$, and so are its conjugates over a number field k containing the coefficients of f. Since N is minimal, the number of distinct conjugates P^σ is at least $\varphi(N)/[k : \mathbb{Q}]$. On the other hand, by Bezout's theorem, either f and $\tilde{g}$ have some common factor or the number of their common zeros is at most $2\sqrt{N} \deg f$.

In the first case f would divide $\tilde{g}$, because f is irreducible. But the only irreducible factors of $\tilde{g}$ are easily seen to be of the shape of the statement (Exercise 4.1), concluding the argument.

In the second case we obtain $\varphi(N) \leq 2[k : \mathbb{Q}]\sqrt{N} \deg f$. Since however we have the easy well-known estimate $\varphi(N) \gg N/\log N$ (and even more - see Exercise 4.2), we obtain a bound on N, so on the number of sought solutions.

Conversely, it is clear that the special shapes for f give infinitely many solutions in roots of unity: just choose an arbitrary root of unity x and solve for y (or conversely). $\qquad\square$

Remark 4.1. When f is not of the said special shape, one can obtain a bound dependent only on $\deg f$ for the number of torsion points (ζ, η), whereas the above proof involves also the field of definition. A sharp estimate, in terms of the area of the Newton polygon of f, is due to Beukers and Smyth [8]. It is easy to see that a dependence on the field of definition cannot be eliminated if we seek a bound for the maximum order N of a torsion point P, rather than for the number of such points; for this order, the above proof yields a bound $N \ll_{k,\epsilon} (\deg f)^{2+\epsilon}$, in which the dependence on $\deg f$ is nearly best-possible (see Exercise 4.4). Recently, a sharpening of this bound has been found provided the curve $f = 0$ has 'small' genus. See [33].

Exercise 4.1. Prove that, for coprime positive integers r, s, $X^r Y^s - 1$ is irreducible over $\overline{\mathbb{Q}}$, and similarly for $X^r - Y^s$. (One of the many possible proofs comes from a substitution $X \mapsto t^s X$, $Y \mapsto t^{\pm r} Y$.) Further, find the complete factorization for general r, s.

Exercise 4.2. Prove that $\varphi(n) \geq n/k$ where φ is the Euler function and k is the largest integer with $k! \leq n$. (Hint: use $\varphi(n) = n \prod_{p|n}((p - 1)/p)$

where p runs through the prime divisors of n; observe that if there are $k - 1$ such primes the product is $\geq \prod_{2 \leq m \leq k}((m-1)/m) = 1/k$.) This yields $\varphi(n) \gg n \log \log n / \log n$; better results are known, e.g. the best-possible $\varphi(n) \gg n / \log \log n$, with a little subtler proofs.

Exercise 4.3. Obtain another proof of Lang's statement as follows (Lang [53] attributes this argument to Serre and Tate): say for simplicity that $f \in \mathbb{Q}[X, Y]$ and let P be a torsion point as above. If l is a prime not dividing N, we obtain by conjugation that P is also a zero of $f(X^l, Y^l)$. Now we may apply Bezout, which will work if l is 'small'. (Some work must however be done to recover the exceptional shapes; for this, supposing that $f(X^l, Y^l)$ is divisible by $f(X, Y)$, find other factors on multiplying X, Y by l-th roots of unity.)

Exercise 4.4. Show that for arbitrarily large N there exist absolutely irreducible polynomials $f \in \mathbb{Q}[X, Y]$, not of special shape, such that $f(\zeta, \eta) = 0$ for roots of unity ζ, η of exact common order $N > (\deg f)^2/4$. (Hint: for prime $p > 2$, let $\zeta = e^{2\pi i/p}, R := [\sqrt{p}] + 1$. For $n \in [0, p-1]$, divide n by R: $n = qR + r$. Form the polynomial $f = \sum_{n=0}^{p-1} X^q Y^r$. Prove $\deg f < 2\sqrt{p}$ and $f(\zeta^R, \zeta) = 0$. Now prove that f is irreducible over $\overline{\mathbb{Q}}$, hence has no special factors, as follows: setting $p - 1 = QR + U, 0 \leq U < R$, we have $f(X, Y) = A(Y)(1 + X + \ldots + X^{Q-1}) + (1 + Y + \ldots + Y^U)X^Q$, where $A(Y) = 1 + Y + \ldots + Y^{R-1}$. Note that $\gcd(U + 1, R) = 1$, because p is prime; hence $A(Y)$ and $1 + Y + \ldots + Y^U$ are coprime. Now apply Eisenstein criterion.)

Note that Lang's statement may be read in geometrical terms as follows: the equation $f(x, y) = 0$ defines an irreducible curve $\mathcal{C}$ in $\mathbb{G}_m^2$, on which we are seeking the torsion points. The special shapes for f correspond to $\mathcal{C}$ being a *translate by a torsion point of an algebraic subgroup of* $\mathbb{G}_m^2$. In fact, the special shapes are of the form $X^r Y^s = \zeta$, with $r, s \in \mathbb{Z}$ and ζ a root of unity. [2] Also, it is immediate to verify that the curve $X^r Y^s = 1$ defines in $\mathbb{G}_m^2$ a subgroup H, which is algebraic by definition; the equation $X^r Y^s = \zeta$ defines a translate of H by a torsion point, e.g. by any point $(1, \zeta^{1/s})$. Lang's statement then implies that *If $\mathcal{C}$ has infinitely many torsion points then it is a translate of an algebraic subgroup, by a torsion point.* (The converse implication also holds because a connected algebraic subgroup of $\mathbb{G}_m^n$ is divisible, so in particular torsion points are Zariski dense on any algebraic subgroup; this may be shown for instance using the fact, to be proved soon, that an algebraic subgroup may always be defined by binomial equations as above.)

Let us now consider a different example, seeking this time points of nonzero small height. The line $x + y = 1$ in $\mathbb{G}_m^2$ contains the torsion points $(e^{\pm \pi i/3}, e^{\mp \pi i/3})$ and no others (as noted above). On the other hand, Zagier has shown [95] that any algebraic non-torsion point on it has height at

---

[2] Note that allowing negative exponents for X, Y leaves us with regular functions on $\mathbb{G}_m^2$.

least $\frac{1}{2}\log\frac{1+\sqrt{5}}{2}$, which is best-possible. [3] We now outline Zagier's proof principle, by sketching a proof of a positive lower bound for the height of non-torsion points on the line.

To start with, recall that by Proposition 3.6 for a nonzero algebraic number ξ of degree d we have $h(\xi) \geqslant \frac{1}{d}\sum_\sigma \log^+|\xi^\sigma|$, the sum being taken over a set of distinct conjugates ξ^σ. Together with the equality $h(\xi)=h(1/\xi)$, we obtain $h(\xi)=(h(\xi)+h(1/\xi))/2 \geqslant \frac{1}{2d}\sum_\sigma|\log|\xi^\sigma||$. As already observed in the previous chapter, this yields that *if the height is small, a 'big' percentage of conjugates of ξ must lie near the unit circle*. We exploit this fact for an algebraic point $P=(\xi, 1-\xi)$ with $\hat{h}(P)\leq\epsilon$; we obtain

$$\sum_\sigma (|\log|\xi^\sigma||+|\log|1-\xi^\sigma||) \leq 2\epsilon d.$$

This implies that for at least $d/2$ conjugates ξ^σ we have $|\log|\xi^\sigma|| + |\log|1-\xi^\sigma|| \leq 4\epsilon$. For these conjugates, both $x_\sigma := \xi^\sigma$ and $y_\sigma := 1-x_\sigma = 1-\xi^\sigma$ are near the unit circle. Then $\overline{x_\sigma}$ is about x_σ^{-1} and similarly for y_σ. Taking the complex conjugate of $x_\sigma+y_\sigma = 1$ we obtain that $x_\sigma^{-1}+y_\sigma^{-1}$ is nearly 1, so $x_\sigma y_\sigma$ is nearly 1, yielding that $\alpha^\sigma := (\xi^\sigma)^2-\xi^\sigma+1$ is nearly 0, specifically $|\alpha^\sigma| \leq c_1\epsilon$ for an absolute constant c_1. So, if $\alpha = \xi^2 - \xi + 1 \neq 0$, we have, if say $2c_1\epsilon < 1$, $h(1/\alpha) \geq d^{-1}\sum_\sigma \log^+(|\alpha^\sigma|^{-1}) > (\log|1/c_1\epsilon|)/2$. But $h(1/\alpha) = h(\alpha) \leq 3h(\xi)+\log 3 \leq 3\epsilon + \log 3$. Hence ϵ is bounded below absolutely.

Remark 4.2. Inspection of the argument shows that it depends on the following facts:
  (i) a number with small height has many conjugates near the unit circle;
  (ii) complex conjugation acts as the map $x \mapsto x^{-1}$ on the unit circle;
  (iii) the curve $x + y = 1$ is sent to $xy = x + y$ by the map $P \mapsto P^{-1}$;
  (iv) the curves $x + y = 1$ and $xy = x + y$ have only the above two torsion points in common. So at bottom the argument works because $xy = x + y$ and $x + y = 1$ are distinct curves.

Now, a similar argument can give something for varieties far more general than the line $x + y = 1$ (Zagier observed this for curves and Schmidt further developed it), but it fails for varieties stable under the map $P \mapsto P^{-1}$, which can have a Zariski dense set of complex points with coordinates in the unit circle. A simple example of this is the curve $(x+y)(xy+1) = xy$ in $\mathbb{G}_m^2$. (Note that this curve is not a translate of an algebraic subgroup.) We shall soon see methods which work for general varieties.

Exercise 4.5. Prove that the curve defined in $\mathbb{G}_m^2$ by $(x+y)(xy+1) = xy$ has a Zariski dense set of points (x, y) with $|x| = |y| = 1$. (Hint: if $x = e^{i\theta}, y = e^{i\phi}$, $\theta, \phi \in \mathbb{R}$, the equation amounts to $\cos\theta + \cos\phi = 1/2$.)

---

[3] Curiously, this coincides with the lower bound of Schinzel for the height of totally real numbers $\neq 0, \pm 1$.

Another example of absolute lower bound for the height comes from the curve defined by $X^r Y^s = c$, where r, s are given integers (not both 0) and $c \in \overline{\mathbb{Q}}^*$ is not a root of unity; this curve is a translate of an algebraic subgroup (the one defined by $X^r Y^s = 1$), but not a translate by a torsion point. For an algebraic point $P = (x, y)$ on it, we have $0 < h(c) = h(x^r y^s) \le |r|h(x) + |s|h(y) \le (\max(|r|, |s|))\widehat{h}(P)$, proving that the height of P is bounded below by a fixed number > 0.

These examples provide some evidence for what we shall recognize to be a general fact: the height of algebraic points on a subvariety of $\mathbb{G}_m^n$ is bounded below by a number > 0 if we stay out of the translates by torsion points of algebraic subgroups. (To exclude these varieties turns out to be necessary, as in the case of Lang's assertion.) This is essentially the said theorem by Zhang; before presenting a proof of it, we shall discuss the structure of algebraic subgroups of $\mathbb{G}_m^n$, which not only will be useful in the arguments but is a topic of independent interest; closely related to it is the theory of sublattices in $\mathbb{Z}^n$, which we shall start with.

4.2. Lattices and algebraic subgroups

4.2.1. Lattices in $\mathbb{Z}^n$

By a *lattice* in $\mathbb{Q}^n$ we shall mean a finitely generated subgroup, not necessarily of rank n. Here we shall work only with sublattices of $\mathbb{Z}^n$. Since $\mathbb{Q}^n$ is torsion-free, by the structure theorem for finitely generated abelian groups, every lattice has a finite basis over $\mathbb{Z}$.[4] The *rank* of the lattice Λ, denoted $rk(\Lambda)$ is the number of elements in a $\mathbb{Z}$-basis, *i.e.* the dimension of $\mathbb{Q}\Lambda$ over $\mathbb{Q}$.

To every lattice $\Lambda \subset \mathbb{Z}^n$ we associate the *saturated* lattice $\tilde{\Lambda} := \mathbb{Q}\Lambda \cap \mathbb{Z}^n \supset \Lambda$.[5] We say that Λ is *primitive* if $\tilde{\Lambda} = \Lambda$. Note that $\mathbb{Q}\tilde{\Lambda} = \mathbb{Q}\Lambda$, so $\tilde{\tilde{\Lambda}} = \tilde{\Lambda}$ and $\tilde{\Lambda}$ is primitive. Note that $\rho(\Lambda) := [\tilde{\Lambda} : \Lambda]$ is finite, because clearly Λ and $\tilde{\Lambda}$ have the same rank, equal to $\dim_{\mathbb{Q}} \mathbb{Q}\Lambda$. The following proposition provides some equivalent conditions for primitivity.

Proposition 4.2. Let Λ be a sublattice of $\mathbb{Z}^n$, and for a prime p let $\Lambda(p)$ be the image of Λ in $\mathbb{F}_p^n$ by the reduction modulo p in $\mathbb{Z}^n$. The following properties are equivalent:

1. Λ is primitive, i.e. $\rho(\Lambda) := [\tilde{\Lambda} : \Lambda] = 1$.
2. $\mathbb{Z}^n / \Lambda$ is torsion-free.

---

[4] In the supplements we shall see a self-contained proof of this assertion, actually in the more general case of discrete subgroups of $\mathbb{R}^n$.

[5] Note that $\tilde{\Lambda}$ is indeed a lattice, as a subgroup of the finitely generated group $\mathbb{Z}^n$.

3. *There exists a lattice Λ_c with $\Lambda \oplus \Lambda_c = \mathbb{Z}^n$ [6], i.e., a(ny) basis of Λ extends to a basis of $\mathbb{Z}^n$.*
4. $\dim_{\mathbb{F}_p} \Lambda(p) = rk(\Lambda)$ *for every prime p.*
5. *Letting $\underline{a}_1, \ldots, \underline{a}_r \in \mathbb{Z}^n$ be a basis of Λ, the gcd of the determinants of the $r \times r$ minors of the matrix of the $\underline{a}_i$ is 1.*

Remark 4.3. For instance, if Λ has rank 1, the equivalence between nos. 3 and 4 (or 5) says that an integer vector $(a_1, \ldots, a_n) \in \mathbb{Z}^n$ is the first row of an integer matrix with determinant ± 1 if and only if $\gcd a_i = 1$. This 'unimodular' property holds in every principal ideal ring; it also holds in polynomial rings over a field, which is a celebrated theorem of Quillen and Suslin (after a conjecture of Serre).

Proof.
(1⇒2) Let $v \in \mathbb{Z}^n$ be torsion modulo Λ, so $mv \in \Lambda$ for some integer $m > 0$. Then $v \in \mathbb{Q}\Lambda$, so $v \in \tilde{\Lambda}$ and if Λ is primitive $v \in \Lambda$.
(2⇒1) Just reverse the previous argument.
(3⇒2) If $\Lambda \oplus \Lambda_c = \mathbb{Z}^n$, then $\mathbb{Z}^n/\Lambda \cong \Lambda_c$ is torsion-free.
(2⇒3) The group $\mathbb{Z}^n/\Lambda$ is abelian and finitely generated. If it is torsion free, it is isomorphic to $\mathbb{Z}^s$, for some s. Let then $g_1, \ldots, g_s \in \mathbb{Z}^n$ be representatives of a basis for $\mathbb{Z}^n/\Lambda$ and define $\Lambda_c := \mathbb{Z}g_1 + \ldots + \mathbb{Z}g_s$. Denoting by $\bar{v}$ the projection modulo Λ of $v \in \mathbb{Z}^n$, we can write $\bar{v} = m_1\bar{g}_1 + \ldots + m_s\bar{g}_s$, finding $v - (m_1 g_1 + \ldots + m_s g_s) \in \Lambda$, so $\mathbb{Z}^n = \Lambda + \Lambda_c$. Since $\bar{g}_1, \ldots, \bar{g}_s$ is a basis for $\mathbb{Z}^n/\Lambda$, we have $\Lambda \cap \Lambda_c = \{0\}$, whence $\mathbb{Z}^n = \Lambda \oplus \Lambda_c$.
(1⇒4) The $\mathbb{F}_p$-vector space $\Lambda(p)$ is generated by the reductions of a $\mathbb{Z}$-basis $\underline{b}_1, \ldots, \underline{b}_r$ of Λ. If these reductions are linearly dependent over $\mathbb{F}_p$, we have some equation $p\underline{v} = m_1\underline{b}_1 + \ldots + m_r\underline{b}_r$, were $\underline{v} \in \mathbb{Z}^n$ and where not all the coefficients $m_i \in \mathbb{Z}$ are divisible by p. Then $\underline{v}$ belongs to $\tilde{\Lambda}$ but not to Λ.
(4⇒1) If a prime p divides ρ (which is finite, as we have noted), then there exists a vector $\underline{v} \in \tilde{\Lambda}$ of order p modulo Λ. Let as above $\underline{b}_1, \ldots, \underline{b}_r$ be a basis of Λ, so a $\mathbb{Q}$-basis of $\mathbb{Q}\Lambda$; then we may write $\underline{v} = c_1\underline{b}_1 + \ldots + c_r\underline{b}_r$ with rationals c_i, not all in $\mathbb{Z}$ and such that $pc_i \in \mathbb{Z}$. Then $(pc_1)\underline{b}_1 + \ldots + (pc_r)\underline{b}_r = p\underline{v}$ is in $p\mathbb{Z}^n$, saying that the reductions of the $\underline{b}_i$ are linearly dependent over $\mathbb{F}_p$, hence $\dim_{\mathbb{F}_p} \Lambda(p) < r$.

Finally, condition 5 is plainly equivalent to 4, because a sequence of r vectors in κ^n (κ a field) is linearly independent if and only if there is some $r \times r$ nonsingular minor. $\qquad\square$

---

[6] As an inner direct sum, *i.e.* $\Lambda \cap \Lambda_c = \{0\}$ and $\Lambda + \Lambda_c = \mathbb{Z}^n$.

4.2.2. Algebraic subgroups

In the sequel we shall work with an algebraically closed ground field κ, which in our applications will be $\overline{\mathbb{Q}}$.

We have already introduced the algebraic group $\mathbb{G}_m$ (see Section 3.4) and its powers $\mathbb{G}_m^n$. We have noted that $\mathbb{G}_m^n$ is affine and we can recover it also from its ring of regular functions (over κ), which is $\kappa[\mathbb{G}_m^n] = \kappa[X_1^{\pm 1}, \ldots, X_n^{\pm 1}]$, namely the ring of Laurent polynomials in n variables.

4.2.3. Some definitions

By *algebraic subgroup* of $\mathbb{G}_m^n$ we mean a closed algebraic subvariety defined over κ, stable under the group operations. We say that an algebraic subgroup of $\mathbb{G}_m^n$ is a *torus* if it is irreducible as an algebraic variety; for instance, $\mathbb{G}_m^n$ is clearly a torus. We may derive from a general property that every algebraic subgroup is a finite disjoint union of translates of a torus (see Exercise 4.6). In our special context, we shall recover independently this property as a consequence of the description below. Since an irreducible algebraic variety is connected (both in the Zariski and in the complex topology) it follows for instance that, for an algebraic subgroup (*e.g.* over $\mathbb{C}$), being irreducible and connected are equivalent properties. We shall refer to a translate of a torus also as a *torus coset*, which shall be called a *torsion coset* if it is a translate of a torus by a torsion point.

To discuss algebraic subgroups of $\mathbb{G}_m^n$, the following notation will be very useful: namely, for a point $P = (x_1, \ldots, x_n) \in \mathbb{G}_m^n$ and an integral vector $\underline{a} = (a_1, \ldots, a_n) \in \mathbb{Z}^n$, we set $P^{\underline{a}} := x_1^{a_1} \cdots x_n^{a_n}$.
Note thay any m-tuple of vectors $\underline{a}_1, \ldots, \underline{a}_m \in \mathbb{Z}^n$ defines a regular map $\varphi: \mathbb{G}_m^n \to \mathbb{G}_m^m$ by $\varphi(P) := (P^{\underline{a}_1}, \ldots, P^{\underline{a}_m})$. This map is plainly an algebraic group homomorphism, called *monoidal*. When $m = n$, the homomorphism φ is invertible if and only if $\Delta_\varphi := \det(\underline{a}_1, \ldots, \underline{a}_m) = \pm 1$; in this case it is called a *monoidal automorphism* of $\mathbb{G}_m^n$.

Exercise 4.6. Let G be an algebraic group. Prove that its irreducible components are pairwise disjoint, that the component G_0 containing the identity is a closed normal subgroup of finite index in G, and that the components are the cosets of G_0 in G. (Hint: observe that left translation is an automorphism and so permutes the irreducible components. Hence, if some element belongs to more than one component, every other element has the same property. But certainly each component contains some element which does not belong to any other. So the components are pairwise disjoint. Now, if $x \in G_0$, then $x^{-1}G_0$ is a component containing the identity, hence $= G_0$, proving that G_0 is a closed subgroup. The rest follows by a similar argument, on considering the conjugation $x^{-1}G_0x$.)

Exercise 4.7. Notation as in the previous exercise, suppose that G is commutative and that $G_0(\kappa)$ is divisible (for a field κ). Prove that for all $x \in G(\kappa)$, xG_0

is a translate of G_0 by a torsion point. (Hint: observe that some nonzero power of x is in $G_0(\kappa)$.)

Exercise 4.8. Prove that a monoidal homomorphism φ is not injective when the determinant $\Delta_\varphi \neq \pm 1$. (Hint: if p is a prime divisor of Δ, find a point of order p in the kernel. In general, the kernel consists of torsion points of order $|\Delta_\varphi|$.)

Given any lattice $\Lambda \subset \mathbb{Z}^n$ we define

$$H_\Lambda := \{P \in \mathbb{G}_m^n(\overline{\mathbb{Q}}) \mid P^\lambda = 1 \ \forall \lambda \in \Lambda\}.$$

This is an algebraic subgroup of $\mathbb{G}_m^n$, since the equations $P^\lambda = 1$ are algebraic and since every map $P \mapsto P^\lambda$ is a group homomorphism. Moreover, the correspondence $\Lambda \mapsto H_\Lambda$ is inclusion-reversing.

Let $\underline{a}_1, \ldots, \underline{a}_r$ be a basis for the saturated lattice $\tilde{\Lambda}$ (defined in the previous subsection); by Proposition 4.2, this basis can be extended to a basis $\underline{a}_1, \ldots, \underline{a}_n$ of $\mathbb{Z}^n$. Without loss of generality, we may perform the monoidal automorphism associated to this basis and consider the coordinates $y_i = \underline{x}^{\underline{a}_i}$, $i = 1, \ldots, n$. In this coordinate system, $\tilde{\Lambda}$ is identified with $\mathbb{Z}^r$, $H_{\tilde{\Lambda}}$ is defined by $y_1 = \ldots = y_r = 1$ and we have, by projection over the last coordinates, that $H_{\tilde{\Lambda}} \cong \mathbb{G}_m^{n-r}$ is a torus.

Now, we have $y_i^\rho = 1$ on H_Λ for $i = 1, \ldots, r$, where $\rho = \rho(\Lambda)$, because $\rho \underline{a}_i \in \Lambda$. Hence, on every irreducible component of H_Λ, all coordinates $y_i, i = 1, \ldots, r$, are constant (equal to ρ-th roots of 1). Since $H_{\tilde{\Lambda}} \subset H_\Lambda$ and $\dim H_{\tilde{\Lambda}} = \dim H_\Lambda$, we have that H_Λ is a finite union of irreducible cosets $P \cdot H_{\tilde{\Lambda}}$; also, every such coset is clearly a torsion coset, because it depends only on the first r coordinates of P, which are roots of unity of exponent ρ. It also follows that such torsion cosets are the components of H_Λ.

From Proposition 4.2 we may further deduce that H_Λ *is irreducible if and only if Λ is primitive*. In particular, the proposition also yields a helpful irreducibility criterion. To prove this claim, recall that by the above monoidal trasformation we may identify $\tilde{\Lambda}$ with $\mathbb{Z}^r$ and so by replacing r with n we are reduced to the case when Λ has finite index ρ in $\mathbb{Z}^n$ and H_Λ is finite. We must prove that H_Λ has just one point if and only if $\rho = 1$. The 'if' part is clear. Conversely, supposing $\rho > 1$, let p be a prime such that $\Lambda(p)$ has rank $< n$ (it exists by n. 4 of the said proposition). Then there is an integral vector $(a_1, \ldots, a_n) \in \mathbb{Z}^n \setminus p\mathbb{Z}^n$, whose scalar product with any vector of Λ lies in $p\mathbb{Z}$. Finally, the point $(\zeta^{a_1}, \ldots, \zeta^{a_n})$, where ζ is a primitive p-th root of unity is easily seen to be a nontrivial point in H_Λ. An alternative argument appears in the following exercise.

Exercise 4.9. Prove that $[H_\Lambda : H_{\tilde{\Lambda}}] = [\tilde{\Lambda} : \Lambda]$ and deduce that the map $\Lambda \to H_\Lambda$ is 1-to-1. (Hint: we have seen that by a monoidal automorphism we may reduce to the case when $\tilde{\Lambda}$ is generated by the first r vectors of a canonical basis of $\mathbb{Z}^n$; this also allows to reduce to the case $r = n$. Now use the theorem of elementary divisors to find a basis $B_1, \ldots, B_n$ of $\mathbb{Z}^n$ such that $m_1 B_1, \ldots, m_n B_n$ is a basis for Λ, for suitable nonzero integers m_i.)

The family H_Λ just defined *a priori* does not exhaust all the possibilities for the algebraic subgroups. But we shall see that this is indeed the case. This result will follow from a general description of the irreducible algebraic subgroups contained in a given algebraic subvariety $\mathcal{X}$ of $\mathbb{G}_m^n$; such a description is 'effective', in the sense that it allows to calculate equations for such subgroups, provided we are given 'effectively' a system of defining equations for $\mathcal{X}$.

Let then $\mathcal{X}$ be defined in $\mathbb{G}_m^n$ by the (finite) system $f_i(\underline{X}) = \sum a_{i,\underline{\lambda}} \underline{X}^{\underline{\lambda}} = 0$. Since we are working in $\mathbb{G}_m^n$, the f_i here may be assumed to be Laurent polynomials, namely we allow negative exponents of the variables. The idea is to view the (Laurent) monomials $\underline{X}^{\underline{\lambda}}$ as homomorphisms from $\mathbb{G}_m^n(\kappa)$ to $\mathbb{G}_m(\kappa)$, *i.e.* characters.[7] So we start by recalling the theorem of Artin on linear independence of characters.

Theorem 4.3 (E. Artin). *Let G be any group and let $\sigma_1, \ldots, \sigma_m$ be pairwise distinct characters (homomorphisms) of G in κ^*. Then they are linearly independent over κ.*

Proof. By induction we may assume that $m > 1$ and that every subsequence of $m - 1$ of our characters is linearly independent. Assume that the characters are linearly dependent and write a nontrivial linear relation $\sum_{i=1}^m c_i \sigma_i = 0$, where necessarily all the coefficients $c_i \in k$ are nonzero.

For every $g, h \in G$ we have $0 = \sum_{i=1}^m c_i \sigma_i(hg) = \sum_{i=1}^m c_i \sigma_i(h) \sigma_i(g)$. Viewing h as fixed, we then conclude that the function $\sum_{i=1}^m c_i (\sigma_i(h) - \sigma_1(h)) \sigma_i$ vanishes on G. But this equals $\sum_{i=2}^m c_i (\sigma_i(h) - \sigma_1(h)) \sigma_i$, whence by induction we have $\sigma_i(h) = \sigma_1(h)$ for each $i = 2, \ldots, m$. Since this holds for all $h \in G$, we conclude that the characters are not distinct, a contradiction which proves the theorem. $\quad\square$

Let now H be an algebraic subgroup of $\mathbb{G}_m^n$ entirely contained in $\mathcal{X}$. Define $\Omega = \Omega_H := \{\omega \in \mathbb{Z}^n \mid \underline{x}^\omega = 1 \ \forall \underline{x} \in H\}$.[8] Note that Ω is plainly a lattice in $\mathbb{Z}^n$, and defines an algebraic subgroup H_Ω. Observe that $H_\Omega \supset H$.

---

[7] This follows M. Laurent's approach; we note that he has not to be confounded with the 'Laurent' giving the name to series and polynomials.

[8] In this argument we may work with points defined over an algebraically closed field κ, such that H is defined over κ. Only *a posteriori* we know that H is defined over $\mathbb{Q}$.

We can now fix a system R of representatives for $\mathbb{Z}^n/\Omega$ and write the equations that define $\mathcal{X}$ in the form $0 = \sum_{\lambda_0 \in R} \underline{x}^{\lambda_0} \left(\sum_{\lambda \in \lambda_0 + \Omega} a_{i,\lambda} \underline{x}^{\lambda - \lambda_0} \right) =: \sum_{\lambda_0 \in R} \underline{x}^{\lambda_0} M_{i,\lambda_0}(\underline{x})$, where we need to consider only finitely many λ_0.

Note that all the terms $M_{i,\lambda_0}(\underline{x}) = \sum_{\lambda \in \lambda_0 + \Omega} a_{i,\lambda} \underline{x}^{\lambda - \lambda_0}$ are constant on the whole H by definition; in fact, since the terms $\underline{x}^{\lambda - \lambda_0}$ are 1 on H, we have $M_{i,\lambda_0}(\underline{x}) = \sum_{\lambda \in \lambda_0 + \Omega} a_{i,\lambda}$ on H. Also, for distinct $\lambda_0 \in R$, the functions $\underline{x}^{\lambda_0}$ induce, by restriction to $H(\kappa)$, characters of H which are pairwise distinct. But then each of the above equations on $\mathcal{X}$ yields a linear relation among these characters, which must be trivial by Artin's theorem. Therefore all the coefficients $\sum_{\lambda \in \lambda_0 + \Omega} a_{i,\lambda}$ must vanish.

Conversely, assume that this holds, and define Λ as the lattice generated by the differences $\lambda - \lambda_0$ which actually appear in some of the (finitely many) equations that we have chosen to define $\mathcal{X}$. Then $\Lambda \subset \Omega$ and thus $H_\Lambda \supset H_\Omega \supset H$. Moreover, we may reverse the above calculation and find that in fact the whole H_Λ is contained in $\mathcal{X}$; hence, if H is maximal (as an algebraic subgroup contained in $\mathcal{X}$), we must have $H = H_\Lambda$. Also, since there only finitely many H_Λ which can arise in this way, each H is contained in a maximal H_Λ.[9] We resume these conclusions in the following

Proposition 4.4. *If H is an algebraic subgroup contained in $\mathcal{X}$, then H is contained in a maximal such subgroup. Each of these maximal subgroups is of the shape H_Λ for some lattice Λ. Moreover such a lattice Λ may be generated by differences $\lambda_1 - \lambda_2$ of exponent vectors which appear in a prescribed finite system of equations defining $\mathcal{X}$, and in particular Λ and H_Λ have only finitely many possibilities for a given $\mathcal{X}$.*

Remark 4.4.
  (i) **Effectivity.** Note that the above procedure allows to 'construct' in finitely many steps all the maximal algebraic subgroups contained in $\mathcal{X}$. It suffices: (a) to take all possible subsets of differences $\lambda_1 - \lambda_2$ as in the proposition (there are only finitely many of them), (b) to consider the corresponding lattices Λ, systems R of representatives and associated decompositions of the equations defining $\mathcal{X}$, and (c) to check whether the terms M_{i,λ_0} vanish.
  (ii) **Cosets contained in $\mathcal{X}$.** On applying the results to a translated variety $P^{-1}\mathcal{X}$ (for some $P \in \mathbb{G}_m^n$) we deduce that every maximal algebraic coset $P \cdot H$ contained in $\mathcal{X}$ is such that H corresponds to a lattice as above. Thus, in particular there are only finitely many algebraic groups H associated to maximal cosets $P \cdot H$ contained in $\mathcal{X}$.

---

[9] If there were infinitely many H_Λ, the union of a chain could *a priori* be non-algebraic; but see Exercise 4.10.

In particular, if H is any algebraic subgroup of $\mathbb{G}_m^n$, we may take $\mathcal{X} = H$ in the proposition to obtain, also recalling a previous remark for the last conclusion, the following

Corollary 4.5. *Every algebraic subgroup of $\mathbb{G}_m^n$ is of the shape H_Λ for some lattice $\Lambda \subset \mathbb{Z}^n$. H_Λ is irreducible if and only if Λ is primitive.*

Exercise 4.10. Prove directly that every algebraic subgroup contained in $\mathcal{X}$ is contained in a maximal such subgroup. (Hint: the problem is to show that a maximal subgroup is algebraic. For this, prove that the Zariski closure of a subgroup is a subgroup.)

Exercise 4.11. Find the algebraic subgroups contained in the variety defined in $\mathbb{G}_m^5$ by $X_1 + \ldots + X_5 = 1$.

Exercise 4.12. Let $\mathcal{X}$ be the variety defined in $\mathbb{G}_m^n$ by $X_1 + \ldots + X_n = 1$. Prove that if a point $P = (\xi_1, \ldots, \xi_n) \in \mathcal{X}$ lies in some algebraic subgroup $H \subset \mathcal{X}$ with $\dim H > 0$, then some subsum of the ξ_i vanishes. (Hint: apply the above criteria for detecting algebraic subgroups contained in $\mathcal{X}$.)

Exercise 4.13.   (Parametrization of algebraic subgroups) Let H be an irreducible algebraic subgroup of $\mathbb{G}_m^n$ of dimension d. Prove that there exists a parametrization $x_i = t_1^{a_{i,1}} \cdots t_d^{a_{i,d}}$, with the $a_{i,j}$ in $\mathbb{Z}$. (Hint: consider the lattice Λ associated to H and its *orthogonal* in $\mathbb{Z}^n$. Alternatively, use a monoidal transformation.) Note that the result says that there is a homomorphism (in the algebraic group sense) from $\mathbb{G}_m^d$ onto H.

Exercise 4.14. Let G be a nonempty subset of $\mathbb{G}_m^n(\kappa)$ closed under multiplication. Prove that the Zariski closure of G is an algebraic subgroup.

Exercise 4.15. Let $P \in \mathbb{G}_m^n(\kappa)$. Prove that the Zariski closure of the set of powers $\{P^m \mid m \in \mathbb{Z}\}$ is an algebraic subgroup H and that $\dim H$ is the multiplicative rank of the subgroup of κ^* generated by the coordinates of P. (Hint: for the last question, use the structure of algebraic subgroups of $\mathbb{G}_m^n$.)

4.2.4. A characterization of torsion cosets

In this subsection we shall discuss torsion cosets and we shall characterize them as those subvarieties of $\mathbb{G}_m^n$ closed under a nontrivial multiplication by m-map. This shall be used in our proof of the theorem of Zhang, but is a fact of independent interest.

As above, we shall work with the points over an algebraically closed field κ, supposed here to be of characteristic zero. It will be $\overline{\mathbb{Q}}$ in all our applications.

By the results just proved, any torus H may be defined by a system of equations $\underline{X}^\lambda = 1$ for λ running through a primitive sublattice $\Lambda = \Lambda_H$ of $\mathbb{Z}^n$; naturally we may consider only the equations corresponding to λ in a basis of Λ. Since Λ is primitive, we may extend a basis of it to a basis of $\mathbb{Z}^n$. Further, by using a monoidal automorphism corresponding to such a basis of $\mathbb{Z}^n$, we may assume that a basis of Λ consists of the first r vectors of the canonical basis of $\mathbb{Z}^n$. In these coordinates H will be defined by the equations $X_i = 1$ for $i = 1, \ldots, r$ and will be identified with $\mathbb{G}_m^{n-r}$ by projection on the last $n - r$ coordinates. In such representation, a coset gH of H will be defined by $X_i = g_i, i = 1, \ldots, r$, for suitable $g_i \in \kappa$. This coset will be a torsion-coset if and only if $g_1, \ldots, g_r$ are roots of unity.

Torsion points in algebraic cosets

This representation also shows that torsion points are Zariski-dense in any torus H [10] and they are Zariski-dense in a coset gH if and only if gH is a torsion coset. (In turn this happens precisely if gH contains at least a torsion point.)

It will be a consequence of Zhang's theorem, proved in the next section, that a converse holds; see Corollary 4.13, which also characterizes torsion cosets. In part for this reason, the torsion cosets are also called *torsion varieties*.[11]

Exercise 4.16. Let v be a finite place of $\mathbb{Q}$ and extend it to $\overline{\mathbb{Q}}$. Prove that the set of torsion points of $H(\overline{\mathbb{Q}})$ is discrete for the v-adic topology. (Hint: reduce to $H = \mathbb{G}_m^r$ by a monoidal transformation. Then show that the distance $|\zeta - 1|_v$ is bounded below for ζ a root of unity, $\zeta \neq 1$. For this, raise an equation $\zeta = 1 + (\zeta - 1)$ to a power and use the binomial theorem.)

The multiplication maps [m]

As is customary, for a nonzero integer m we denote by $[m]$ the endomorphism $P \mapsto P^m$ of $\mathbb{G}_m^n$. Let us study a little this map. It is plainly surjective and with finite kernel; a homomorphism with these properties is also called an *isogeny*. The kernel of $[m]$ has $|m|^n$ elements.[12] This map is finite; viewing it on complex points, it is a finite covering map. In particular, for a subvariety $\mathcal{X}$ of $\mathbb{G}_m^n$, the image $[m]\mathcal{X}$ of $\mathcal{X}$ under the map $[m]$ is always a closed subvariety; it is irreducible if $\mathcal{X}$ is irreducible. Naturally, the inverse image $[m]^{-1}\mathcal{X}$ is also closed, but may well be reducible. If $\mathcal{X}$ is defined by equations $f_i(\underline{X}) = 0$, then $[m]^{-1}\mathcal{X}$ is defined by the equations $f_i(\underline{X}^m) = 0$.

Exercise 4.17. For a subvariety $\mathcal{X}$ of $\mathbb{G}_m^n$, prove that $[m]^{-1}[m]\mathcal{X} = \cup_{g^m=1} g\mathcal{X}$.

---

[10] In the complex topology their closure is a product of circles (see the exercise below for the p-adic case), and thus is not dense.

[11] Another motivation for this terminology is that gH becomes a torsion point of the quotient $\mathbb{G}_m^n/H$.

[12] In positive characteristic p things are different and depend on whether or not p divides m.

Exercise 4.18. Prove that if $\mathcal{X}$ is irreducible the irreducible components of $[m]^{-1}\mathcal{X}$ are permuted transitively by translation by the points of order m. (Hint: Observe that $[m]$ is onto $\mathcal{X}$ on each component.)

Exercise 4.19. Prove that $\dim[m]\mathcal{X} = \dim \mathcal{X} = \dim[m]^{-1}\mathcal{X}$.

Exercise 4.20. Let $\mathcal{X} \subset \mathbb{G}_m^2$ be a curve, defined by $f(X, Y) = 0$. Find some equation defining $[m]\mathcal{X}$. (Hint: consider $\prod_{\zeta^m=\eta^m=1} f(\zeta X, \eta Y)$.) Generalize the result.

Exercise 4.21. For an algebraic subgroup H, prove that $[m]H = H$ if and only if $\gcd([H : H_0], m) = 1$, where H_0 is the component of the identity in H. (Hint: observe that H is a union of translates zH_0, for z in a group of order $[H : H_0]$.)

Exercise 4.22. Let X, Y be irreducible algebraic groups and $\varphi : X \to Y$ be a homomorphism with dense image. (i) Prove that φ is surjective. (Hint: let $y \in Y$. Observe that $y\varphi(X)$ and $\varphi(X)$ have nonempty intersection.) (ii) Observe that φ need not be closed. (Hint: consider *e.g.* the projection from $\mathbb{A}^2$ to $\mathbb{A}^1$.) (iii) Prove that if φ has finite degree then it is closed. (Hint: use (i).)

Plainly, if $\mathcal{X} = H$ is an algebraic subgroup of $\mathbb{G}_m^n$, we have $[m]H \subset H$, with equality if H is irreducible. (See Exercise 4.21.) We also have $[m](gH) \subset gH$ for a torsion point g of order dividing $m - 1$. Our object is to prove a converse.

Theorem 4.6. *Let $\mathcal{X}$ be a nonempty irreducible subvariety of $\mathbb{G}_m^n$ such that $[m]\mathcal{X} \subset \mathcal{X}$ for some integer $m > 1$. Then $\mathcal{X}$ is a torsion coset.*

For example, if $n = 1$ the nonempty irreducible varieties are the whole $\mathbb{G}_m$ and single points $\alpha \in \mathbb{G}_m(\kappa)$, in which case the assumption $\alpha^m = \alpha$ implies that α is a torsion point. If $n = 2$, either $\mathcal{X}$ is a point or the whole $\mathbb{G}_m^2$ as before, or $\mathcal{X}$ is an irreducible curve in $\mathbb{G}_m^2$, say defined by $f(X, Y) = 0$. Now the assumption says that $f(x^m, y^m) = 0$ whenever $f(x, y) = 0$. This forces $f(X, Y)$ to divide $f(X^m, Y^m)$, and it is a pleasant exercise to prove directly that this happens only if f has the special shape that we have met in connection with Lang's problem, namely when $\mathcal{X}$ is a translate of an algebraic subgroup. (Given this conclusion, it is very easy to deduce that the translate is actually torsion.) This method generalizes to hypersurfaces $\mathcal{X}$ in $\mathbb{G}_m^n$, for any n.

Exercise 4.23. Let $f \in \kappa[X, Y]$ be irreducible, not a monomial, and suppose f is a factor of $f(X^m, Y^m)$, for an integer $m > 1$. Prove that f is a binomial. Generalize this to any number of variables. (Hint: observe that $f(\zeta X, \eta Y)$ also divides $f(X^m, Y^m)$, for ζ, η any m-th roots of unity. Then consider degrees and substitutions sending f to a multiple of itself; this will be simpler for prime m.)

This approach appears *e.g.* in [21]: combining the case of hypersurfaces with projections, one can then prove Theorem 4.6 generally, by induction. Here we shall use a different argument, which seems to be new. It relies on the complex points, through the following lemma.

Lemma 4.7. *Let* $\psi_1, \ldots, \psi_r$ *be non-constant rational functions on an irreducible algebraic variety* $\mathcal{X}$ *defined over* $\mathbb{C}$. *Then there is a point* P *in* $\mathcal{X}(\mathbb{C})$ *such that* $|\psi_i(P)| \neq 0, 1$ *for* $i = 1, \ldots, r$.

Proof. Since the ψ_i are nonconstant, $\mathcal{X}$ is nonempty and actually of positive dimension. Then, by cutting with hyperplanes we can assume that $\mathcal{X}$ is a curve. Let P_0 be a smooth point in $\mathcal{X}(\mathbb{C})$, not a pole of any of the ψ_i, and let z be a local parameter at P_0. Then on a suitably small neighborhood of P_0 every ψ_i can be viewed as an analytic function of z. Each ψ_i thus becomes an open mapping of z, and the conclusion follows. $\square$

Remark 4.5. A slightly different way to conclude is as follows: write ψ_i as a series $\sum_{n \geq 0} a_{in} z^n$. No function $|\psi_i|^2$ is constant because of Parseval formula, hence $\prod_i |\psi_i| (1 - |\psi_i|^2)$ cannot vanish identically, and the claim again follows.

Proof of Theorem 4.6. Since we are working in characteristic zero, we may assume that the ground field is $\mathbb{C}$: in fact, $\mathcal{X}$ is defined by finitely many equations, and then it suffices to embed in $\mathbb{C}$ the field generated over $\overline{\mathbb{Q}}$ by the coefficients.

We now use induction on n. We have seen above the easy case $n = 1$, when $\mathcal{X}$ is either a point or $\mathbb{G}_m$. Hence let us assume $n > 1$ and the conclusion true up to $n - 1$.

Let then $\mathcal{X}$ satisfy the assumptions. If $\mathcal{X} = \mathbb{G}_m^n$ we are done, so let us be given a non-zero polynomial $f \in \mathbb{C}[X_1, \ldots, X_n]$ vanishing on $\mathcal{X}$. We can write it as a sum of distinct monomials $f(\underline{X}) = \sum_{i=1}^s c_i M_i$, with $c_1 \cdots c_s \neq 0$; the monomial M_i defines by restriction a regular function $\varphi_i \in \mathbb{C}[\mathcal{X}]^*$.

Assume first that no ratio φ_i / φ_j with $i \neq j$ is constant on $\mathcal{X}$. Then, by Lemma 4.7 applied with the ψ_μ equal to these ratios, there exists a point $P \in \mathcal{X}(\mathbb{C})$ such that the $|\varphi_i(P)|$ are pairwise distinct. Let i_0 be such that $|\varphi_{i_0}(P)|$ is maximal. Since $\mathcal{X}$ is supposed to be stable by the map $[m]$, we have in particular that all the points $[m]^t P = P^{m^t}$ lie in $\mathcal{X}$, for $t = 1, 2, \ldots$. Then, since f vanishes on $\mathcal{X}$, we have $\sum_{i=1}^s c_i \varphi_i(P^{m^t}) = 0$ for every integer $t > 0$. However $|\varphi_i(P^{m^t})| = |\varphi_i(P)|^{m^t}$, whence (since $m > 1$) the term with $i = i_0$ grows faster in t than any other term, a contradiction.

We deduce that there exist distinct indices $i \neq j$ so that φ_i / φ_j is constant, whence there exist a nontrivial (Laurent) monomial $\underline{X}^{h\underline{b}}$ constant on $\mathcal{X}$, for an integer $h \neq 0$ and a primitive vector $\underline{b} \in \mathbb{Z}^n$. Since $\mathcal{X}$ is irreducible, $\underline{X}^{\underline{b}}$ must itself be constant on $\mathcal{X}$. Since $\underline{b}$ is primitive, it can be completed to a basis of $\mathbb{Z}^n$, so we may view it as the first basis vector. On performing the corresponding monoidal automorphism, we may assume that $X_1 = \alpha$ is constant on $\mathcal{X}$. Since $\mathcal{X}$ is assumed to be stable by $[m]$, we conclude that $\alpha^m = \alpha$, *i.e.* α is an $(m - 1)$-th root of unity. Also, the projection of $\mathcal{X}$ on the last $n - 1$ coordinates induces an isomorphism of $\mathcal{X}$ with a subvariety $\mathcal{Y}$ of $\mathbb{G}_m^{n-1}$, also stable by the map $[m]$. Finally, we may apply the induction hypothesis to $\mathcal{Y}$ to conclude the proof. $\square$

Remark 4.6. Inspection shows that for this proof-pattern the crucial point consists in drawing informations from a vanishing $\sum_{i=1}^{s} c_i \alpha_i^{m^t} = 0$ for all integers $t > 0$, where the c_i, α_i are nonzero complex numbers. In the above proof we have managed so that the relevant α_i have pairwise distinct absolute values, so there is a 'dominant' term and the vanishing cannot actually occur for all t. However other devices are suitable. For instance, by taking the α_i to be values of the φ_i at a κ-generic point $P \in \mathcal{X}(\mathbb{C})$ (where $\kappa \subset \mathbb{C}$ is a field of definition for $\mathcal{X}$), one can also argue by using derivations of the function field extension $\kappa(\mathcal{X})/\kappa$ (see [20, 24] or [96]). Still alternatively, one may use an interesting theorem of Skolem, later extended by Mahler and Lech, implying that *if* $\sum_{i=1}^{s} c_i \alpha_i^u = 0$ *for infinitely many integers u then some ratio α_i/α_j is a root of unity.* The usual proof of this result uses some theory of p-adic analytic functions, but is substantially elementary. (See the supplements and [101] for a sketch and for references.)

Exercise 4.24. Prove that the above theorem does not hold in positive characteristic without suitable new assumptions. (Observe for instance that any variety defined over $\mathbb{F}_p$ is stable by the map $[p]$.)

4.3. Heights on subvarieties of $\mathbb{G}_m^n$

4.3.1. The theorem of Zhang

We are now ready to state and prove the main result of this chapter, the theorem of Zhang. As announced, it bounds from below by a fixed number > 0 the height of algebraic points on an algebraic subvariety $\mathcal{X} \subset \mathbb{G}_m^n$, with the exception of those points lying on a certain exceptional set. To state this precisely, a further definition will be convenient.

Definition 4.8. We let $\mathcal{X}^*$ be the complement in $\mathcal{X}$ of the union of all the torsion cosets entirely contained in $\mathcal{X}$.

Theorem 4.9 (Shou-Wu Zhang). *Let $\mathcal{X}$ be an algebraic subvariety of $\mathbb{G}_m^n$. Then:*

(i) *$\mathcal{X} \setminus \mathcal{X}^*$ is a finite union of torsion cosets, so in particular $\mathcal{X}^*$ is Zariski-open in $\mathcal{X}$.*
(ii) *There exists a number $c = c(\mathcal{X}) > 0$ such that $\widehat{h}(P) \geqslant c$ for all $P \in \mathcal{X}^*(\overline{\mathbb{Q}})$.*

The original proof [107] was sophisticated, depending on Arakelov Theory. Since then, other proofs of various nature have been given. Here we shall present a rather elementary proof due to Bombieri and Zannier (see [21]). In principle, it resembles Zagier's approach for the curve $x + y = 1$, that we have sketched above. However, instead of the map $P \mapsto P^{-1}$ (which on points of small height acts 'nearly' as the complex conjugation), it uses a 'Frobenius' map $P \mapsto P^l$, which may be read as an l-adic analogue. This variation has the crucial advantage that the prime l may be chosen arbitrarily, so that the method eventually applies to all varieties. On the contrary, recall our previous remark that Zagier's approach would not work for certain varieties stable by $P \mapsto P^{-1}$. It is also to be observed that the Frobenius map was used by Dobrowolski to establish his lower bound for the height of algebraic numbers, mentioned in the previous chapter.

To outline the principle at bottom, note that the height of an algebraic number ξ^{-1} acquires a contribution from the divisibility of ξ by a prime ideal, say above a prime l; if we prove a high divisibility, we shall obtain a good lower bound for the height $h(\xi) = h(\xi^{-1})$. To obtain the divisibility, we may consider, thinking of Fermat's Little Theorem, $\xi = \alpha^{l^r} - \alpha$ for suitable r. In the present context, this use of the Frobenius map appears in the following basic lemma.

Lemma 4.10 ([21]). *Let k be a number field, Galois over $\mathbb{Q}$, put $d = [k : \mathbb{Q}]$ and let $f \in k[X_1, \ldots, X_n]$. If $l > l(f, k)$ is a sufficiently large prime, there exists $c = c(f, l, d) > 0$ with the following property: if $f(P) = 0$, where $P \in \mathbb{G}_m^n(\overline{\mathbb{Q}})$, then either $\widehat{h}(P) \geq c$ or $f(P^{l^d}) = 0$.*

Proof. We can assume that $f \in \mathcal{O}_k[\underline{X}]$, i.e. that the coefficients of f are algebraic integers. If l is a prime number unramified in k, for any $\omega \in \mathcal{O}_k$ we have $\omega^{l^d} \equiv \omega \pmod{l}$, as is easy to verify. Together with Fermat's congruence, this implies $f^{l^d}(X_1, \ldots, X_n) = f(X_1^{l^d}, \ldots, X_n^{l^d}) + lg(X_1, \ldots, X_n)$, where $g \in \mathcal{O}_k[\underline{X}]$; we clearly have $\deg g \leq l^d \deg f$.

Evaluating at P we find $f(P^{l^d}) = -lg(P) =: \beta$; we have $\beta \in k(P) =: K$. If $\beta = 0$ one of the alternatives is verified and we are done; if not, we proceed to prove a lower bound for $\widehat{h}(P)$ by applying the product formula $\prod_{v \in M_K} \|\beta\|_v = 1$ (with the same normalizations used to compute the height on K).

We distinguish two cases.

When $v | l$, we use the equality $\beta = -lg(P)$, which yields $\|\beta\|_v = \|l\|_v \|g(P)\|_v$. Now observe that since the coefficients are algebraic integers in k and since the monomials in g have degree at most $\deg g$, we have $\|g(P)\|_v \leq \sup(1, \|P\|_v)^{\deg g}$, where $\|P\|_v$ is the v-adic sup norm. Taking logarithms we get $\log \|\beta\|_v \leqslant \log \|l\|_v + l^d \deg f \log \sup(1, \|P\|_v)$.

When $v \nmid l$, we use $\beta = f(P^{l^d})$. Then, letting N be the number of terms in f and γ_j denote its coefficients, we have $\log \|\beta\|_v \leqslant \log \sup(1, \|N\|_v) + l^d \deg f \log \sup(1, \|P\|_v) + \log \sup_j \|\gamma_j\|$.

Finally, writing $P = (\xi_1, \ldots, \xi_n)$ and using $h(1 : \xi_1 : \ldots : \xi_n) \leq h(\xi_1) + \ldots + h(\xi_n) = \widehat{h}(P)$, we obtain

$$0 = \sum_v \log \|\beta\|_v \leqslant \sum_{v|l} \log \|l\|_v + l^d \deg f \sum_v \log \sup(1, \|P\|_v)$$
$$+ \sum_{v \nmid l} \log \sup(1, \|N\|_v) + \sum_{v \nmid l} \log \sup_j \|\gamma_j\|_v$$
$$\leqslant -\log l + l^d (\deg f)\widehat{h}(P) + c_1,$$

where c_1 depends only on f, and may be taken $h(N) + h(\gamma_1 : \ldots : \gamma_N)$. Taking a prime $l \geqslant e^{2c_1}$, we obtain $\frac{1}{2}\log l \leqslant \log l - c_1 \leqslant l^d (\deg f)\widehat{h}(P)$. Thus $\widehat{h}(P) \geqslant \dfrac{\log l}{2l^d \deg f} =: c(f, l, d)$, as wanted.   □

The lower bound coming from this argument is very small, but it does not depend on the chosen point and will be amply sufficient for our purposes; also, observe that by choosing l as the smallest prime unramified in k and larger than the number e^{2c_1}, we obtain a lower bound > 0, effectively computable in terms only of f and k.

The lemma leads to the following statement, essentially a rephrasing of Zhang's theorem.

Theorem 4.11. *Let $\mathcal{X}$ be a subvariety of $\mathbb{G}_m^n$. Then there exist a constant $\gamma = \gamma(\mathcal{X}) > 0$ and a finite union $T \subset \mathcal{X}$ of torsion cosets such that every algebraic point $P \in \mathcal{X} \setminus T$ satisfies $\widehat{h}(P) \geq \gamma$.*

Proof. We argue by induction on $\delta := \dim \mathcal{X}$. In proving this theorem we may plainly assume that $\mathcal{X}$ is irreducible and defined over $\overline{\mathbb{Q}}$.[13] If $\mathcal{X}$ has dimension 0, it is a point $P \in \mathbb{G}_m(\overline{\mathbb{Q}})$. Either P is a torsion point, or $\widehat{h}(P) > 0$, which proves the conclusion in this case. Hence we can assume $\delta > 0$ and the result true for varieties of dimension up to $\delta - 1$.

Further, if $\mathcal{X}$ is a torsion coset, we are done by taking the said union to consist just of $\mathcal{X}$. (The statement on the lower bound now becomes empty.) Hence in the sequel we also assume that $\mathcal{X}$ is not a torsion coset. Let $\mathcal{X}$ be defined by the finitely many polynomials $f_1, \ldots, f_m \in k[X_1, \ldots, X_n]$, where k is a number field Galois over $\mathbb{Q}$, of degree d. Let us apply the lemma to each of the f_i (with a sufficiently large prime

---

[13] In fact, observe that the Zariski closure of any set of algebraic points is a variety defined over $\overline{\mathbb{Q}}$.

l, chosen once and for all, the same for all i), by taking P to be any algebraic point on $\mathcal{X}$, so in fact $f_i(P) = 0$. If $c = c(f_1, \ldots, f_m, l, d) > 0$ is the minimum of the numbers $c(f_i, l, d)$ appearing in the lemma, we conclude that either $\widehat{h}(P) \geq c$ or $f_i(P^{l^d}) = 0$ for $i = 1, \ldots, m$. In other words, either we have a lower bound of the required type or $[l^d]P \in \mathcal{X}$.

Since $\mathcal{X}$ is assumed not to be a torsion coset, we may apply Theorem 4.6 to conclude that the variety $[l^d]\mathcal{X}$ is not contained in $\mathcal{X}$, so $[l^d]^{-1}\mathcal{X}$ does not contain $\mathcal{X}$. Then the intersection $\mathcal{Y} := [l^d]^{-1}\mathcal{X} \cap \mathcal{X}$ has lower dimension than $\mathcal{X}$ and we may apply to it the inductive assumption. Let $\gamma(\mathcal{Y}) > 0$ be a corresponding constant and let T' be a corresponding finite union of torsion cosets. On defining $\gamma(\mathcal{X}) := \min(\gamma(\mathcal{Y}), c)$, this union T' of torsion cosets works for $\mathcal{X}$ as well: firstly, T' is contained in $\mathcal{X}$; secondly, if $P \in \mathcal{X}(\overline{\mathbb{Q}})$ does not belong to T', its height is bounded below either by $\gamma(\mathcal{Y})$ or by c, according as $[l^d]P \in \mathcal{X}$ (which implies $P \in \mathcal{Y}$) or not.

This concludes the proof.   □

Proof of Theorem 4.9. Let us apply Theorem 4.11 to $\mathcal{X}$, preserving the notation; certainly $\mathcal{X} \setminus T$ contains $\mathcal{X}^*$ by definition. Hence we obtain the required lower bound for the height on $\mathcal{X}^*(\overline{\mathbb{Q}})$ and we have just to show that $\mathcal{X} \setminus \mathcal{X}^*$ is Zariski-closed and actually a finite union of torsion cosets; we shall show that it equals this union T. In fact, let gH be a torsion coset contained in $\mathcal{X}$. As we have seen in 4.2.4, the torsion points are Zariski dense in gH (*e.g.*, use a suitable monoidal transformation to identify H with some $\mathbb{G}_m^r$). By Theorem 4.11 these torsion points must be contained in T, because they have zero height. Therefore $gH \subset T$, as asserted.   □

A further rephrasement of Zhang's theorem is the following, apparently more general but in fact equivalent; by saying that a set of algebraic points has *height tending to* 0 we mean (as usual) that for any $\epsilon > 0$ we have $\widehat{h}(P) \leq \epsilon$ for all but finitely many P in the set.[14]

Theorem 4.12. *Let* Σ *be any set of algebraic points in* $\mathbb{G}_m^n$, *with height tending to* 0. *The Zariski closure of* Σ *is a union of a finite set with a finite union of torsion cosets.*

Proof. Let $\mathcal{X}$ be the Zariski closure of Σ and let $c = c(\mathcal{X}) > 0$ be the constant appearing in Theorem 4.9. Since $\widehat{h}(P) \geq c$ for $P \in \mathcal{X}^*(\overline{\mathbb{Q}})$, we

---

[14] This is the usual notion of *tending to* 0 *outside the compact sets*, when our set is given the discrete topology.

have that $\Sigma_0 := \Sigma \cap \mathcal{X}^*$ is a finite set. By Theorem 4.9 we have a disjoint union $\mathcal{X} = \mathcal{X}^* \cup T$, for a certain finite union T of torsion cosets, and the complementary set $\Sigma_1 := \Sigma \setminus \Sigma_0$ is contained in T. Since Σ is Zariski dense in $\mathcal{X}$, we conclude that $\mathcal{X}$ is the union of Σ_0 and the closure of Σ_1. Since this closure is contained in T, which is contained in $\mathcal{X}$, it must be T, proving what we need. □

Exercise 4.25. Deduce Theorem 4.9 from Theorem 4.12. (Hint: enumerate the torsion cosets in a sequence $T_1, T_2, \ldots$ and for all m pick, if possible, a point $P_m \in \mathcal{X} \setminus (T_1 \cup \ldots \cup T_m)$ with $\hat{h}(P_m) \le 1/m$. See also the deduction of Zhang's theorem from Bilu's theorem in the next section.)

Since the torsion points have zero height, we now obtain that those lying on $\mathcal{X}$ are confined in a finite union of torsion cosets contained in $\mathcal{X}$. Of course this yields a broad sharpening of Lang's statement of Proposition 4.1. We explicitly state such result in the fashion of the last theorem:

Corollary 4.13. *Let Σ be any set of torsion points in $\mathbb{G}_m^n$. The Zariski closure of Σ is a finite union of torsion cosets. In particular, the torsion points on an irreducible variety $\mathcal{X}$ are Zariski-dense if and only if $\mathcal{X}$ is a torsion coset.*

Proof. The first part is immediate by the previous theorem, since any torsion point is a torsion coset (of dimension 0), and this yields the 'only if' in the last part. For the 'if', it suffices to recall that the torsion points are Zariski-dense in any torsion coset, as has been observed above. □

A further corollary (essentially equivalent) concerns the special case of torsion points on a hyperplane $\mathcal{X} : \{a_1 X_1 + \ldots + a_n X_n = 1\}$. We want to study the torsion points on it, which in turn is a simple case of the *generalized S-unit equation* mentioned in the previous chapter. We say that a point $P = (x_1, \ldots, x_n) \in \mathcal{X}$ is *degenerate* if there exists a proper subset I of $\{1, \ldots, n\}$ such that $\sum_{i \in I} a_i x_i = 0$.

Corollary 4.14. *Let $\mathcal{X} : \{a_1 X_1 + \ldots + a_n X_n = 1\}$, where $a_1, \ldots, a_n \in \overline{\mathbb{Q}}^*$. Then there are at most finitely many non-degenerate torsion points on X.*

Proof. By Corollary 4.13 it will be sufficient to show that any torsion coset of strictly positive dimension in $\mathcal{X}$ contains no non-degenerate points.

For every torsion coset $gH \subset \mathcal{X}$, with $g = (g_1, \ldots, g_n)$, the torus H is a subvariety of $g^{-1}\mathcal{X}$, which is defined by the equation $g_1 a_1 X_1 + \ldots + g_n a_n X_n = 1$. As in a previous proof, consider the characters $x_0 = 1, x_1, \ldots, x_n$, defined as the restrictions to H of the coordinate functions, and let $B_1, \ldots, B_r$ be a partition of $\{0, 1, \ldots, n\}$ such that $x_i = x_j$ if and only if i and j belong to a same subset B_s. Note that, if $\dim H > 0$, then $r > 1$, so there exists an s such that B_s is nonempty and contained in $\{1, \ldots, n\}$. Now, by the theorem of Artin, we have $\sum_{i \in B_s} g_i a_i x_i = 0$ (on H); so, if a point of $\mathcal{X}$ lies in gH, this equation proves that it is degenerate.   □

Remark 4.7.
 (i) Zhang [107] actually proved more precise results. For an irreducible variety $\mathcal{X} \subset \mathbb{G}_m^n$ over $\overline{\mathbb{Q}}$, he defined the *essential minimum* $\mu(\mathcal{X})$ as the infimum of the real numbers c such that the set of P with $\widehat{h}(P) \leq c$ is Zariski-dense in $\mathcal{X}$. (The next exercise says that $\mu(\mathcal{X}) < \infty$.) The results proved above may be rephrased by saying that $\mu(\mathcal{X}) = 0$ *if and only if $\mathcal{X}$ is a torsion variety* (*i.e.* a torsion coset). Zhang used a certain notion of height $\widehat{h}(\mathcal{X})$ of the variety $\mathcal{X}$ and proved an inequality implying that $\deg(\mathcal{X})\mu(\mathcal{X}) \leq \widehat{h}(\mathcal{X}) \leq (\dim \mathcal{X} + 1) \deg(\mathcal{X})\mu(\mathcal{X})$; also, $\widehat{h}(\mathcal{X})$ turns out to be zero just for torsion varieties. For more on this we refer to the Appendix below by Amoroso, where several explicit quantitative results are mentioned.
 (ii) **Effectivity.** Inspection of the above arguments shows that the constants in Lemma 4.10 and in Theorem 4.9 may be effectively computed in terms of a set of defining equations for the variety $\mathcal{X}$. They would depend on the heights of the coefficients of such equations and on the degree of a field of definition. Moreover, from these data one may also compute the finitely many torsion cosets in Theorem 4.11, which in turn leads to an effective form of Corollary 4.13. Naturally, in view of Section 4.2, the maximal torsion cosets gH contained in $\mathcal{X}$ correspond to only finitely many tori H, which can be found also with the method outlined therein. Any given torus H can be identified with $\mathbb{G}_m^r$ by means of a monoidal automorphism, and then the relevant g correspond, by projection on the last $n - r$ coordinates, to torsion points on the projected variety, which will be 'isolated', apart for a proper subvariety; this procedure reduces the computation to the case of *isolated torsion points*; these may be also described as maximal torsion cosets of dimension 0 and may be also found with independent methods. (See Exercise 4.29 below.)
 (iii) **Discreteness of height.** By applying Zhang's theorem to a translated variety $Q^{-1}\mathcal{X}$, one may prove a lower bound for the height $\widehat{h}(PQ^{-1})$, where $P, Q \in \mathcal{X}(\overline{\mathbb{Q}})$ are such that $P \in (Q^{-1}\mathcal{X})^*$. As recalled earlier, $\widehat{h}(PQ^{-1})$ may be seen as a kind of '(semi)distance' between P, Q considered up to torsion [15] so the alluded lower bound asserts that the algebraic points are somewhat 'discrete' with respect to the height, not merely 'away from torsion points'. However some remarks are in order. First, the lower bound depends also on Q, so we do not obtain uniform discreteness with this approach. Secondly, also the exceptional set depends on Q: it is the union of translates QgH contained in $\mathcal{X}$, for H a torus and g a torsion point.

---

[15] Recall also the last section of the previous chapter.

Alternatively, one may apply Zhang's theorem to $\mathcal{X} \cdot \mathcal{X}^{-1} \subset \mathbb{G}_m^n$ defined as the set of products PQ^{-1} for $P, Q \in \mathcal{X}$. This is not necessarily a closed subvariety, but (being the image of the regular map $(P, Q) \mapsto PQ^{-1}$) it contains an open dense set in its Zariski closure $\mathcal{Y}$. Zhang's theorem yields a bound $\widehat{h}(PQ^{-1}) \geq c(\mathcal{Y}) > 0$ independent of P, Q, provided however $PQ^{-1} \in \mathcal{Y}^*$. To study this set $\mathcal{Y}^*$ we must study the torsion translates gH contained in the closure of $\mathcal{X} \cdot \mathcal{X}^{-1}$; this new set of translates can be very big: for instance if $\mathcal{X}$ is a curve in $\mathbb{G}_m^2$, the variety $\mathcal{Y}$ will be 'generally' the whole $\mathbb{G}_m^2$; in such case this method will not give anything interesting; this is no surprise because in fact in this case we could prescribe arbitrarily small height for $\widehat{h}(PQ^{-1})$ with no restrictions of algebraic nature on P, Q. But when $\mathcal{Y}$ is in some sense small, we may gain uniform discreteness. See Exercise 4.33 below for a simple example.

(iv) **Uniformity.** We shall soon state a uniform version of Zhang's theorem, namely with a lower bound for $\widehat{h}(P)$ depending only on the degree of $\mathcal{X}$ and the dimension of the ambient space (with suitable new restrictions on P and up to finitely many exceptional points). In particular, this will yield a uniform lower bound for the height on $Q^{-1}\mathcal{X}$ and thus provides another approach to a lower bound for $\widehat{h}(PQ^{-1})$, i.e. 'discreteness' of the height on $\mathcal{X}(\overline{\mathbb{Q}})$.

Exercise 4.26. Let $\mathcal{X}$ be any subvariety of $\mathbb{G}_m^n$, defined over $\overline{\mathbb{Q}}$. Prove that there exists $c = c(\mathcal{X})$ such that the points $P \in \mathcal{X}(\overline{\mathbb{Q}})$ with $\widehat{h}(P) \leq c$ are Zariski-dense in $\mathcal{X}$. (Hint: for instance use a birational correspondence of $\mathcal{X}$ with a hypersurface and use Exercise 3.6.)

Exercise 4.27. Let $a_1\zeta_1 + \ldots + a_n\zeta_n = 0$, where the a_i are given rational numbers, the ζ_i are roots of unity and no proper subsum vanishes. Show that the common order of the ζ_i/ζ_j divides the product of all primes up to n. (Hint: consider the irreducibility of the p^l-th cyclotomic polynomial over cyclotomic fields.)

Exercise 4.28. Let $a_1\zeta_1 + \ldots + a_n\zeta_n = 0$, where the a_i are given algebraic numbers and the ζ_i are roots of unity. Prove that the minimal order of ζ_i/ζ_j for $i \neq j$ is bounded only in terms of n and the a_i. (Hint: one can reduce to the case when no subsum of the $a_i\zeta_i$ vanishes. Then, as in the previous exercise, if p divides the order of some ratio and p is large one obtains a contradiction with the irreducibility of the p-th cyclotomic polynomial.)

Exercise 4.29. Using the result of the previous exercise, give another proof of Corollary 4.13. (Hint: view the equations holding on $\mathcal{X}$, calculated at torsion points, as linear relations among roots of unity, with given coefficients.)

Exercise 4.30. Work out effectively the above proof of Zhang's theorem for the curve $x + y = 1$ treated by Zagier. Find an explicit lower bound for the height of algebraic nontorsion points on it. (The choice $l^d = 5$ suffices in Lemma 4.10.)

Exercise 4.31. Let $\mathcal{X}$ be an irreducible curve in $\mathbb{G}_m^2$. Prove that either $\mathcal{X} \cdot \mathcal{X}^{-1}$ is Zariski dense in $\mathbb{G}_m^2$ or $\mathcal{X}$ is a translate of an algebraic subgroup. (Hint: prove that if the first alternative is not verified, all the curves $Q^{-1}\mathcal{X}$, $Q \in \mathcal{X}$, are equal.)

Exercise 4.32. Let $\mathcal{X}$ be a line $Y = aX+b, ab \neq 0$, in $\mathbb{G}_m^2$. Prove that $\mathcal{X} \cdot \mathcal{X}^{-1}$ is Zariski-dense in $\mathbb{G}_m^2$ but does not contain the lines (algebraic subgroups) $Y = 1$, $X = 1, X = Y$.

Exercise 4.33. Let $\mathcal{X} \subset \mathbb{G}_m^3$ be the line $X_2 = X_1 + 1, X_3 = X_1 - 1$.

 (i) Prove that $\mathcal{X} \cdot \mathcal{X}^{-1}$ is Zariski-dense in the hypersurface $\mathcal{Y}$ defined by $2X_2X_3 - X_1X_2 - X_1X_3 - X_2 - X_3 + 2X_1 = 0$. More precisely, find the complement $\mathcal{Y} \setminus \mathcal{X} \cdot \mathcal{X}^{-1}$. (It consists of the union of the subgroups $X_1 = X_2 = 1, X_1 = X_3 = 1, X_2 = X_3 = 1$ deprived of the origin.)

 (ii) Find the torsion cosets contained in $\mathcal{Y}$ and prove a lower bound for $\widehat{h}(P)$ for P in $\mathcal{Y}^*(\overline{\mathbb{Q}})$. (To speed up calculations Exercises 4.28 and 4.29 may be helpful.)

(iii) Deduce a lower bound for $\widehat{h}(PQ^{-1})$, for $P, Q \in \mathcal{X}(\overline{\mathbb{Q}}), P \neq Q$.

We conclude this subsection by mentioning a uniform version of Zhang's theorem, which first appeared in the paper [21]. Note that the lower bound for $\widehat{h}$ on $\mathcal{X}^*$ provided by the above approach depends rather heavily on various data associated with the variety $\mathcal{X}$: the number of variables and the degrees in a system of defining equations, but also the degree (over $\mathbb{Q}$) and heights of the involved coefficients. Actually, it can be easily seen that these dependencies cannot be eliminated. (Consider for instance non-torsion translates of algebraic subgroups.) On the other hand, it turns out that if we work in a subset of $\mathcal{X}$ a priori smaller than $\mathcal{X}^*$, one can prove a lower bound for the height which depends only on the degree of $\mathcal{X}$ and on the dimension n of the ambient space. To state this precisely, we need another definition.

Definition 4.15. For a subvariety $\mathcal{X}$ of $\mathbb{G}_m^n$, we define $\mathcal{X}^\circ$ as the complement in $\mathcal{X}$ of the union of all the torus cosets gH of dimension > 0 which are contained in $\mathcal{X}$ (not merely the torsion cosets). By degree of $\mathcal{X}$ we mean the degree with respect to the natural embedding $\mathbb{G}_m^n \subset \mathbb{P}_n$.

Theorem 4.16. * Let $\mathcal{X}$ be an algebraic subvariety of $\mathbb{G}_m^n$ of degree at most d. Then:

 (i) $\mathcal{X} \setminus \mathcal{X}^\circ$ is Zariski-open in $\mathcal{X}$ and the number and degrees of the irreducible components of $\mathcal{X} \setminus \mathcal{X}^\circ$ are bounded (effectively) only in terms of n, d.

 (ii) There exist (effective) numbers $N = N(n, d), c = c(n, d) > 0$ such that $\widehat{h}(P) \geqslant c$ for all $P \in \mathcal{X}^\circ(\overline{\mathbb{Q}})$ with the exception of at most N points.

Note that $\mathcal{X}^*$ and $\mathcal{X}^\circ$ are related sets. To obtain the first one we throw away from $\mathcal{X}$ only the torsion cosets (of any dimension) whereas for the second one we throw away all the torus cosets of *strictly positive* dimension. So $\mathcal{X}^* \cup F \supset \mathcal{X}^\circ$, where F is a finite set of torsion points. For a 'general' variety we certainly expect $\mathcal{X}^* = \mathcal{X}^\circ = \mathcal{X}$.

Note that the finite set in (ii) can be effectively found (for a variety $\mathcal{X}$ given 'effectively'). In fact, if a point $P \in \mathcal{X}(\overline{\mathbb{Q}})$ satisfies $\widehat{h}(P) < c$, then the same holds for the conjugates of P over a number field k of definition for $\mathcal{X}$. Therefore the number of conjugates cannot exceed N by the theorem, so $[k(P) : k] \leq N$. On following the proof of Northcott Theorem, we conclude that P may be found.

As noted in Remark 4.7, (iv), this statement yields a lower bound for $\widehat{h}(PQ^{-1})$, $P, Q \in \mathcal{X}^\circ(\overline{\mathbb{Q}})$, uniform in P, Q (although with a finite set E_Q of exceptional P for every given Q).

We finally remark that this theorem, although concerning only $\mathcal{X}^\circ$, allows an inductive procedure for dealing with the non-torsion cosets in $\mathcal{X}$, of dimension > 0, in practice with $\mathcal{X}^* \setminus \mathcal{X}^\circ$. In fact, we have already noticed in studying algebraic subgroups that the maximal torus cosets in $\mathcal{X}$ are of the shape gH for a torus H which has only (effectively) finitely many possibilities. For a given torus H of codimension $r < n$, we can use a monoidal automorphism to suppose that H is defined by $X_1 = \ldots = X_r = 1$, so gH is defined by $X_i = g_i$ for $i = 1, \ldots, r$. Let $f(X_1, \ldots, X_n)$ be a polynomial vanishing on $\mathcal{X}$; if $gH \subset \mathcal{X}$ we have that $f(g_1, \ldots, g_r, X_{r+1}, \ldots, X_n)$ vanishes identically. This gives equations for $(g_1, \ldots, g_r)$. So the relevant $(g_1, \ldots, g_r)$ form a variety $\mathcal{X}_H$ in a space of dimension $r < n$. Note that for a point $P \in gH$ we have $\widehat{h}(P) \geq \widehat{h}(g_1, \ldots, g_r)$. Hence we may apply again the result, this time to $\mathcal{X}_H$, to obtain a lower bound and so on.

We do not prove this uniform Theorem 4.16 here, but refer to [21] or [17]. We only say that the proof is based on the following principle (applied also by Schlickewei in the context of torsion points). Consider points $P_1, \ldots, P_m \in \mathcal{X}$. For any polynomial f vanishing on $\mathcal{X}$, we have $f(P_1) = \ldots = f(P_m) = 0$. If f has l monomial terms, we may view these equations as an $l \times m$ system of linear equations where the unknowns are the coefficients of f and the entries are the monomials appearing in f, evaluated at the points P_i. The equations show that the matrix of this system has rank $< l$, which for $m \geq l$ gives certain determinantal equations on the P_i. The point is that these equations do not anymore depend on the coefficients of f; they define a variety $\mathcal{Y}$ in $\mathbb{G}_m^{nm}$ to which one can apply the theorem of Zhang, obtaining a lower bound which now depends only on n and the monomials involved in f. Naturally to carry out this program it is necessary to study $\mathcal{Y}^*$, which requires some detail and some induction step; but this may be actually done and leads to the above result.

We shall see below an explicit example of this technique, in a 'concrete' case which shall be used in the next chapter. See also the Appendix by Amoroso for references on more precise quantitative estimations of the mentioned numbers $c(n, d)$, $N(n, d)$, gotten by other methods.

By applying this uniform version one obtains in particular an estimate for the number of non degenerate solutions in Corollary 4.14 which depends only on n, not on the coefficients a_i.

4.3.2. Bilu's approach through equidistribution

A completely different proof of Zhang's theorem 4.9 was given by Bilu in [10]. This relied on a result of his, on the distribution of the Galois conjugates of algebraic numbers with small height, of considerable interest in itself. Roughly speaking, he showed that such conjugates 'tend' to be uniformly distributed 'around' the unit circle. A precise statement is as follows.

Theorem 4.17 (Bilu). *Let ψ be a continuous function on $\mathbb{C}$, with compact support, and let $(\xi_m)_{m \in \mathbb{N}}$ be a sequence in $\overline{\mathbb{Q}}$ with $h(\xi_m) \to 0$ and $[\mathbb{Q}(\xi_m) : \mathbb{Q}] \to \infty$. Then, letting ξ_m^σ be the distinct conjugates of ξ_m, the 'mean' $[\mathbb{Q}(\xi_m) : \mathbb{Q}]^{-1} \sum_\sigma \psi(\xi_m^\sigma)$, tends to $\int_0^1 \psi(e^{2\pi i t}) dt$ as $m \to \infty$.*

If we take for instance $\psi = 1$ on an arc $\{e^{2\pi i \theta} \mid |\theta - \theta_0| < \epsilon\}$, $0 \le \epsilon \le 1/2$, and $\psi = 0$ outside a small neighborhood I of the arc, we deduce that asymptotically the percentage of conjugates of ξ_m contained in I is 2ϵ. Still in other words, we may say that as $m \to \infty$ most conjugates of ξ_m tend to have absolute value near 1 and uniformly distributed arguments. An interesting instance occurs when ξ is a root of unity. If m is its exact order, its conjugates are the ξ^b, for b an integer coprime to m. The total number of conjugates is $\phi(m)$ and the number of those in the arc $\{e^{2\pi i \theta} \mid 0 \le \theta \le \lambda\}$, $0 < \lambda < 1$, is $\#\{b \mid 0 \le b \le \lambda m, \ (b, m) = 1\}$. By Bilu's theorem, this quantity is asymptotically $\lambda \phi(m)$, which is just what one would expect. Of course, it is not difficult to recover this last statement directly (even with error term estimates - see exercise below) but it is significant that this follows from considerations of heights.

We shall not give a detailed proof of Bilu's statement, nor of the deduction of Zhang Theorem from it, but only sketch the main points of both arguments.

Sketch of deduction of Zhang Theorem from Theorem 4.17

The idea is that, for points P of small height, the value at P of any given monomial also has small height; suppose now that there is a fixed linear relation, say with rational coefficients, among a finite set of monomials, valid at P. Then this relation continues to hold at the conjugates of P. But then the values of the monomials at the conjugates P cannot be independently equidistributed. Now an application of Bilu's theorem will say that some ratio of the monomials, evaluated at P, has bounded degree over R. On the other hand, by Northcott Theorem we infer that this ratio will be constant on an infinite sequence of points P, which will give equations for a torsion coset containing such points.

We now add some detail to this program. We argue by induction on n. We assume that the relevant irreducible variety $\mathcal{X}$ is not $\mathbb{G}_m^n$ and we take a nontrivial equation $f(P) = 0$ valid for $P \in \mathcal{X} \subset \mathbb{G}_m^n$, where $f \in \overline{\mathbb{Q}}[\mathbb{G}_m^n]$ is a Laurent polynomial. By replacing f with the product of its conjugates over $\mathbb{Q}$, we may assume it has coefficients in $\mathbb{Q}$ and on dividing by a monomial we may also assume that its constant coefficient is not zero. Then we may write the equation in the form $1 + \sum_{\lambda \in M} a_\lambda P^\lambda = 0$, where the sum runs over a finite set M of distinct nonzero vectors $\lambda \in \mathbb{Z}^n$ and the a_λ are in $\mathbb{Q}$.

Claim. *For every infinite sequence S of points $P_m \in \mathcal{X}$ with $\widehat{h}(P_m) \to 0$, there is a $\lambda \in M$ and an infinite subsequence S' of S such that P_m^λ is constant for P_m in S'.*

We prove this claim by contradiction, denoting by $d_m = [\mathbb{Q}(P_m) : \mathbb{Q}]$ the degree of P_m; note that by Northcott Theorem we may assume $d_m \to \infty$. Let us fix $\lambda \in M$ for the moment and set $\xi_m = \xi_m(\lambda) := P_m^\lambda$. We have $h(\xi_m) \to 0$ and $[\mathbb{Q}(\xi_m) : \mathbb{Q}] \to \infty$ because otherwise ξ_m would be constant on some infinite subsequence (Northcott Theorem again). Hence we may apply Theorem 4.17 to the sequence (ξ_m). For this, we choose $R > r > 1$ and we consider a smooth function ψ that coincides with the identity inside a circle of radius r and vanishes outside a circle of radius R. Note that for $m \to \infty$ the set $C_\lambda(m)$ of conjugates P_m^σ of P_m such that $|\xi_m^\sigma| > r$ contains only $o(d_m)$ elements, for otherwise $h(\xi_m)$ would not tend to 0. Hence $C(m) := \cup_{\lambda \in M} C_\lambda(m)$ also contains only $o(d_m)$ elements for $m \to \infty$.

Finally, using the above equation for $P = P_m$, conjugating P_m in all possible ways outside $C(m)$ and summing over the conjugates outside $C(m)$, which we indicate with a star, we obtain

$$0 = \sum_\sigma{}^* \left(1 + \sum_{\lambda \in M} a_\lambda (P_m^\lambda)^\sigma \right) = (d_m - \#C(m)) + \sum_{\lambda \in M} a_\lambda \left(\sum_\sigma{}^* (P_m^\lambda)^\sigma \right).$$

It is now an easy matter to see that $\sum_{\sigma}^{*}(P_m^{\lambda})^{\sigma}$ is asymptotic to $\sum_{\sigma}^{*}\psi((P_m^{\lambda})^{\sigma}) = \sum_{\sigma}^{*}\psi(\xi_m^{\sigma})$, which however is $o(d_m)$ by Theorem 4.17.[16]

This gives a contradiction which proves the Claim.

Let us now enumerate all the varieties of the shape $\underline{X}^{\lambda} = \alpha$ ($\lambda \in M, \alpha$ a root of 1) in a sequence $T_1, T_2, \ldots$. We contend that *there is a finite union* $T = T_1 \cup \ldots \cup T_r$ *such that every sequence* (Q_j) *in* $\mathcal{X}(\overline{\mathbb{Q}})$ *with* $\hat{h}(Q_j) \to 0$ *is contained eventually in* T. For otherwise we could construct an infinite sequence $S := (P_m)$ by taking P_m to be a point not in $T_1 \cup \ldots \cup T_m$ and with $\hat{h}(P_m) < 1/m$. This sequence would plainly violate the claim.

In turn, this implies that $\mathcal{X} \setminus \mathcal{X}^{*}$ is contained in T: for otherwise, since the torsion points are Zariski dense in any torsion coset contained in $\mathcal{X} \setminus \mathcal{X}^{*}$, we could construct an infinite sequence S of torsion points in $\mathcal{X}$ and not contained eventually in T.

Note that there is a $c > 0$ such that every $P \in \mathcal{X}(\overline{\mathbb{Q}})$ with $\hat{h}(P) < c$ lies in T; the argument is as above: if this is not the case there exists $P_m \notin T$ and $\hat{h}(P_m) < 1/m$, violating the above conclusions.

To conclude it now suffices to apply induction to each of the varieties $\mathcal{X} \cap T_i, i = 1, \ldots, r$. The variety T_i can split into a finite union of torsion cosets when the corresponding λ is not primitive, but by replacing T_i with each individual element in this finite union (and increasing r) we may assume that each T_i is a torsion coset. Let us fix i; by applying a monoidal automorphism, this variety $\mathcal{X} \cap T_i$ can viewed as the intersection of $\mathcal{X}$ with $X_1 = \alpha$, where α is a root of unity. Projection to the last $n - 1$ coordinates then gives an isomorphism with a variety $\mathcal{Y}_i \subset \mathbb{G}_m^{n-1}$ to which we can apply induction. We leave the remaining easy details to the interested reader.

Sketch of proof of Theorem 4.17

It suffices to prove the result for functions of type $\psi(re^{i\theta}) = r^h e^{il\theta}$ (where $h, l \in \mathbb{Z}$) for $r < R$ and 0 for $r \geq R + 1$. In fact, by using these functions wth fixed suitably large R we may approximate uniformly for $r \leq R$ any continuous function with support contained in the circle of center 0 and radius R.

Let us put $d_m := [\mathbb{Q}(\xi_m) : \mathbb{Q}] \to \infty$. Using Proposition 3.6, it is very easy to see that if $h(\xi_m) \to 0$ only $o(d_m)$ conjugates of ξ can lie outside an annulus $1 - \epsilon < r < 1 + \epsilon$, for any fixed $\epsilon > 0$; in practice, the

---

[16] We are using here the conjugates of P_m rather than of ξ_m, hence some conjugate of ξ_m could be repeated; however the mean value of Bilu's theorem does not change.

absolute value is near 1 for most conjugates of ξ_m, and hence we easily reduce to prove the conclusion only for the functions $e^{il\theta}$, $l \neq 0$.

Dropping the suffix m, let us consider the discriminant $D :=$ $a^{2d-2} \prod_{\sigma \neq \tau} (\xi^\sigma - \xi^\tau) \in \mathbb{Z}$, where d is the degree of ξ, a is the leading coefficient of the minimal polynomial of ξ over $\mathbb{Z}$ and ξ^σ, ξ^τ are the distinct conjugates. We have $|D| \geq 1$, whence $0 \leq (2d-2)\log|a| + \sum_{\sigma \neq \tau} \log|\xi^\sigma - \xi^\tau|$. However $\log|a| \leq dh(\xi)$ by Proposition 3.6, so applying this with $\xi = \xi_m$ we get for $m \to \infty$,

$$\sum_{\sigma \neq \tau} \log|\xi^\sigma - \xi^\tau| \geq o(d^2).$$

We now put $\xi^\sigma = r_\sigma e^{i\theta_\sigma}$ with real θ_σ and $r_\sigma \geq 0$. As above, since $h(\xi) \to 0$, we may assume that, for every fixed $\epsilon > 0$, for all but $o(d)$ conjugates, $|r_\sigma - 1| < \epsilon$. If ξ_σ, ξ_τ are among these conjugates we may use the inequality, valid for small ϵ,

$$\log|\xi_\sigma - \xi_\tau| \leq \log|1 - \rho e^{i(\theta_\sigma - \theta_\tau)}| + O(\epsilon),$$

where $\rho := 1 - 3\epsilon$. Summing over all the relevant pairs σ, τ, we find complex conjugates pairs, so we may forget about the absolute value in the right term, and by expanding $\log(1-z)$ we find

$$\sum_{\sigma \neq \tau} \log|\xi^\sigma - \xi^\tau| \leq O(\epsilon d^2) - \sum_{\sigma \neq \tau} \sum_{l=1}^{\infty} \frac{\rho^l e^{il(\theta_\sigma - \theta_\tau)}}{l}$$

$$\leq O(\epsilon d^2) - \sum_{l=1}^{\infty} \frac{\rho^l}{l} \left(\left| \sum_\sigma e^{il\theta_\sigma} \right|^2 - d \right).$$

The sum over the remaining pairs is also estimated $o(d^2)$. Combining with the above we find $d^{-1}| \sum_\sigma e^{il\theta_\sigma}| = O(\sqrt{\epsilon}) + O(\sqrt{|\log \epsilon|/d}) + o(1)$, whence the sought result on letting $d \to \infty$ and then $\epsilon \to 0$.

The details may be found in Bilu's quoted paper.

Exercise 4.34. Let $m > 0$ be an integer, let $0 < \lambda < 1$ and put $\phi(\lambda, m) :=$ $\#\{b \mid 0 < b \leq \lambda m, \ (b, m) = 1\}$. Prove that for fixed λ, $\phi(\lambda, m) \sim \lambda \phi(m)$ as $m \to \infty$. (Hint: the sum $\sum_{d|b,m} \mu(d)$, where μ is the Möbius function, is 1 or 0 according as b is coprime to m or not. Hence $\phi(\lambda, m) = \sum_{b \leq \lambda m} \sum_{d|b,m} \mu(d) = \sum_{d|m} \mu(d) \sum_{d|b \leq \lambda m} 1$. The inner sum is $= \lambda m/d + O(1)$ and the result follows. This also leads to error term estimates of the shape $O(\#\{d : d|m\})$.)

4.4. An application to the S-unit equation

In the last chapter we have considered the S-unit equation $x + y = 1$, to be solved with x, y in the group of S-units $\mathcal{O}_S^*$ of some number field, or, more or less equivalently, with (x, y) in a finitely generated group $\Gamma \subset \mathbb{G}_m^2(\overline{\mathbb{Q}})$. We have also seen that this equation, apparently very special, is in fact rather interesting: not only it embodies the general genus 0-case of Siegel's theorem on integral points (*i.e.* integral points for $\mathbb{P}_1 \setminus \{0, 1, \infty\}$), but it is also directly linked with several classical diophantine equations of higher genus, like the Thue-Mahler's or the hyperelliptic ones.

We have noticed that Roth Generalized Theorem (or the Thue-Mahler finiteness theorem) implies the finiteness of the set of solutions of the S-unit equation. We have not proved completely this finiteness statement here, *i.e.* Theorem 3.13, apart from a rather special case in the supplements to Chapter 2. However, as announced above, a complete proof will be given in the next chapter; this proof will actually give more, *i.e.* a remarkable upper bound for the number of solutions, depending only on the rank of Γ (but neither on a number field of definition nor on the height of a system of generators). Such quantitative proof consists of two steps:

Step (a). Estimating the number of solutions of 'large' height: this is the main and more difficult step, even leaving aside the quantitative viewpoint; we do not say more on it now.

Step (b). Estimating the number of solutions of 'small' height: for this, if we are merely interested in a finiteness result, the easy Northcott Theorem suffices. However if we seek good and uniform bounds, this step too becomes important, and cannot be treated in a satisfactory way with the method used for the 'large' solutions. Instead, it turns out that Zhang's theorem (especially the uniform version) is a quite powerful weapon for this task.

In this section we shall perform in detail this program of applying Zhang's theorem to Step (b).

In the sequel we let Γ denote a finitely generated subgroup of $\mathbb{G}_m^2(\overline{\mathbb{Q}}) = (\overline{\mathbb{Q}}^*)^2$, of rank r. We shall also consider solutions in a larger group, *i.e.* the *division group* of Γ defined by $\Gamma' := \{P \mid \exists n > 0, P^n \in \Gamma\}$. Note that Γ' is neither finitely generated nor contained in any number field (for instance, it contains all the torsion points in $\mathbb{G}_m^2$) but has the same rank r as Γ. We are going to prove the following theorem.

Theorem 4.18. *There exist absolute constants* $N, \gamma > 0$ *such that the number of solutions* $(x, y) \in \Gamma'$ *to* $x + y = 1$ *with* $h(x, y) \leq t$ *is at most* $N(1 + \frac{2t}{\gamma})^r$.

The significance is that (for given t) this bound depends on r but not otherwise on Γ. One can give explicit values for γ, N but we shall not insist on this and only say a little more on it in the sequel. We start by proving a fundamental lemma, derived from Zhang's theorem 4.9. The proof will provide a good illustration of the method used in [21] for the uniform version of Zhang's theorem, *i.e.* Theorem 4.16 above, which we have stated but not proved.

Lemma 4.19. *There exist absolute constants* $N_1, \gamma_1 > 0$ *such that for every* $a, b \in \overline{\mathbb{Q}}^*$ *the number of solutions* $(x, y) \in \Gamma'$ *to* $ax + by = 1$ *with* $\widehat{h}(x, y) < \gamma_1$ *is at most* N_1.

Proof. An application of Theorem 4.16 to the curve $\mathcal{X} = \mathcal{X}_{a,b}$ defined by $aX + bY = 1$ would immediately yield the conclusion. However we have not proved that theorem here, so we give a direct argument, following the methods of [21].

Consider the matrix $\mathbf{M} = \mathbf{M}(Z_1, W_1, Z_2, W_2) := \begin{pmatrix} 1 & 1 & 1 \\ Z_1 & W_1 & 1 \\ Z_2 & W_2 & 1 \end{pmatrix}$ and the variety $\mathcal{X} \subset \mathbb{G}_m^4$ defined by $\det(\mathbf{M}) = Z_1 W_2 - Z_2 W_1 - Z_1 - W_2 + Z_2 + W_1 = 0$. This equation may be rewritten as $(Z_1 - 1)(W_2 - 1) = (Z_2 - 1)(W_1 - 1)$.

For every three solutions $P_i = (x_i, y_i) \in \Gamma'$ to $ax_i + by_i = 1$, $i = 1, 2, 3$, set $z_i = x_i/x_3$ and $w_i = y_i/y_3$, $i = 1, 2$. Specializing $Z_i = z_i$, $W_j = w_j$, the vector $\mathbf{v} = (ax_3, by_3, -1)^t$ is a non-trivial solution to $\mathbf{M}(z_1, w_1, z_2, w_2)\mathbf{v} = 0$, so the point $(z_1, w_1, z_2, w_2) = (P_1 P_3^{-1}, P_2 P_3^{-1})$ belongs to $\mathcal{X}$.[17]

The Theorem 4.9 of Zhang now provides an absolute constant $\gamma_2 > 0$ such that every point $P \in \mathcal{X}^*(\overline{R})$ has height $\widehat{h}(P) \geq \gamma_2$. This implies that, if $(P_1 P_3^{-1}, P_2 P_3^{-1})$ belongs to $\mathcal{X}^*$, then $\widehat{h}(P_1) + \widehat{h}(P_2) + 2\widehat{h}(P_3) > \gamma_2$, so at least one of our three points has height $\widehat{h}(P_i) \geqslant \gamma_1 := \gamma_2/4$.

To deal with the remaining cases, we shall now describe $\mathcal{X} \setminus \mathcal{X}^*$: we shall show that every torsion coset in $\mathcal{X}$ is contained in one of the six following tori of dimension 2:

$$\{Z_1 = Z_2, W_1 = W_2\}, \quad \{Z_1 = Z_2 = 1\}, \quad \{W_1 = W_2 = 1\},$$
$$\{Z_1 = W_1, Z_2 = W_2\}, \quad \{Z_1 = W_1 = 1\}, \quad \{Z_2 = W_2 = 1\}.$$

This list might be obtained by performing explicitly the above proof of Zhang's theorem, or also using the method outlined in Remark 4.7 (ii), through Exercises 4.27, 4.28 and 4.29. As this may be lengthy, we give another *ad hoc* argument, related to Zagier's method. [18]

---

[17] It would have been more natural to consider the matrix in six variables, with rows $(P_i, 1)$ for $i = 1, 2, 3$, but the present substitution eliminates two variables and simplifies the formulas.

[18] We note however that this explicit list would not be essential for the method to work, leaving aside the explicit values of γ_1 and N_1. In fact, following the proofs of Theorem 4.16 in [21] or [17] one realizes that it is possible to work by induction on the dimension, without calculating at each step the structure of the relevant $\mathcal{X}^*$.

Let $P = (z_1, w_1, z_2, w_2)$ be a torsion point on $\mathcal{X}$, so $(z_1 - 1)(w_2 - 1) = (z_2 - 1)(w_1 - 1)$. Conjugating this equation and recalling $\bar{\zeta} = \zeta^{-1}$ for a root of unity ζ, we find $(1 - z_1)(1 - w_2)z_2 w_1 = (1 - z_2)(1 - w_1)z_1 w_2$.

Suppose first that no coordinate of P equals 1. Then comparison of the two equations so obtained gives $z_1 w_2 = z_2 w_1$. Inserting this into $(z_1 - 1)(w_2 - 1) = (z_2 - 1)(w_1 - 1)$ we find $z_1 + w_2 = z_2 + w_1$. Hence the pair $\{z_1, w_2\}$ coincides with $\{z_2, w_1\}$. In this case P lies in the union of the two tori above listed in the first column.

If on the other hand some coordinate equals 1, then again the equation $(z_1 - 1)(w_2 - 1) = (z_2 - 1)(w_1 - 1)$ shows that P lies in the union of the four remaining tori.

In conclusion, certainly any torsion point on $\mathcal{X}$ lies in the union of the above six tori. On the other hand the above tori are plainly contained in $\mathcal{X}$; since the torsion points are Zariski-dense in any torsion coset, we deduce that every torsion coset contained in $\mathcal{X}$ must be contained in the union of the above tori, which completes the verification that $\mathcal{X} \smallsetminus \mathcal{X}^*$ equals the union of the above six tori.

Now, recall the constraints $ax_3 + by_3 = 1$ and $ax_3 z_i + by_3 w_i = 1$, and $ax_3, by_3 \neq 0$. They entail that for each $i = 1, 2$ the three equalities $z_i = 1$, $w_i = 1$, and $z_i = w_i$ are pairwise equivalent. Thus our three points P_1, P_2, P_3 satisfy $(P_1 P_3^{-1}, P_2 P_3^{-1}) \in \mathcal{X} \smallsetminus \mathcal{X}^*$ if and only if two of them coincide. In other words, for any three pairwise distinct points P_1, P_2, P_3 as above, we have $(P_1 P_3^{-1}, P_2 P_3^{-1}) \in \mathcal{X}^*$, so, as we have seen, $\max \widehat{h}(P_i) \geq \gamma_1$. In turn, this implies that there are at most $N_1 = 2$ solutions $(x, y) \in \Gamma'$ to $ax + by = 1$ with $\widehat{h}(x, y) < \gamma_1$. $\square$

Remark 4.8.

(i) In this proof Zhang's theorem appears as a kind of 'gap principle' for small solutions. It is crucial that the estimate does not depend on a, b.

(ii) The special shape of the above tori has allowed a quick conclusion, moreover with the 'good' value $N_1 = 2$. For varieties more general than the line $aX + bY = 1$, we can imitate the above procedure and construct a suitable 'universal' determinantal variety $\mathcal{X}$. Then, working on $\mathcal{X} \setminus \mathcal{X}^*$, we may decrease the dimension, which allows an inductive procedure; one may show that this works generally, leading eventually to Theorem 4.16; see [21] or [17] for details.

(iii) Explicit values for γ_1 and N_1 were found by Schlickewei and Wirsing [69] and then sharpened by Beukers and Zagier [9]; to obtain their results, all these authors considered the variety $\mathcal{X}^{-1}$ in the notation of Lemma 4.19. Their approach makes use of Zagier's method (see Section 1): the equality $h(P) = h(P^{-1})$ gives a lower bound for $h(\xi)$ for any non-torsion point P. The key for these proofs is the peculiar fact that, for this special $\mathcal{X}$, $\mathcal{X} \cap \mathcal{X}^{-1}$ consists only of torsion cosets. (If this were not the case, their method could not be iterated for the points in $\mathcal{X} \cap \mathcal{X}^{-1}$, because this kind of variety is stable by the inversion map; see Remark 4.2. So, since a 'general' variety $\mathcal{X}$ does not contain any torsion coset, we may consider this event as a little 'piece of good luck'.)

(iv) A different way to obtain suitable numerical values for γ_1 and N_1, in the general case, is via the proof of the Theorem of Zhang given above, with careful inspection of the various steps, in particular of Lemma 4.10.

Explicit bounds. A slight variation, which may lead to better numerical values, is to consider different 'good' primes l_i in Lemma 4.10; each of them gives a lower bound $c(l_i) > 0$ for the height of any point in $\mathcal{X}$, with the exception of those lying in $[l_i^d]^{-1} \mathcal{X}$. This yields a lower bound $\min\{c(l_i)\}$ for the height of algebraic points outside the intersection $\mathcal{Y}$ of all such varieties; the more primes we consider, the lower the dimension of $\mathcal{Y}$ will be, until $\mathcal{Y}$ consists of a finite number of points. Using resultants to eliminate one variable at each step, one can effectively compute all the data involved in this process.

In the case of interest for us we shall not produce this kind of actual computations, but merely a final estimate, obtained without calculating any resultant. We start by observing that if f and g are polynomials of separate degrees resp. at most m and n with respect to variables X_i, then any term in the resultant $\mathrm{Res}_{X_1}(f, g)$ will be a product of at most n monomials of f and at most m monomials of g (with respect to the remaining variables); thus its degree will be at most $2mn$ with respect to any of the remaining variables. We apply this to the above determinantal variety $\mathcal{X}$, using subsequently Lemma 4.10 with four primes $l_1 < l_2 < l_3 < l_4$, eliminating one variable each time by means of a resultant. (The resultants will be nonzero in $\mathcal{X}^*$.) Using the estimates for degrees just given, by inspection of the proof of the lemma, we end up with a polynomial in one variable of degree $2^7 \cdot l_1^4 \cdot l_2^2 \cdot l_3 \cdot l_4$ and with $N_1 = 2^{11} \cdot l_1^8 \cdot l_2^4 \cdot l_3^2 \cdot l_4$; we also obtain positive lower bounds $c(l_i)$ for the relevant heights, and we may retain only the minimum of these, corresponding to the largest of the primes. Taking, for instance, the primes 7, 11, 13, 17, we obtain $\gamma_1 \approx 10^{-3}$ and $N_1 \approx 10^{18}$.

Note that, although this method provides only rough values, it can be applied in a more general context, for it uses no *ad hoc* arguments.

Proof of Theorem 4.18. We have to estimate the number of points $(x, y) \in \Gamma'$ with $x + y = 1$ and $\widehat{h}(x, y) \leqslant t$.

Let $P := (x, y)$ and $P' := (x', y')$ be two such points and assume first that $P' = ZP$ for a torsion point Z, i.e. $x' = \zeta x$ and $y' = \theta y$ with roots of unity ζ, θ. Then we have both $x + y = 1$ and $\zeta x + \theta y = 1$. For every fixed (x, y) there can be at most two solutions (ζ, θ) of this system, corresponding to the intersections of the circles in $\mathbb{C}$ parametrized by $x e^{\alpha i}$ and $1 - y e^{\beta i}$, $\alpha, \beta \in \mathbb{R}$. This implies that we can count the points in question up to torsion, multiplying at the end by 2 the resulting estimate.

As in Section 3.4 of the previous chapter, we may associate to the group Γ of rank r a norm $\|\cdot\|$ on $\mathbb{R}^r$ and a homomorphism $\psi: \Gamma' \to \mathbb{Q}^r \hookrightarrow \mathbb{R}^r$ with kernel Γ'_{tors} and such that $\|\psi(P)\| = \widehat{h}(P)$ for every point $P \in \Gamma'$.

Let then Ω be a set of solutions to $x + y = 1$ in Γ', pairwise inequivalent modulo torsion and with $\widehat{h}(x, y) \leqslant t$. Their images $\psi(P)$, $P \in \Omega$ in $\mathbb{R}^r$ are pairwise distinct and lie in the ball $B(t) = \{x \in \mathbb{R}^r \mid \|x\| \leqslant t\}$.

To take advantage from the last lemma, we try to include the points in $\psi(\Omega)$ in the smallest possible number of translates of $B(\gamma)$, where $\gamma := \gamma_1/2$ and γ_1 is the constant of the lemma. The following is a rather common method to deal with this matter. Let us consider a maximal set of pairwise disjoint translates of $B(\gamma/2)$, centered at some of the points $\psi(P)$, $P \in \Omega$; let $B_1, \ldots, B_l$ be such translated balls. Since no other ball of radius $\gamma/2$ centered at some $\psi(P)$, $P \in \Omega$, can be added without intersecting some of the B_j, each of the $\psi(P)$, $P \in \Omega$, has distance $\leq \gamma/2$ from some ball B_j, hence distance at most γ from its center. Defining now B'_j as the ball with the same center as B_j but double radius (*i.e.* radius γ), we have obtained that $B'_1 \cup \ldots \cup B'_l$ contains $\psi(\Omega)$.

To estimate l, note that since the balls B_j are pairwise disjoint and contained in $B(t + (\gamma/2))$, we have $l \leqslant \frac{\text{vol} B(t + \gamma/2)}{\text{vol} B(\gamma/2)} = (\frac{t + (\gamma/2)}{\gamma/2})^r = (1 + \frac{2t}{\gamma})^r$.

We are left with the task of estimating the number of points P' in Ω such that $\psi(P')$ lies in a fixed ball B'_j. Choose $P = (a, b) \in \Omega$ so that $\psi(P)$ is the center of B'_j, as in the previous construction. If $\psi(P')$, $P' \in \Omega$, lies in B'_j, then, by definition of the norm, the point $(x, y) := P'P^{-1} \in \Gamma'$ has height $\widehat{h}(x, y) \leq \gamma < \gamma_1$. Also, $P' = P \cdot (x, y) = (ax, by)$, so $ax + by = 1$. By the above lemma, the equation $ax + by = 1$ has at most N_1 solutions in Γ' with height $< \gamma_1$. A *fortiori*, the number of points $Q \in \psi(\Omega) \cap B'_j$ is $\leq N_1$ (including $Q = \psi(P)$), and finally we find that the total number m of our points in Ω satisfies $m \leq N_1 l \leq N_1(1 + \frac{2t}{\gamma})^r$.

We have already observed that doubling this estimate gives an upper bound for the total number of the sought solutions. Since N_1 is an absolute constant, this concludes the proof.   □

Supplements to Chapter 4

Lattices and closed subgroups of $\mathbb{R}^n$

Since we have mentioned and used lattices and discrete subgroups, we pause here to give a brief account of some basic theory, including also general closed subgroups; in particular, this has applications to diophantine approximation.

Discrete subgroups of $\mathbb{R}^n$

We say that a subgroup Λ of $\mathbb{R}^n$ is *discrete* if it inherits the discrete topology from $\mathbb{R}^n$. This means that each point of Λ is isolated and in particular the origin is isolated, *i.e.* there is $\delta > 0$ such that the ball $B_\delta := \{x \in \mathbb{R}^n : |x| \leq \delta\}$ intersects Λ only at the origin. Hence for any two distinct points $P, Q \in \Lambda$, the difference $P - Q$ lies outside that ball, which means that P, Q have distance $> \delta$. In particular, Λ has no accumulation points in $\mathbb{R}^n$ and so each bounded region in $\mathbb{R}^n$ contains only finitely many elements of Λ. Conversely, if Λ has these last properties it is plainly discrete. A discrete subgroup of $\mathbb{R}^n$ is also called a *lattice* (in $\mathbb{R}^n$). Natural examples of lattices are $\mathbb{Z}^n$ and its subgroups.

Exercise 4.35. Prove that a lattice in $\mathbb{R}^n$ has $\mathbb{Q}$-rank $\leq n$. Deduce that it is finitely generated. (Hint: let $v_1, \ldots, v_m \in \Lambda$ be linearly independent over $\mathbb{Q}$. Then the map $(a_1, \ldots, a_m) \mapsto \sum a_i v_i$ from $\mathbb{Z}^m$ to Λ is injective, but sends the m-ball of radius R into the n-ball of radius $\leq cR$. If $m > n$, for large R one obtains a contradiction with discreteness. For the second assertion, prove that $\Lambda / \sum_{i=1}^m \mathbb{Z}v_i$ is finite if m is maximal with the v_i independent over $\mathbb{Q}$.)

Since any subgroup of $\mathbb{R}^n$ is torsion-free, the result of this exercise implies, through the structure theorem for finitely generated abelian groups, that any lattice is isomorphic to $\mathbb{Z}^r$, for some r. However this conclusion may be refined, with a self-contained proof, as in the following

Theorem 4.20. *Every discrete subgroup Λ of $\mathbb{R}^n$ has a finite $\mathbb{Z}$-basis $\lambda_1, \ldots, \lambda_r$ of vectors linearly independent over $\mathbb{R}$. The number r equals both $\dim_\mathbb{Q} \mathbb{Q}\Lambda$ and $\dim_\mathbb{R} \mathbb{R}\Lambda$.*

Proof. We may suppose that Λ contains some nonzero element λ and we argue by induction on n. In doing this we shall implicitly apply the induction hypothesis with $\mathbb{R}^n$ replaced with $\mathbb{R}$-vector spaces of dimension $< n$, identifying them with appropriate powers of $\mathbb{R}$. Let first $n = 1$; by discreteness, we may choose λ to have minimal nonzero absolute value. We may write any element $x \in \mathbb{R}$ as $x = q\lambda + \rho$, where $q \in \mathbb{Z}$ and $|\rho| < |\lambda|$. If $x \in \Lambda$, we deduce $\rho = x - q\lambda \in \Lambda$, and minimality entails $\rho = 0$, so $\Lambda = \mathbb{Z}\lambda$, proving the conclusion in this case. Let now $n > 1$ and consider the line $L := \mathbb{R}\lambda$; note that L intersects Λ in a nonzero subgroup which must be discrete; hence by the result for $n = 1$ we may assume that $\Lambda \cap L$ is generated by some nonzero element, which may be taken as λ. Let $V \cong \mathbb{R}^{n-1}$ be the orthogonal complement of L in $\mathbb{R}^n$ and let $\pi : \mathbb{R}^n \to V$ be the orthogonal projection. We contend that the subgroup $\Gamma := \pi(\Lambda)$ of V is discrete; in fact, let $B \subset V$ be a bounded region and define $B' := B + [0, 1]\lambda$, so B' is a bounded region in $\mathbb{R}^n$. Let $y \in B \cap \Gamma$, so $y = \pi(x)$ for some $x \in \Lambda$ of the form $x = y + t\lambda$, $t \in \mathbb{R}$. By subtracting from x an integral multiple of λ we may suppose that $0 \leq t \leq 1$ without changing $y = \pi(x)$; but then $x \in B'$ and therefore x has only finitely many possibilities. Then the same holds for y, proving that Γ is in fact discrete. By induction, there exists a finite $\mathbb{Z}$-basis $\gamma_1, \ldots, \gamma_s$ of Γ, with the γ_i linearly independent over $\mathbb{R}$, and we may choose $\lambda_1, \ldots, \lambda_s \in \Lambda$ with $\pi(\lambda_i) = \gamma_i$ for $i = 1, \ldots, s$. Let now $x \in \Lambda$; then $\pi(x) \in \Gamma$ so we may write $\pi(x) = m_1\gamma_1 + \ldots + m_s\gamma_s = \pi(m_1\lambda_1 + \ldots + m_s\lambda_s)$ with integers $m_1, \ldots, m_s$. Then $\pi(x - (m_1\lambda_1 + \ldots + m_s\lambda_s)) = 0$, so $x - (m_1\lambda_1 + \ldots + m_s\lambda_s) \in L \cap \Lambda = \mathbb{Z}\lambda$. This proves that $\lambda, \lambda_1, \ldots, \lambda_s$ generate Λ over $\mathbb{Z}$. Also, they are plainly linearly independent over $\mathbb{R}$, because the γ_i are; hence they are a basis of $\mathbb{R}\Lambda$ over $\mathbb{R}$, and thus their number is just the dimension of $\mathbb{R}\Lambda$ over $\mathbb{R}$, and also the dimension of $\mathbb{Q}\Lambda$ over $\mathbb{Q}$, which concludes the proof. $\qquad \square$

Closed subgroups of $\mathbb{R}^n$

Lattices in $\mathbb{R}^n$ are in particular *closed* subgroups. In general, for a closed subgroup G of $\mathbb{R}^n$ we define its *rank* $r = r(G)$ as its $\mathbb{R}$-rank, *i.e.* the maximum number of elements of G linearly independent over $\mathbb{R}$ (*i.e.* $r(G) = \dim_{\mathbb{R}} \mathbb{R}G$) and its *local rank* $l = l(G)$ as the maximum integer h such that every ball B_ϵ ($\epsilon > 0$) around the origin contains $\geq h$ elements of G linearly independent over $\mathbb{R}$. Note that $l(G) = 0$ if and only if G is discrete. We prove

Theorem 4.21. *Every closed subgroup G of $\mathbb{R}^n$ is of the shape $W + \Lambda$, where W is a vector subspace of dimension $l = l(G)$ and Λ is a lattice of rank $r - l$ with $W \cap \mathbb{R}\Lambda = \{0\}$. The space W is the maximal $\mathbb{R}$-vector space contained in G. We may choose Λ to be orthogonal to W, and then W, Λ are unique. Conversely, every subgroup of the shape $W + \Lambda$, W a vector subspace, Λ a lattice with $W \cap \mathbb{R}\Lambda = \{0\}$, is closed.*

Proof. For $\epsilon > 0$, let $W(\epsilon)$ be the vector space generated by $G \cap B_\epsilon$ over $\mathbb{R}$. By definition of $l = l(G)$, we have $\dim W(\epsilon) \geq l$, with equality for small enough ϵ. Since $W(\epsilon) \supset W(\delta)$ for $\epsilon \geq \delta > 0$, we see that for small ϵ the space $W(\epsilon)$ does not depend on ϵ, and we call it W. Let now $w \in W$; if $v_1, \ldots, v_l$ is a basis for W contained in $G \cap B_\epsilon$, we can write $w = (m_1 + t_1)v_1 + \ldots + (m_l + t_l)v_l$, where m_i are integers and $0 \leq t_i < 1$ and hence there exists $g = g(\epsilon) \in G$ with $|w - g| \leq l\epsilon$. Since ϵ can be taken arbitrarily small, this proves that w lies in the closure of G, hence lies in G. We have thus proved that $W \subset G$. Note that this implies that W is uniquely determined as the maximal $\mathbb{R}$-vector space contained in G.

Let now V be the orthogonal complement to W in $\mathbb{R}^n$ and let π be the associated projection onto V. Note that $G = W + \pi(G)$, because $W \subset G$, and note that this sum is orthogonal. If ϵ is small enough, the above shows that $G \cap B_\epsilon \subset W$, hence $\pi(G) \cap B_\epsilon$ reduces to the origin of V. Hence $\pi(G)$ is a discrete subgroup of V. It now suffices to define $\Lambda = \pi(G)$ to obtain the first part.

For the converse, let Λ be a lattice in $\mathbb{R}^n$ such that $W \cap \mathbb{R}\Lambda = \{0\}$ and let V be a subspace containing $\mathbb{R}\Lambda$ and with $W \cap V = \{0\}$, $W + V = \mathbb{R}^n$. Let $\pi : \mathbb{R}^n \to V$ be the associated projection. If $\gamma \notin G := W + \Lambda$, we have $\pi(\gamma) \notin \Lambda$. Since Λ is discrete there is a neighborhood I of $\pi(\gamma)$ disjoint from Λ. Then $W + I$ is a neighborhood of γ disjoint from G, proving that G is closed. This concludes the proof. $\qquad\square$

The next result concerns closed subgroups containing $\mathbb{Z}^n$; it admits a nice application to diophantine approximation, given in the subsequent corollary. By saying that a vector subspace V of $\mathbb{R}^n$ is 'defined over $\mathbb{Q}$' we mean that it may be defined by linear equations with rational coefficients. Equivalently, it has a basis in $\mathbb{Q}^n$.

Corollary 4.22. *Let G be a closed subgroup of $\mathbb{R}^n$ containing $\mathbb{Z}^n$. Then $G = W + \Gamma$ where W is a vector subspace defined over $\mathbb{Q}$, Γ is a lattice such that $[\Gamma : \Gamma \cap \mathbb{Z}^n]$ is finite, and $W \cap \mathbb{R}\Gamma = \{0\}$, $W + \mathbb{R}\Gamma = \mathbb{R}^n$.*

Proof. By the last theorem, $G = W + \Lambda$, where W, Λ are as in that statement, with Λ orthogonal to W. Let V be the orthogonal complement to W, so $V \supset$

$\mathbb{R}\Lambda$, and let $\pi : \mathbb{R}^n \to V$ be the orthogonal projection. Note that $\pi(\mathbb{Z}^n) \subset \pi(G) \subset \Lambda$, so $\pi(\mathbb{Z}^n)$ is discrete and thus has a $\mathbb{Z}$-basis $v_1, \ldots, v_s$, where the $v_i \in V$ are linearly independent, so $s \le \dim V$. If $z_i \in \mathbb{Z}^n$ are such that $\pi(z_i) = v_i$, we have $\mathbb{Z}^n \subset \Omega + W$, where we have put $\Omega := \mathbb{Z}z_1 + \ldots + \mathbb{Z}z_s \subset \mathbb{Z}^n$. In particular $\mathbb{Z}^n \cap W$ contains at least $n - s$ linearly independent vectors. On the other hand $s \le r(\Lambda) \le \dim V = n - \dim W$, whence we must have $r(\Lambda) = \dim V = s$ and $\dim W = n - s$; so W is spanned by integral vectors and is thus defined over $\mathbb{Q}$. Note also that $\pi|_{\mathbb{R}\Omega}$ is injective, because the $\pi(z_i) = v_i$ form a basis of V. Now we set $\Gamma := \pi^{-1}(\Lambda) \cap \mathbb{R}\Omega$. Note that Γ is a lattice of rank s and $\Gamma \cap \mathbb{Z}^n$ contains Ω, which is in $\mathbb{Z}^n$ and also has rank s, so $[\Gamma : \Gamma \cap \mathbb{Z}^n]$ is indeed finite. Also, we have $\pi(\Gamma) = \Lambda$, because $\pi(\mathbb{R}\Omega) = V$. Finally, since $\pi(\Gamma) = \Lambda$ and since $W \subset G$ we have $G = W + \Gamma$, whereas $W \cap \mathbb{R}\Gamma \subset W \cap \mathbb{R}\Omega = W \cap V = \{0\}$. (For the last part of the proof we could also apply n. 3 of Proposition 4.2 to the primitive lattice $\mathbb{Z}^n \cap W$, on defining Γ so that $\mathbb{Z}^n = (\mathbb{Z}^n \cap W) \oplus \Gamma$.) $\qquad\square$

Corollary 4.23 (Kronecker's Theorem). *Let* $1, \xi_1, \ldots, \xi_n \in \mathbb{R}$ *be linearly independent over* $\mathbb{Q}$ *and* $\beta_1, \ldots, \beta_n \in \mathbb{R}$. *For every* $\epsilon > 0$ *there are integers* $q, p_1, \ldots, p_n$ *such that* $|q\xi_i - \beta_i - p_i| < \epsilon$ *for* $i = 1, \ldots, n$.

Proof. Let G be the closure of the subgroup of $\mathbb{R}^n$ generated by $\mathbb{Z}^n$ and by $v := (\xi_1, \ldots, \xi_n)$. Let us write $G = W + \Gamma$ as in the last corollary, which we may do because G is closed and contains $\mathbb{Z}^n$. To obtain the conclusion it is plainly sufficient to prove that $W = \mathbb{R}^n$. Suppose the contrary; then, since W is defined over $\mathbb{Q}$, there is a nontrivial linear form L with rational coefficients and vanishing on W. Since $v \in G$ we may write $v = w + \gamma$ where $w \in W$ and $\gamma \in \Gamma$. Since $\Gamma \cap \mathbb{Z}^n$ has finite index in Γ, we have $m\gamma \in \mathbb{Z}^n$ for some integer $m > 0$. Then $L(v) = L(w) + L(\gamma) = L(\gamma) \in \mathbb{Q}$. However, this contradicts the linear independence over $\mathbb{Q}$ of $1, \xi_1, \ldots, \xi_n$. $\qquad\square$

This corollary says that under the independence assumption the fractional parts of the $q\xi_i$ are dense in the *unit cube* $[0, 1)^n$, for $q \in \mathbb{Z}$. Compare it with Lemma 3.20, which implies that, even dropping the independence assumptions, we can make all the $q\xi_i$ near to integers.

Exercise 4.36. Let $\xi \in \mathbb{R} \setminus \mathbb{Q}$ and let $\beta \in [0, 1)$. Prove that there exist infinitely many integers p, q such that $|q\xi - \beta - p| \le q^{-1}$. (Hint: apply Dirichlet Lemma 1.1 to approximate ξ with a rational r/s up to $(sQ)^{-1}$. Now write $\beta = (h+\theta)/s$ with integer h and $|\theta| < 1$ and find $q \le s$ with $rq \equiv h \pmod{s}$.)

Exercise 4.37. Show that the independence assumption in the last corollary cannot be eliminated.

Exercise 4.38. Dropping the independence assumption in the last corollary, prove that the conclusion holds for a $(\beta_1, \ldots, \beta_n) \in \mathbb{R}^n$ (and every $\epsilon > 0$) provided every linear form taking integral values at $(\xi_1, \ldots, \xi_n)$ and on $\mathbb{Z}^n$ necessarily takes integral values at $(\beta_1, \ldots, \beta_n)$. (Hint: mimic the above proof.)

Exercise 4.39. (Volumes of lattices) Let Λ be a lattice of rank r in $\mathbb{R}^n$, with basis $\lambda_1, \ldots, \lambda_r$. We define its *volume* $\mathrm{vol}(\Lambda)$ as the r-dimensional euclidean volume of the region $F_\Lambda := \{\sum_{i=1}^r t_i\lambda_i : 0 \le t_i < 1\}$, also called a *fundamental domain* for Λ; it is a system of representatives for the quotient group $\mathbb{R}\Lambda/\Lambda$.

(i) Prove that $\mathrm{vol}(\Lambda)$ is indeed well defined, *i.e.*, exists and is independent of the chosen basis.

(ii) Let $\mathcal{B}$ be an 'orthogonal box' in $\mathbb{R}\Lambda$ of sides $L_1 \geq L_2 \geq \ldots \geq L_r \geq 1$. Prove that $\#(\mathcal{B} \cap \Lambda) = L_1 \cdots L_r(\mathrm{vol}(\Lambda))^{-1} + O(L_1 \cdots L_{r-1})$. (Hint: estimate the number of fundamental domains in $\mathcal{B}$.)

(iii) Let Λ be a sublattice of $\mathbb{Z}^n$ and define be the *orthogonal lattice* $\Lambda' :=$ $\mathbb{Z}^n \cap V$, where V is the orthogonal complement of $\mathbb{R}\Lambda$ in $\mathbb{R}^n$. Prove that if Λ is primitive, we have $\mathrm{vol}(\Lambda) = \mathrm{vol}(\Lambda')$. (Hint: among the possible proofs, one of the simplest is as follows, supposing for simplicity that $\Lambda = \mathbb{Z}v$ has rank 1. Pick once for all a vector $z \in \mathbb{Z}^n$ with $(z, v) = 1$; it exists by primitivity. For $L > 1$, let B_L be the cube in V of side L centered at the origin. Fix an integer $c > 0$; the vectors $w \in B_L + [0, c]v$, satisfy $0 \leq (v, w) \leq c|v|^2 =: C$. Now, all the integer vectors x with $(x, v) = l$ are of the shape $x = lz + \lambda$ where $\lambda \in \Lambda'$. By (ii), the number N_l of those in $B_L + lv/|v|^2$ equals $L^{n-1}(\mathrm{vol}(\Lambda'))^{-1} + O(L^{n-2})$. Hence the number $N = \sum_{l=0}^{C} N_l$ of integer vectors in $B_L + [0, c]v$ equals $c|v|^2 L^{n-1}(\mathrm{vol}(\Lambda'))^{-1} + O(cL^{n-2})$. But, again by (ii), this also equals $\mathrm{vol}(B_L + [0, c]v) + O(L^{n-1} + cL^{n-2}) = c|v|L^{n-1} + O(L^{n-1} + cL^{n-2})$. Dividing by cL^{n-1}, letting $L \to \infty$ and then $c \to \infty$ we get the sought conclusion.)

Exercise 4.40. Let V be a vector subspace of $\mathbb{R}^n$ such that $V + \mathbb{Z}^n$ is dense in $\mathbb{R}^n$. Prove that there exists a line $L \subset V$ such that $L + \mathbb{Z}^n$ is also dense. (Hint: Use Corollary 4.22 to show that the minimal subspace W containing V and defined over $\mathbb{Q}$ is $\mathbb{R}^n$. Then cut V with a suitable hyperplane.)

Exercise 4.41. Let V be a vector subspace of $\mathbb{R}^n$ such that $V + \mathbb{Z}^n$ is dense in $\mathbb{R}^n$. Prove that for any open ball B centered at 0, $V + B$ contains a set of generators for $\mathbb{Z}^n$. (Hint: Let Λ be the lattice generated by $(V + B) \cap \mathbb{Z}^n$. Observe first that also $V + \Lambda$ is dense in $\mathbb{R}^n$: in fact, the closure of $V + \mathbb{Z}^n$ contains $B/2$, whence the closure of $V + \Lambda$ must contain $B/2$. Now note that for any $l \in \mathbb{Z}^n$, $l - B$ intersects $V + \Lambda$, whence $l + \Lambda$ intersects $V + B$ and hence $l \in \Lambda$. An alternative approach comes from the previous exercise, to reduce to the case of lines.)

The Skolem-Mahler-Lech Theorem and a generalization

This result concerns zeros of *linear recurrence sequences*, *i.e.* sequences $\{f(n)\}_{n \in \mathbb{N}}$ of complex numbers such that there exist $a_0, \ldots, a_{r-1} \in \mathbb{C}$ ($r \geq 1$, $a_0 \neq 0$) with $f(n+r) = a_0 f(n) + \ldots + a_{r-1} f(n+r-1)$ for all $n \in \mathbb{N}$. The minimum integer $r \geq 1$ with this property is called the *order* of the recurrence f.

It is easily seen that a formal power series $\sum_{n=0}^{\infty} f(n)X^n$ is the Taylor series of a rational function if and only if $f(n)$ coincides with a linear recurrence for all large n. In turn, the partial fraction decomposition for rational functions easily shows that every linear recurrence sequence may be represented uniquely as an exponential polynomial

$$f(n) = \sum_{i=1}^{s} c_i(n)\rho_i^n, \qquad (4.1)$$

where the $c_i \in \mathbb{C}[X]$ are nonzero and the $\rho_i \in \mathbb{C}^*$ are pairwise distinct. We say it is *over k* if $c_i \in k[X]$, $\rho_i \in k$. The ρ_i are usually called the *roots* of the recurrence. (They are roots of the polynomial $X^r - a_{r-1}X^{r-1} - \ldots - a_0$.) The recurrence is called *simple* if the c_i are constant and *nondegenerate* if no ratio $\rho_i/\rho_j, i \neq j$, is a root of unity. We refer to [101] and the related bibliography for more on recurrences.

By *zeros* of the recurrence f we mean integers $n \in \mathbb{N}$ with $f(n) = 0$ and we are interested in describing them. The expression (4.1) puts our problem in the context of zeros of entire functions. In general, there may be infinitely many complex zeros [19], but here we consider integer zeros. Recall that this problem has been relevant in the context of Theorem 4.6 (see especially Remark 4.6), but we shall see also a few other diophantine applications.

For recurrences over $\mathbb{R}$ a simple argument depending on Rolle's theorem bounds the number of real zeros by $\sum \deg c_i + s - 1$ (see Exercise 4.43 below). Another easy case occurs when there is a place v (*e.g.* a complex one) such that there exists a single root ρ_i having maximal absolute value $|\rho_i|_v$; one then speaks of a *dominant root*, whose term in fact dominates the whole sum, so there are only finitely many zeros. However, in general the problem of zeros is nontrivial. Note that when $\rho_i = \rho \zeta_i$ for roots of unity ζ_i we have that $f(n)/\rho^n$ is periodic, so may well have an infinity of zeros, which form a union of arithmetic progressions. That this description is essentially general was established by Skolem for recurrences over $\overline{\mathbb{Q}}$ and later extended by Mahler and Lech to fields of zero characteristic. The final result is known as the theorem of Skolem-Mahler-Lech. We shall give a brief account of Skolem's p-adic proof, limiting ourselves to the algebraic case. (See the exercises for an extension to $\mathbb{C}$.)

Theorem 4.24. *Let $f(n)$ be a linear recurrence over $\overline{\mathbb{Q}}$. Then the set of zeros of f is a union of a finite set with a finite union of arithmetic progressions. If f is nondegenerate it is a finite set.*

Proof. We argue with the representation (4.1); the idea is to view the right side suitably as a p-adic analytic function. For this, let k be a number field containing all the roots ρ_i and all coefficients of the polynomials c_i. Further, we choose once for all a finite place v of k trivial at all the roots ρ_i, and we embed k in the completion k_v. The residue field will be a finite field $\mathbb{F}_q$, where q is a power p^m of a prime p. Note that $\rho_i^{q-1} \equiv 1$ in the residue field, so for a sufficiently large power p^h of p we shall have $\xi_i := \rho_i^{(q-1)p^h} = 1 + p^2 \gamma_i$, where γ_i is v-integral. For $n \in \mathbb{N}$ we may write

$$\xi_i^n = (1 + p^2 \gamma_i)^n = \sum_{r \geq 0} p^{2r} \gamma_i^r \binom{n}{r} = \sum_{r \geq 0} p^{2r} \gamma_i^r \frac{n(n-1)\cdots(n-r+1)}{r!}.$$

Now, observe that $p^r/r!$ is p-integral for all $r \geq 0$, since the power of p dividing $r!$ is $[r/p] + [r/p^2] + \ldots \leq r/(p-1) \leq r$. Therefore the polynomials $p^{2r}\binom{x}{r}$ converge uniformly to 0 in the ring $k_v[[x]]$ equipped with the metric coming from the sup-norm of the coefficients. Taking into account that γ_i is v-integral,

---

[19] It may be shown that the only case of finitely many complex zeros occurs for $s = 1$.

it follows that the series

$$\psi_i(x) := \sum_{r \geq 0} p^{2r} \gamma_i^r \binom{x}{r}$$

is well-defined in $k_v[[x]]$, converges uniformly in the v-adic disk $\mathcal{O}_v := \{x \in k_v : |x|_v \leq 1\}$, $i.e.$ in the valuation ring of k_v, and represents an analytic function there. Moreover, by the above formula we have $\psi_i(n) = \xi_i^n$ for any integer $n \in \mathbb{N}$. (For this analytic representation of the exponential functions ξ_i^n we could have used equivalently the easy theory of p-adic exponential and logarithmic functions to write $\xi_i^n = \exp(n \log(1 + p^2 \gamma_i))$; see $e.g.$ [23] or [40].)

Now, we split $\mathbb{N}$ as a finite disjoint union of the progressions $\{a + n(q-1)p^h : n \in \mathbb{N}\}$, for $0 \leq a < (q-1)p^h$, and we restrict our exponential polynomial f to each progression. On writing $g_a(n) := f(a + n(q-1)p^h)$, we have

$$g_a(n) = \sum_{i=1}^{s} c_i(a + n(q-1)p^h)\rho_i^a \xi_i^n = \sum_{i=1}^{s} c_i(a + n(q-1)p^h)\rho_i^a \psi_i(n).$$

Hence each function g_a is the restriction to $\mathbb{N}$ of an analytic function $g_a(x) := \sum_{i=1}^{s} c_i(a + x(q-1)p^h)\rho_i^a \psi_i(x)$ on $\mathcal{O}_v$. Since $\mathcal{O}_v$ is compact and open, each function g_a either has only finitely many zeros in $\mathcal{O}_v$ or is identically zero therein. [20] Reading back these conclusions for $f(n)$, we deduce that on each of the said progressions f has either only finitely many zeros or vanishes identically, which proves the first part.

For the second part, suppose that f is nondegenerate, but vanishes on a whole progression $a + b\mathbb{N}$, $b > 0$. Then $\sum_{i=1}^{s} c_i(a + bn)\rho_i^a \eta_i^n = 0$ for all n, where $\eta_i := \rho_i^b$. Nondegeneracy implies that the η_i are pairwise distinct, but then it is easy to obtain a contradiction (see $e.g.$ Exercise 4.42 below).   □

Exercise 4.42. Let $f(n)$ be given by (4.1), with distinct nonzero ρ_i. Prove directly that if $f(n) = 0$ for all large $n \in \mathbb{N}$ then $c_i(n) = 0$ identically for all $i = 1, \ldots, s$. (Hint: argue for instance by induction on $\sum(\deg c_i + 1)$, using $g(n) := f(n+1) - \rho_1 f(n)$ for the induction step.)

Exercise 4.43. Let $\alpha_1 > \ldots > \alpha_s > 0$ be real numbers and let $\beta_1, \ldots, \beta_s \in \mathbb{R}[X]$. Prove that the number of zeros of the function $\sum_{i=1}^{s} \beta_i(n)\alpha_i^n$ on $\mathbb{N}$ (or even on $\mathbb{R}$) is at most $-1 + \sum_{i=1}^{s}(\deg \beta_i + 1)$. (Hint: use induction, on applying Rolle's theorem to the function $\sum_{i=1}^{s} \beta_i(x)(\alpha_i/\alpha_1)^x$.)

Exercise 4.44. Obtain a partially different proof of Theorem 4.6 by using the Skolem-Mahler-Lech theorem. (Hint: see Remark 4.6.)

Exercise 4.45. Obtain a proof of Theorem 4.24 for arbitrary fields of characteristic 0. (Hint: it suffices to work with function fields, and one may easily reduce to the case of curves $\mathcal{C}/\overline{\mathbb{Q}}$. Then, one may further easily reduce the theorem to the case of a nondegenerate recurrence. The ratios ρ_i/ρ_j, $i \neq j$ become thus

---

[20] A very easy proof of this fact may be obtained as in the complex case.

functions on the curve, not constantly equal to roots of unity. By Exercise 3.9 there are points $P \in \mathcal{C}(\overline{\mathbb{Q}})$ where all of these functions have large height or are constant; in particular, none of the values at P can be a root of unity. Now it suffices to apply Theorem 4.24 to the recurrence specialized at P.)

Exercise 4.46. Use Theorem 4.24 to prove that the equation $2^n = x^2 + 7$ has only finitely many integer solutions. (Hint: factoring in the ring of integers of $\mathbb{Q}(\sqrt{-7})$ easily leads to $\alpha_+^m - \alpha_-^m = \pm\sqrt{-7}$, where $\alpha_\pm := (1 \pm \sqrt{-7})/2$, $m = n - 2$. Since $\alpha_+/\alpha_- = (-3 + \sqrt{-7})/4$ is not a root of unity, Theorem 4.24 applies immediately. Actually, a detailed analysis along the lines of above proof, choosing *e.g.* $p = 11$, leads to the determination of all solutions, given by $n = 3, 4, 5, 7, 15$. See [26].)

An application to Thue Equations

Theorem 4.24 has a nice application to cubic Thue equations, found by Skolem. We sketch a simple example for the equation $X^3 - dY^3 = c$, d, c nonzero integers, for which we seek solutions $(p, q) \in \mathbb{Z}^2$. Letting δ be a real cube root of d, assumed irrational, we write the equation as $N(p - \delta q) = c$, where N is the norm from $\mathbb{Q}(\delta)$. Simple arguments of factorization in number fields show that $p - \delta q$ is of the shape $\alpha\mu$ where α runs through a finite set and μ is a unit. But the units of such a cubic field have rank 1, thus $p - \delta q = \beta\nu^n$ where β runs through a finite set, ν is fixed and $n \in \mathbb{Z}$. For fixed β, conjugating in three ways and eliminating p, q leads to an equation $\gamma_1\nu_1^n + \gamma_2\nu_2^n + \gamma_3\nu_3^n = 0$. By Theorem 4.24 either there are only finitely many possible integers n, or some ratio ν_i/ν_j is a root of unity, which can be easily excluded. Working backwards, one recovers the finiteness of the integral solutions.

This proof essentially resembles the deduction, seen in Chapter 3, of the Thue-Mahler Theorem 3.12 from the S-unit Theorem 3.13. In fact, the case of *simple* recurrences in Theorem 4.24 (*i.e.* constant coefficients c_i) may be interpreted in the 'S-unit context' as the study of the intersection of the hyperplane $\sum_{i=1}^{s} c_i X_i = 0$ with the cyclic subgroup $\{(\rho_1, \ldots, \rho_s)^n : n \in \mathbb{N}\} \subset \mathbb{G}_m^s$.

Exercise 4.47. Prove that in fact the ratio of conjugates $\nu_i/\nu_j, i \neq j$, appearing in the last proof, cannot be a root of unity. (Hint: Otherwise a power of ν would be in $\mathbb{Q}$, whence...)

A generalization to algebraic groups

There is an attractive reformulation of Theorem 4.24 in the context of algebraic groups, which we now briefly describe; this also leads to a significant generalization. All the ideas go back to Chabauty and Skolem (see [77, Section 5], or [23]), but we do not have an explicit reference for what we are going to discuss.

As a motivation, let us start by considering the following easy result, which we formulate as an exercise (a generalization of Exercise 4.15):

Exercise 4.48. Let Γ be an algebraic group, let $g \in \Gamma$ and let G be the Zariski-closure of the set $\{g^n : n \in \mathbb{Z}\}$ of powers of g.

(i) Prove that G is a commutative algebraic subgroup of Γ.
 Now, define $G(b)$ as the closure of the set $\{g^{bn} : n \in \mathbb{Z}\}$, so $G = G(1)$.

(ii) Prove that the connected component G_0 of the identity in G equals $G(b)$ for a suitable integer b.

(iii) Finally, prove that for $q \neq 0$, $G(q)$ is a finite union of cosets of G_0. (Hint: for the last question, one may replace g with g^b to assume that G is connected. Then, G is a finite union of cosets of $G(q)$, whence $G = G(q)$.)

Now, the question arises of what can be said about the Zariski-closure of a subset of all the powers of g, namely of a set $\{g^{a_n}, n \in \mathbb{N}\}$ where $(a_n)_{n \in \mathbb{N}}$ is a sequence of integers. Theorem 4.24 fits into this question as follows. Suppose for simplicity that $f(n)$ is a *simple* recurrence over $\mathbb{C}$, namely a purely exponential polynomial of the shape $f(n) = \sum_{i=1}^{s} c_i \rho_i^n$. We may view $f(n)$ as the value of a fixed linear form $L(x_1, \ldots, x_s)$ at the point g^n where $g := (\rho_1, \ldots, \rho_s) \in \Gamma := \mathbb{G}_m^s$. The zeros of $f(n)$ correspond to the integers n with $L(g^n) = 0$. If we write these zeros as a sequence (a_n) we see that the Zariski-closure of $\{g^{a_n}\}$ in Γ is contained in the hyperplane $L = 0$, falling thus in the above context.

We shall sketch a proof of the following general result, containing Theorem 4.24 as a special case.

Theorem 4.25. *Let Γ be an algebraic group over an algebraically closed field κ of characteristic zero, let $g \in \Gamma$ and let $(a_n)_{n \in \mathbb{N}}$ be a sequence of integers. The Zariski-closure of $\{g^{a_n} : n \in \mathbb{N}\}$ is a finite union of points and cosets of the connected component of the identity of the Zariski-closure of $\{g^n : n \in \mathbb{Z}\}$.*

Proof. We only sketch the argument, as a fully detailed proof would take us too far. By Exercise 4.48 (i) we can replace Γ with the Zariski closure G of $\{g^n : n \in \mathbb{Z}\}$, which is a commutative algebraic group; we denote by G_0 the connected component of the identity in G and use from now on an additive notation. Further, by partitioning G into (finitely many) cosets of G_0, we may assume, on replacing g with a suitable power of it, that $G = G_0$ is connected. We prove that if $\{a_n\}$ is infinite then $\{a_n g\}$ is Zariski-dense in G; this plainly leads at once to the theorem. Then suppose by contradiction that there is a regular nonconstant section f on G, defined at infinitely many points $a_n g$, so that $f(a_n g) = 0$ for all n.

Now, G, g and f are defined over a finitely generated subfield κ_1 of κ, and it is well known that this may be embedded in some finite extension κ_2 of some field $\mathbb{Q}_p$ (see [77, page 61]). Let $\mathcal{O}$ be the valuation ring of κ_2; by [75, Corollary 4 to Theorem 2, page 151], $G(\kappa_2)$ has an open subgroup H analytically isomorphic to $\mathcal{O}^d$, where $d = \dim G$. By taking p very large, we may assume that G, g have good reduction at p. Since the residue field of κ_2 is finite, it follows that a suitable multiple lg lies in H. Then, by replacing (a_n) with a suitable infinite subsequence we may assume that the a_n are pairwise congruent modulo l, so we may write $a_n = c + b_n l$ with a fixed integer c and integers b_n. Applying the isomorphism $H \cong \mathcal{O}^n$ to $lg \in H$ and to $f(c + x)$ we obtain a $\gamma \in \mathcal{O}^n$ and an analytic section ϕ on $\mathcal{O}^n$ such that $\phi(b_n \gamma) = 0$ for all n. The section $\phi(z\gamma)$ induces a local analytic function on the compact set $\mathcal{O}$ with infinitely many zeros therein, so it must vanish identically. But then $\phi(n\gamma) = 0$ for all integers n, whence $f((c + n)lg) = 0$ for all n, and we have a contradiction because $\{nlg : n \in \mathbb{Z}\}$ is Zariski-dense (*e.g.* by Exercise 4.48). $\square$

Exercise 4.49. Using also Exercise 4.48, deduce Theorem 4.24 from Theorem 4.25, by the arguments before the statement. (The case of general recurrences is similar by taking $\Gamma = \mathbb{G}_m^s \times \mathbb{G}_a$.)

Exercise 4.50. Let E be an elliptic curve over a number field k, let $f \in k(E^n)$ be a rational function and let $P \in E^n(k)$. Prove that the set of integers n such that $f(nP) = 0$ is the union of a finite set and a finite number of arithmetical progressions. (The same holds for any abelian variety in place of E^n; the context has been chosen for its analogy to Theorem 4.24 .)

An open question

Theorem 4.24 describes the set of zeros of linear recurrences; as we have remarked, this translates into a description of the possible sets of zeros for the coefficients a_n of the Taylor series $T(X) = \sum_n a_n X^n$ of a rational function $T(X) \in \overline{\mathbb{Q}}(X)$. It seems natural to ask for a similar description when the series $T(X) \in \overline{\mathbb{Q}}[[X]]$ is *algebraic* over $\overline{\mathbb{Q}}(X)$. Is it still true that the set of zeros of the coefficients differs by a finite set from the union of arithmetic progressions?

It seems not easy to answer this question, neither by reproducing the above method of proof nor by other known methods valid for rational functions (see the Notes below). Already for algebraic functions of degree 2, like $\sqrt{1+X} + r(X)$, $r \in \overline{\mathbb{Q}}(X)$, the answer seems not easy to obtain. This corresponds to describe the zeros of a sequence $\binom{2n}{n} + f(n)$, where $f(n)$ is a linear recurrence. It should be not too difficult to prove, in the general case, that the set of zeros differs by a set of density zero from a finite union of arithmetical progressions.

Notes to Chapter 4

Lang's problem is mentioned for instance in [53], p. 201. An analogue for the case of abelian varieties was the object of a conjecture of Manin and Mumford, predicting finiteness for the set of torsion points on a curve of genus ≥ 2 embedded in its Jacobian. This conjecture was first proved by Raynaud in [63].

An account of lattices and algebraic subgroups of $\mathbb{G}_m^n$, more complete than the present one, is in [17], from which we have borrowed freely. See also [71].

The characterization of torsion cosets given by Theorem 4.6 follows implicitly from the methods of [21]. Essentially, one reduces by projection to the case of codimension 1, and then uses a result of Gourin, reproduced in [21] , Lemma 3. (This also appears as Exercise 4.23 above.)

Zhang's Theorem appeared first in the case of curves [106], then in the general case [107]. After the paper [21], a similar argument for Zhang's theorem was given by Schmidt [71]; this interesting variation led to better quantitative bounds. This paper also avoids Theorem 4.6: by appealing to Lemma 4.10 for several primes, it reduces the problem to varieties which are invariant for several multiplications, and then a Vandermonde argument suffices. Previous results by Schmidt used a method similar to Zagier's for other algebraic varieties, leading again to explicit bounds. The whole subject was motivated by a conjecture of Bogomolov, prior to Zhang's Theorem; this conjecture predicted an absolute lower bound for the canonical height of algebraic points on a subvariety of an abelian variety, not on a torsion coset. This abelian context led to deeper difficulties compared to the case of $\mathbb{G}_m$. After some intermediate steps a proof was obtained by Ullmo in 1997. See [88].

The set $\mathcal{X}^\circ$ was introduced in [21], where Theorem 4.16 first appeared. See also [17] for a treatment. In particular, this uniform version of Zhang's theorem yields a uniform bound for the number of nondegenerate solutions of linear equations in n roots of unity. This corollary has been proved independently also by Schlickewei (see [68]); to date, the best quantitative estimate $(n + 1)^{3(n+1)^2}$ is due to Evertse [45].

Excellent quantitative bounds for the constants appearing in Zhang's Theorem and its uniform version are due to Amoroso, David, Philippon. See the Appendix by Amoroso for more on this.

Algebraic subgroups are relevant also concerning the distribution of points in the intersection of a subvariety $\mathcal{X}$ of $\mathbb{G}_m^n$ with a finitely generated group Γ, or its division group Γ'. The S-unit Theorem 3.13 represents the case $n = 2$; a generalization to higher dimensions is due to Laurent and depends on the deep Subspace Theorem of Schmidt, in the formulations by Schlickewei for arbitrary finite sets of places (see [17] or [101]). The result predicts that the Zariski closure of $\mathcal{X} \cap \Gamma$ is a finite union of cosets. This was a conjecture of Lang (solved by him for plane curves); another conjecture of Lang was the analogous fact for abelian varieties. This deeper problem was later solved by Faltings; a special case is the celebrated Mordell conjecture, also solved by Faltings earlier with a different method (see [17] for the Bombieri-Vojta proof of the Mordell conjecture and for references).

The results in Exercises 4.28 and 4.29 are essentially due to Mann; see [39] for proofs, improvements and references. These results bound the order in minimal vanishing sums of roots of unity, and may be considered as counterparts of the mentioned results on the counting of the solutions.

Bilu's theorem has been interpreted by Bombieri in terms of capacity theory, which led to a quick proof. (See *e.g.* [17].) An effective treatment of Bilu's equidistribution theorem has been given by Petsche [62]. Similar results have been proved by Szpiro, Ullmo, Zhang, for abelian varieties, leading eventually to the first proof of the above mentioned conjecture of Bogomolov.

A treatment of discrete and of closed subgroups of $\mathbb{R}^n$ can be found for instance in [80]. This also contains the simple elegant theory of continuous characters, which leads to the theorem of Kronecker, stated above as Corollary 4.23. We have not inserted this theory for brevity, but our arguments at bottom contain similar principles. For other proofs of Kronecker's theorem see [49].

For the diophantine theory of linear recurrences see [44, 101] and [103]. Skolem used his p-adic method, which we have essentially reproduced, also for other diophantine problems; here we have given just a simple illustration to the cubic Thue equation, but see [23] for a far more complete account. The general version of the Skolem-Mahler-Lech Theorem for fields of characteristic zero, sketched here in Exercise 4.45 above, can also be reduced to the algebraic case by methods based on derivations; see [102]. The case of positive characteristic is substantially different and has been satisfactorily treated only recently in [36].

Finally, the generalization to algebraic groups that we have presented as Theorem 4.25 does not seem to appear explicitly in the literature, although all the ideas for the proof are implicit in Skolem and Chabauty (see [77]).

The problem of zeros can be dealt with also by deeper methods, using the Subspace Theorem of Schmidt. The uniform versions of this result by Evertse and Schlickewei eventually led Schmidt [72] to solve a longstanding conjecture by proving that the number of zeros of a nondegenerate recurrence is bounded only in terms of its order. (The p-adic method instead leads to a bound depending on the field k of definition and on the least prime p such that the *roots* are p-adic units. Such a bound may be obtained by an explicit estimate for the number of zeros of a p-adic analytic function in terms of its Newton polygon, the corresponding theory appearing *e.g.* in [40].)

From an effective point of view, the situation is not satisfactory because there is no known algorithm for finding all the zeros of a given nondegenerate recurrence over $\overline{\mathbb{Q}}$ (except for order ≤ 3). Skolem's p-adic method in fortunate cases leads to the actual determination.

Chapter 5
The S-unit equation

In this chapter we shall give a complete proof of the S-unit theorem, independent of the General Roth Theorem stated in Chapter 3. Compared to other approaches, this proof is simultaneously simpler and more informative. These arguments, due to Beukers and Schlickewei, are similar in spirit to Thue's. However, compared to Thue's, we shall recognize three main differences: firstly, in this proof the auxiliary function is constructed explicitly, which results in better estimates; secondly, we shall apply the very useful theory of the euclidean norm associated to the height, developed earlier in this book; thirdly, the results on 'small' solutions obtained through Zhang's theorem, given in detail in the last chapter, also allow good and uniform estimates. The output will be an excellent quantitative bound for the number of solutions, depending remarkably only on the rank of the relevant group of S-units.

As we have seen in Chapter 3, the S-unit theorem implies many interesting finiteness diophantine results; correspondingly, the said quantitative bounds lead to rather explicit estimates for the number of solutions in these theorems. In this direction, we shall present a simple application to the Thue-Mahler Equation.

5.1. A quantitative S-unit theorem

In this short section we shall state in precise form the promised quantitative version of the S-unit Theorem 3.13, to be proved later in this chapter. This result consists in an explicit estimate for the number of solutions of $x + y = 1$ with $x, y \in \mathcal{O}_{k,S}$. Actually, we have remarked several times that $\mathcal{O}_{k,S}$ is a finitely generated multiplicative subgroup of $\overline{\mathbb{Q}}^*$. Hence it will suffice to estimate, more generally, the number of solutions of $x + y = 1$ with (x, y) lying in any given finitely generated subgroup Γ of $\mathbb{G}_m^2(\overline{\mathbb{Q}})$; although Γ is anyway a subgroup of some $\mathcal{O}_{k,S}$, of course it may have smaller rank, in which case we shall obtain more precise bounds. It turns out that the methods yield estimates for even more general solutions, namely allowing (x, y) to lie in the *division* group of Γ, *i.e.* the set

$$\Gamma' := \{P \in \Gamma : \exists m \in \mathbb{Z}, m \neq 0, P^m \in \Gamma\}.$$

The group Γ' is formed by all the m-th roots, of any order, of elements of Γ. Note that Γ' has the same rank as Γ, but is not finitely generated. We shall prove the following

Theorem 5.1. *There exist computable positive absolute constants C_1, C_2 with the following properties. Let Γ be a finitely generated subgroup of $\mathbb{G}_m^2(\overline{\mathbb{Q}})$, of rank r, and let Γ' be its division group. Then the number of solutions $(x, y) \in \Gamma'$ to the equation $x + y = 1$ is at most $C_1 C_2^r$.*

Note that this estimate depends only on the rank; it does not depend neither on data related to a number field containing Γ, nor on the height of possible generators for Γ. The proof splits into two parts: (a) estimating the number of solutions of 'large' height and (b) estimating the number of solutions of 'small' height. Actually, the second part has already been done in the last chapter.

The proof of an upper bound for the number of large solutions (part (a)) in this remarkable theorem is based on explicit Padé approximations to the polynomials $(1 - X)^n$. As we shall observe, it resembles in a certain way Thue's method.

In the next section we shall develop some preliminaries concerning Padé approximants; then we shall go on with the proof, formulated in terms of the euclidean norm associated to the height, that we have constructed in Chapter 3, whereas in the last section we shall present some applications of the theorem.

We have not given here admissible values for C_1, C_2, though it will be clear that the arguments lead to a completely explicit result, if desired.

Exercise 5.1. Let $a, b, c, r_0, r_1 \in \overline{\mathbb{Q}}$ and consider the second order linear recurrence $\{r_n\}$ defined for $n \geq 0$ by $r_{n+2} = a r_{n+1} + b r_n$. Prove that either r_n is constant or the number of n such that $r_n = c$ is bounded by an absolute constant. (Hint: obtain a formula for r_n of one of the shapes $\alpha + \beta n + \gamma n^2$, $(\alpha + \beta n)l^n$, $\alpha_1 l_1^n + \alpha_2 l_2^n$ and apply Theorem 5.1. For the second shape, take a conjugate to obtain $(\alpha \beta^\sigma - \alpha^\sigma \beta) = c l^{-n} - c^\sigma (l^\sigma)^{-n}$; or, even more simply, take complex conjugates and multiply the equations, then apply Exercise 4.43 to obtain an elementary sharp estimate independent of Theorem 5.1.)

Exercise 5.2. Observe that Theorem 5.1 yields a similar bound for the number of solutions of $ax + by = 1$, $x, y \in \Gamma$, independent of $a, b \in \overline{\mathbb{Q}}^*$.

Exercise 5.3. Let p be a prime number. Prove that the number of integer solutions of the equation $x^2 + 7 = p^n$ is bounded independently of p. (Compare with Exercise 4.46. (Hint: One may assume that -7 is a quadratic residue of

p, so p splits in the ring of integers of $\mathbb{Q}(\sqrt{-7})$ as $p = \pi\bar{\pi}$. Factoring and eliminating x leads to $\pi^m - \bar{\pi}^m = c$, whence the result by Theorem 5.1.)

5.2. Padé approximations

The auxiliary polynomials that we shall use below come from the theory of the so-called Padé approximations. Now we shall develop this concept in a little more detail, following [17], but thinking mainly of the special case we shall need; this leads to some simplifications. Below, κ will denote a field of characteristic zero.

For positive integers L, M, an (L, M)-*Padé approximation* to a formal power series $f \in \kappa[[x]]$ consists in a pair of polynomials $P, Q \in \kappa[x]$, not both zero, with $\deg P \le L, \deg Q \le M$, such that $P - fQ$ has order at least $L + M + 1$ at zero, *i.e.*

$$P(x) - f(x)Q(x) = x^{L+M+1}R(x) \qquad (5.1)$$

for some $R \in \kappa[[x]]$. Note that if this holds then $Q \ne 0$ for otherwise $\deg P \le L$ implies $P = 0$, which would be excluded. We have already met such Padé approximations as *good rational approximations* to functions; in fact, note that the rational function P/Q approximates $f(x)$ in the topology of $\kappa[[x]]$. For explicit examples, recall for instance the irrationality proofs related to $\exp(x)$ in the Supplements to Chapter 1.

Proposition 5.2. *For every triple* (L, M, f) *there exists an* (L, M)-*Padé approximation to* f. *Also, the ratio* P/Q *is uniquely determined.*

Proof. The triple (L, M, f) determines a system of $L + M + 1$ homogeneous linear equations in the $L + M + 2$ coefficients of P and Q. This grants the existence of a non-trivial solution.

Now, let (P, Q) and $(\tilde{P}, \tilde{Q})$ be two (L, M)-approximations to f. Note that the polynomial $P(x)\tilde{Q}(x) - Q(x)\tilde{P}(x)$ has degree at most $L+M$, but is equal to $x^{L+M+1}(R(x)\tilde{Q}(x) - Q(x)\tilde{R}(x))$, with order at least $L+M+1$ at zero. This implies $P\tilde{Q} = Q\tilde{P}$, that is $P/Q = \tilde{P}/\tilde{Q}$, as required. $\square$

Remark 5.1. In the diophantine applications, when κ is a number field, it is often important to have estimates for the height of the coefficients of P, Q. For this one can use Siegel Lemma (see Chapter 2) but this leads to bad bounds. To obtain better bounds through Siegel's lemma, one can slightly relax the approximation condition (5.1), by requiring a zero of multiplicity smaller than $L + M + 1$ (e.g. $\le \lambda(L + M)$ for a fixed $\lambda < 1$). This is what we have done in the course of the proof of Thue's theorem. However, in the present situation, we shall produce explicitly the Padé approximations; it will turn out that a good estimate for the coefficients is true, and can be proved directly.

We shall use the the Padé approximations for the functions $f(x) = (1 - x)^n$, and we shall find *hypergeometric* formulas for them, expressed in terms of binomial coefficients. Before that, let us note the following fact:

Proposition 5.3. *Let L, M, N be positive integers and let P, Q, R be nonzero polynomials over κ of degrees resp. $\leq L, M, N$, with $P(x) - (1 - x)^{L+N+1} Q(x) = x^{L+M+1} R(x)$. Then P, Q, R are pairwise coprime and none of them has multiple roots or vanishes at $0, 1$. We also have $\deg P = L$, $\deg Q = M$, $\deg R = N$. Further, P, Q, R are uniquely determined by L, M, N up to a constant factor.*

Proof. We apply the abc-theorem for polynomials (Corollary 3.17) to the equation in the statement. We set $A = P(x), B = -(1 - x)^{L+N+1} Q(x)$, and $C = x^{L+M+1} R(x)$. Note that $A + B = C$ and that these polynomials have degrees resp. $\deg P, L+N+1+\deg Q, L+M+1+\deg R$, whereas the product ABC has at most $2 + \deg P + \deg Q + \deg R$ distinct roots. Hence, if A, B, C were coprime, the said abc-theorem would yield the inequality

$$\max(L+N+1+\deg Q, L+M+1+\deg R) \leq 1+\deg P+\deg Q+\deg R,$$

whence, on taking into account the upper bounds for the degrees of P, Q, R, we would in fact have a case of equality and we would also infer that $\deg P = L, \deg Q = M, \deg R = N$.

To take care of a possible nontrivial common divisor of A, B, C, we actually apply the abc-theorem to $a := A/d, b := B/d, c := C/d$, where d is a gcd of A, B, C, so a, b, c are coprime, not all constant (otherwise B would divide A, whence, on looking at degrees, $A = 0$) and $a + b = c$. Note that each root of the product abc lies among $0, 1$ or the roots of $P/(P, d), Q/(Q, d)$ or $R/(R, d)$. The total number of roots is therefore at most $2 + \deg P + \deg Q + \deg R - \deg(P, d) - \deg(Q, d) - \deg(R, d)$. The maximum degree is $\geq L + \deg Q + N + 1 - \deg d$, so by Corollary 3.17 we obtain $L + \deg Q + N + 1 - \deg d \leq (2 + \deg P + \deg Q + \deg R - \deg(P, d) - \deg(Q, d) - \deg(R, d)) - 1$, *i.e.* $(L + N - \deg P - \deg R) + \deg(P, d) + \deg(Q, d) + \deg(R, d) \leq \deg d$. Since d divides P we have $(L + N - \deg P - \deg R) + \deg(Q, d) + \deg(R, d) \leq 0$, *i.e.* $L = \deg P, N = \deg R$ and d is coprime with both R, Q. Hence, d must divide both a power of x and a power of $1 - x$, which implies in fact that d is constant, so A, B, C are coprime.

Then the opening argument applies, and we also deduce that all the above inequalities must be equalities, so PQR may have neither a multiple root nor a root in $\{0, 1\}$, as required.

Uniqueness follows since by Proposition 5.2 the ratio P/Q is uniquely determined and so are P, Q up to constants, because they have just been shown to be coprime. □

Remark 5.2. In Exercise 3.60 we have noted that the cases of equalities in Theorem 3.16 correspond to rational maps unbranched outside $\{0, 1, \infty\}$. In the present context, the proof of this proposition has shown that the identity in the statement yields a case of equality in the abc-theorem for polynomials. Putting all of this together we deduce that the rational map $x^{L+M+1} R(x)/P(x)$ (or, equivalently, $(1 - x)^{L+N+1} Q(x)/P(x)$) is unbranched outside $\{0, 1, \infty\}$. (See [17] for a direct deduction.)

Theorem 5.4. *For positive integers L, M, N, let*

$$Q_{L,M,N}(x) := \sum_{j=0}^{M} \binom{N+j}{N}\binom{L+M-j}{L} x^j; \qquad (5.2)$$

$$P_{L,M,N}(x) := (-x)^L Q_{N,L,M}\left(1 - \frac{1}{x}\right); \qquad (5.3)$$

$$R_{L,M,N}(x) := (-1)^L Q_{L,N,M}(1 - x). \qquad (5.4)$$

Then $P_{L,M,N}$, $Q_{L,M,N}$, $R_{L,M,N}$ are polynomials with integer coefficients and degrees L, M, N respectively. They satisfy

$$P_{L,M,N}(x) - (1 - x)^{L+N+1} Q_{L,M,N}(x) = x^{L+M+1} R_{L,M,N}(x). \quad (5.5)$$

Also, the ℓ_1 norm of $Q_{L,M,N}$ is $Q_{L,M,N}(1) = \binom{L+M+N+1}{M}$.

Proof. First, note that $Q_{L,M,N}$ is a polynomial of degree M with coefficients in $\mathbb{N}$ and such that $Q_{L,M,N}(1) \neq 0$; thus also $P_{L,M,N}$ and $Q_{L,M,N}$ lie in $\mathbb{Z}[X]$ and have the stated degrees.

Now, consider the integral $\int_0^1 t^M (t - 1)^N (t - x)^L dt$; with the substitution $t = xu$, the splitting $\int_0^{1/x} = \int_0^1 - \int_{1/x}^1$, and the further substitution $xu = 1 - (1 - x)v$ in the second piece, we obtain

$$\int_0^1 t^M (t - 1)^N (t - x)^L dt = x^{L+M+1} \int_0^{1/x} u^M (xu - 1)^N (u - 1)^L du$$

$$= x^{L+M+1} \int_0^1 u^M (xu - 1)^N (u - 1)^L du$$

$$+ (-1)^N (1 - x)^{L+N+1} \int_0^1 (1 - (1 - x)v)^M v^N (1 - v)^L dv.$$

Further, put $P(x) := \int_0^1 t^M (1 - t)^N (t - x)^L dt$, $Q(x) := \int_0^1 v^N (1 - v)^L (1 - (1 - x)v)^M dv$, and $R(x) := (-1)^L \int_0^1 u^M (1 - u)^L (1 - xu)^N du$.

Apart from the sign $(-1)^N$, these are the three terms appearing in the last displayed equation, which now reads: $P(x) = (-1)^N (1-x)^{N+L+1} Q(x) + (-1)^L x^{L+M+1} R(x)$.

The coefficients of P, Q, and R can be 'explicitly' determined by expanding the binomials involving x and computing the integrals which arise. These are obtained from Euler's beta function, defined for every pair (a, b) of positive integers as $\beta(a, b) := \int_0^1 t^a(1-t)^b dt$. When $b \neq 0$, integration by parts gives $\beta(a, b) = \frac{b}{a+1}\beta(a+1, b-1)$; thus we obtain by recursion the (standard) formula $\beta(a, b) = \frac{a!b!}{(a+b+1)!}$.

In particular we have

$$Q(x) = \int_0^1 v^N(1-v)^L(1-v+xv)^M dv$$

$$= \sum_{j=0}^M \binom{M}{j} x^j \int_0^1 v^{N+j}(1-v)^{L+M-j} dv$$

$$= \sum_{j=0}^M \frac{M!(N+j)!(L+M-j)!}{j!(M-j)!(L+M+N+1)!} x^j$$

$$= D^{-1} \sum_{j=0}^M \binom{N+j}{N}\binom{L+M-j}{L} x^j$$

where $D := \frac{(L+M+N+1)!}{L!M!N!}$; note that $D \in \mathbb{N}$.

We now define $P_{L,M,N} = DP$, $Q_{L,M,N} = DQ$, and $R_{L,M,N} = DR$. Equation (5.5) is verified, as we have just shown, and we are left with the task of verifying (5.2), (5.3), and (5.4).

The first of these equations holds by definition. The second and third one follow by uniqueness, ensured by Proposition 5.3: one observes that, by the substitution $x \mapsto (1-x)$ (resp. $x \mapsto (1-x^{-1})$), with subsequent clearing of a power of x-denominator) the (L, M)-Padé approximation to $(1-x)^{L+N+1}$ tranforms into the (L, N)-(resp. (M, N)-)Padé approximation to $(1-x)^{L+M+1}$ (resp. $(1-x)^{M+N+1}$). The undetermined constant factor is checked e.g.. by evaluating at $x = 1$.

Finally, since every coefficient of $Q_{L,M,N}$ is ≥ 0, its ℓ_1 norm is just the sum of its coefficients, $Q_{L,M,N}(1)$; by definition we have $Q_{L,M,N}(1) = DQ(1) = D\beta(N, L) = \binom{L+M+N+1}{M}$. □

Remark 5.3. The above substitutions $x \mapsto (1-x)$ and $x \mapsto (1-1/x)$ generate the automorphism group of $\mathbb{P}^1 \setminus \{0, 1, \infty\}$; the appearance of this affine curve is no special surprise, because of its link with the S-unit theorem, as in Remark 3.14. The $\mathbb{Q}(x)$-point $(x^{L+M+1} R(x)/P(x), (1-x)^{L+N+1} Q(x)/P(x)) \in \mathbb{Q}(x)^2$

can be viewed as an 'almost' S-unit point on $X + Y = 1$, *i.e.* an 'almost' integral point for $\mathbb{P}_1 \setminus \{0, 1, \infty\}$ (relative to $k = \mathbb{Q}(x)$).

Unless otherwise specified, from now on we shall omit indices and write simply P, Q, R to denote the polynomials of Padé approximations to $(1 - x)^{L+N+1}$ (and not those used in the above proof), normalized so that $Q = Q_{L,M,N}$ as given by (5.2).

5.3. Proof of Theorem 5.1

We recall from Chapter 3 that the restriction of the height $\widehat{h}$ to a subgroup Γ of $\mathbb{G}_m^n(\overline{\mathbb{Q}})$ of finite rank r corresponds, up to torsion elements, to a norm on the euclidean space $\mathbb{R}^r$, via a homomorphism $\varphi : \Gamma \to \mathbb{Z}^r \subset \mathbb{R}^r$, whose kernel is precisely the set Γ_{tors} of torsion points of Γ. Thus the solutions $\underline{u} = (u_1, u_2) \in \Gamma \subset \mathbb{G}_m^2(\overline{\mathbb{Q}})$ we are looking for, *i.e.* with $u_1 + u_2 = 1$, can be studied in this space, up to torsion elements. (As observed in Chapter 4, there are at most two solutions corresponding to a same point of $\mathbb{R}^r$.)

We shall denote by U the subset of $\mathbb{R}^r$ corresponding to solutions, *i.e.* to points $\underline{u} = (u_1, u_2) \in \Gamma$ with $u_1 + u_2 = 1$. So U in fact is a subset of $\mathbb{Z}^r$.

In the sequel we shall sometimes identify elements of Γ (modulo torsion) with vectors in U, omitting an explicit reference to φ for brevity and for notational convenience. The words 'small', 'near', etc. will then refer to the associated euclidean norm, corresponding the height $\widehat{h}$.

Plan of the proof

The plan of the proof (insofar as we are concerned with the counting of large solutions) parallels in several respects Thue's one and, in analogy with Chapter 2, may be briefly summarized as follows:

0. We suppose to be given two solutions $\underline{u} = (u_1, u_2)$, $\underline{v} = (v_1, v_2) \in \Gamma$ to $X_1 + X_2 = 1$, with sufficiently large height, and $\widehat{h}(\underline{v})$ much larger than $\widehat{h}(\underline{u})$. Thinking of the solutions as vectors in $\mathbb{R}^r$, we shall also assume that $\underline{u}$, $\underline{v}$ have nearly the same direction. This can be achieved just by partitioning the solutions in finitely many subsets corresponding to finitely many cones with small angular width, covering $\mathbb{R}^r$.

Depending on these two solutions we shall:

1. Construct auxiliary identites $r_n(x) + s_n(x) = 1$ ($n \in \mathbb{N}$) where r_n, s_n are 'almost units relative to $\{0, 1, \infty\}$ for the field $\mathbb{Q}(x)$; actually this construction has been already performed in Theorem 5.4, depending on parameters L, M, N, to be chosen conveniently in terms of $\underline{u}$, $\underline{v}$. (Below we shall set $L = M = N = n$, with a convenient n depending on $\underline{u}$, $\underline{v}$.)

Specializing these identites at $x = u_1$ we shall produce other solutions $\underline{u}^{(n)} := (r_n(u_1), s_n(u_1))$, somewhat 'near' to $\underline{u}^n$, i.e. such that $r_n(u_1) = u_1^n \sigma_1$, $s_n(u_1) = u_2^n \sigma_2$, where σ_1, σ_2 have not too large height.

2. Thinking again of the solutions as vectors in $\mathbb{R}^r$, since $\underline{u}$ and $\underline{v}$ have almost the same direction, a suitable power $\underline{u}^n$ will go 'near' $\underline{v}$. (This step, together with the last part of the previous one, corresponds to the 'Upper bound' in Thue's method). Note however that $\underline{u}^n$ will generally not be a solution (but $\underline{u}^{(n)}$ will be).

3. Since $\underline{u}^n$ goes both near $\underline{v}$ and also near the solution $\underline{u}^{(n)}$, the two solutions $\underline{v}$ and $\underline{u}^{(n)}$ are themselves near. However two large solutions cannot be too near: If $v_1 + v_2 = v_1\xi_1 + v_2\xi_2 = 1$, where $\widehat{h}(\underline{\xi})$ is small, then by elimination we also find that $\widehat{h}(\underline{v})$ is small (*unless* $\xi_1 = \xi_2 = 1$). (This step corresponds to the 'Gap principle' or the 'Lower bound' in Thue's method).

4. Deriving a contradiction: this follows on comparing nos. 2,3. However for Step n. 3 to work one needs $(\xi_1, \xi_2) \neq (1, 1)$. To avoid the danger of equality corresponds to the critical 'Nonvanishing' step in Thue's proof. Similarly to that proof, this danger may be overcome on differentiating the auxiliary function, to obtain other similar functions. Here such functions are given explicitly, and it will turn out that a single differentiation suffices. (On the contrary, in Thue's proof we needed higher order differentiations, and a further problem occurred in finding a good upper bound for a suitable order.)

This analogy with Thue's proof that we have presented in a previous chapter, is however a partial one: in fact, here we have to estimate (and moreover in a uniform way) the number of solutions, whereas our version of Thue's Theorem was purely qualitative, namely it merely proved the finiteness of the set of solutions.[1] Partly for this reason, below we shall proceed in a somewhat different order, following [17], to simplify the exposition. It is an instructive exercise for the interested reader to match the steps in the detailed proof with the rough description just given.

5.3.1. Distribution of solutions in euclidean spaces

We start with a lemma playing a role in n. 3 above.

Lemma 5.5. Let $(a, b, c), (a', b', c') \in \overline{\mathbb{Q}}^3$. If $ab' \neq a'b$, then every solution (x_1, x_2) to the system $ax_1 + bx_2 = c$ and $a'x_1 + b'x_2 = c'$

---

[1] See however Remark 2.11, in which we pointed out how Thue's argument can be adapted to estimate the number of solutions.

satisfies

$$h(1 : x_1 : x_2) \leqslant \log 2 + h(a : b : c) + h(a' : b' : c').$$

Proof. Solving for x_1, x_2 we find $x_1 = (cb' - c'b)/(ab' - a'b)$ and $x_2 = (ac' - a'c)/(ab' - a'b)$. We take a number field k containing all the involved quantities; by definition of the height we have

$$h(\underline{x}) = h(ab' - a'b, ac' - a'c, bc' - b'c)$$
$$= \sum_v \log \sup(\|ab' - a'b\|_v, \|ac' - a'c\|_v, \|bc' - b'c\|_v),$$

where the summations are for $v \in M_k$, with the appropriate normalizations. Note that $\|ab' - ba'\|_v \leqslant \max(1, \|2\|_v) \sup(\|a\|_v \|b'\|_v, \|b\|_v \|a'\|_v)$ and similarly for the other terms. Then we obtain

$$h(1 : x_1 : x_2) \leqslant h(2) + \sum_v \log \sup(\|a\|_v, \|b\|_v, \|c\|_v)$$
$$+ \sum_v \log \sup(\|a'\|_v, \|b'\|_v, \|c'\|_v),$$

as required. $\qquad\qquad\qquad\qquad\qquad\qquad\qquad\qquad\qquad\qquad\square$

Corollary 5.6. *Let* $(x_1, x_2), (y_1, y_2) \in \mathbb{G}_m^2(\overline{\mathbb{Q}})$ *be distinct elements such that* $x_1 + x_2 = 1$ *and* $y_1 + y_2 = 1$. *Then* $\widehat{h}(\underline{x}) \leqslant 2 \log 2 + 2\widehat{h}(\underline{x}\underline{y}^{-1})$.

Proof. We apply Lemma 5.5 with $(a, b, c) = (1, 1, 1)$ and $(a', b', c') = (y_1 x_1^{-1}, y_2 x_2^{-1}, 1)$. Note that the requirements of the lemma are satisfied, since $(x_1, x_2) \neq (y_1, y_2)$. It now suffices to recall that $h(1 : x_1 : x_2) \leq \widehat{h}(\underline{x}) := h(x_1) + h(x_2) \leq 2h(1 : x_1 : x_2)$ $\qquad\qquad\square$

Lemma 5.7. *Let* $\underline{x}, \underline{y} \in \mathbb{G}_m^2(\overline{\mathbb{Q}})$ *be such that* $x_1 + x_2 = 1$ *and* $y_1 + y_2 = 1$. *There exists an absolute effective constant* C *such that for every integer* $n \geqslant 2$ *we have*

$$\widehat{h}(\underline{x}) \leqslant C + \frac{2}{n-1} \widehat{h}(\underline{y}\underline{x}^{-2n}).$$

Proof. Using $x_2 = 1 - x_1$, we can apply Theorem 5.4 with $L = M = N = n$ to obtain polynomials $P, Q, R \in \mathbb{Z}[x]$ such that $x_1^{2n+1} R(x_1) + x_2^{2n+1} Q(x_1) = P(x_1)$ and, by differentiation, $x_1^{2n} \tilde{R}(x_1) + x_2^{2n} \tilde{Q}(x_1) = \tilde{P}(x_1)$. Here we have abbreviated

$$\tilde{P}(x) = P'(x),$$
$$\tilde{Q}(x) = (1 - x)Q'(x) - (2n + 1)Q(x),$$
$$\tilde{R}(x) = xR'(x) + (2n + 1)R(x).$$

In Lemma 5.5 we set $(a, b, c) := (y_1 x_1^{-2n}, y_2 x_2^{-2n}, 1)$ and x_1^{2n}, x_2^{2n} in place of x_1, x_2. Also, we set either $(a', b', c') = (x_1 R(x_1), x_2 Q(x_1), P(x_1))$ or $(a', b', c') = (\tilde{R}(x_1), \tilde{Q}(x_1), \tilde{P}(x_1))$: we contend that at least one of these two possibilities satisfies the assumption $ab' \neq a'b$ of that lemma.

To prove this contention amounts to show that

$$\Delta(x_1) := \begin{vmatrix} x_1 R(x_1) & (1-x_1)Q(x_1) \\ \tilde{R}(x_1) & \tilde{Q}(x_1) \end{vmatrix}$$

is not zero. Now, note that $x^{2n}(1 - x)^{2n}\Delta(x)$ is the Wronskian $W := W(x^{2n+1} R(x), (1 - x)^{2n+1} Q(x))$, so it is not identically zero: for otherwise the said polynomials would be linearly dependent whereas they are nonconstant and coprime by Proposition 5.3; on the other hand our basic identity yields $W = W(x^{2n+1} R(x), P(x))$, showing that W has degree $\leq 2n + 1 + \deg R + \deg P - 1 \leq 4n$. Therefore $\Delta(x)$ is a nonzero constant; in particular, it does not vanish at $x = x_1$, concluding the proof of the claim.

Therefore we may apply Lemma 5.5 with one of the above two choices, and we proceed to estimate the right hand side of the inequality provided by the lemma.

In the first place, note that $h(a : b : c) \leq \widehat{h}(y\underline{x}^{-2n})$.

As to $h(a' : b' : c')$, if we make the first choice for (a', b', c'), we have $h(a' : b' : c') = h(x_1 R(x_1) : x_2 Q(x_1) : P(x_1))$. In this case the estimates of Theorem 5.4 easily yield $h(a' : b' : c') \leq (n+1)h(x_1)+(3n+1)\log 2$.

If the first choice is not possible, then we have just proved that we can make the second one, in which case we obtain $h(a' : b' : c') = h(\tilde{R}(x_1), \tilde{Q}(x_1), \tilde{P}(x_1))$. Taking into account the above formulae, again Theorem 5.4 yields $h(a' : b' : c') \leq nh(x_1) + h((4n + 1)2^{3n+1}) \leq nh(x_1) + (6n + 1)\log 2$.

Therefore, Lemma 5.5 gives $2nh(x_1) \leq \log 2 + \widehat{h}(y\underline{x}^{-2n}) + (n + 1)h(x_1)+(6n+1)\log 2$, whence $(n-1)h(x_1) \leq (6n+2)\log 2+\widehat{h}(y\underline{x}^{-2n})$.

By symmetry this holds with x_2 in place of x_1. Summing the resulting inequalities we have finally $(n - 1)\widehat{h}(\underline{x}) \leq 2(6n + 2)\log 2 + 2\widehat{h}(y\underline{x}^{-2n})$, proving the result with $C := \max_{n\geq 2}(12n+4)\log 2/(n-1) = 28\log 2$. □

The next lemma merely rephrases these results in terms of the euclidean space $\mathbb{R}^r$. This yields actually two gap principles. The second one will be used to show that a solution of large norm bounds the norm of any other solution having nearly the same 'direction'; the first one will be used to estimate the number of solutions satisfying these bounds.

Lemma 5.8. *There exist effective absolute constants c_1, c_2 such that for every $u, v \in U$ the following holds:*

$$\|u\| \leqslant c_1 + 2\|v - u\| \qquad \text{for } u \neq v; \qquad (5.6)$$

$$\|u\| \leqslant c_2 + \frac{2}{n-1}\|v - 2nu\| \qquad \text{for every integer } n \geqslant 2. \qquad (5.7)$$

5.3.2. Final arguments

After these technical preliminaries we are ready to proceed with the proof of Theorem 5.1, following the above outlined program and working in the euclidean space $\mathbb{R}^r$, equipped with the norm provided by the height on Γ; when speaking of balls, cones etc. we refer to this norm.

Below $c_1, c_2, \ldots$ will denote positive computable absolute constants.

Proof of Theorem 5.1. We define $\psi(u) := u/\|u\|$ for every $u \in \mathbb{R}^r$, so that $u = \|u\| \psi(u)$ and $\|\psi(u)\| = 1$. For a given u, $\psi(u)$ defines the 'direction' of u in $\mathbb{R}^r$. For given ϵ, we shall consider solutions having the same 'direction', up to ϵ, that is we consider the pairs u, v with $\|\psi(v) - \psi(u)\| < \epsilon$. For a given u, the corresponding set of $v \in \mathbb{R}^r$ plainly constitutes a cone, which we call the ϵ-*cone of* u. We also say that $\psi(u)$ is a *center* of such a cone.

For u, v satisfying this inequality, with the purpose to apply (5.7), we seek an even multiple of u going 'near' v; for this we set $n := \left[\frac{\|v\|}{2\|u\|}\right]$. Note that $2n\|u\| \leq \|v\| < 2(n+1)\|u\|$, i.e. $0 \leq \|v\| - 2n\|u\| < 2\|u\|$. In particular, if $\|v\| > 4\|u\|$ we have $n \geqslant 2$, so by inequality (5.7) we obtain

$$\|u\| \leqslant c_2 + \frac{2}{n-1}\Big\| \|v\| \psi(v) - 2n\|u\| \psi(u) \Big\|$$

$$\leqslant c_2 + \frac{2}{n-1}\Big(\|v\| \Big\|\psi(v) - \psi(u)\Big\| + \|\psi(u)\| \Big(\|v\| - 2n\|u\| \Big) \Big)$$

$$\leqslant c_2 + \frac{2}{n-1}(\|v\|\epsilon + 2\|u\|) \leqslant c_2 + \frac{2}{n-1}((2n+2)\epsilon + 2)\|u\|.$$

For $n \geqslant 14$ and $\epsilon = 1/12$ we obtain $\frac{2}{n-1}((2n+2)\epsilon + 2) \leqslant 4\epsilon + \frac{8\epsilon+4}{n-1} \leqslant 2/3$, so that $\|u\|$ is bounded by a constant $c_3 := 3c_2$. So, with this choice $1/12$ for ϵ, if $\|u\| > c_3$ we must have $n \leq 13$. Thus, recalling $\|v\| < (2n+2)\|u\|$ we have obtained the following:

Intermediate conclusion

If a point $u \in U$ has norm $\|u\| > c_3$, every other point $v \in U$ lying in its $(1/12)$-cone has norm $\|v\| < 28\|u\|$.

This result already implies the finiteness of the set U: on the one hand, since $U \subset \mathbb{Z}^r$, there are only finitely many elements of U with norm $\leq c_3$.[2] On the other hand, we can partition the remaining points into finitely many $(1/24)$-cones and, by the result just proved, any point in such a cone yields a bound for the norm of any other point in the same cone; so every such cone contains only finitely many points of U.

However, this result is not effective, not even in the weak sense of estimating the cardinality of U (leaving aside the more difficult question of determining its elements, so far unsolved with the present methods): although we can estimate (and find) the number of points of norm $\leq c_3$, for the remaining points of U the 'Intermediate conclusion' only yields a bound in each cone depending on the norm of a *possible* point of U lying in the cone; the higher the norm N of such a point, the higher the bound for the norm (at most $28N$) and the number of the other possible points lying in the same cone. The problem is that we do not know anything about the minimal norm of possible points outside the ball of radius c_3.

To overcome this obstacle, we shall use the gap principle expressed by (5.6), which so far has not been invoked. (The fact that $\|v\| / \|u\|$ is absolutely bounded will be crucial here.)

Let us first subdivide $\mathbb{R}^r$ into finitely many $(1/24)$-cones. For every such cone $\mathcal{D}$, let us consider the points in $U \cap \mathcal{D}$ which have norm $> c_3$, and let us pick such a point $u \in U \cap \mathcal{D}$ of minimal norm (if there are any such points). By what we have proved any other point in the cone $\mathcal{D}$ will have norm $\leq 28 \|u\|$.

Now let us fix a number $\lambda > 1$ and let us subdivide such cone $\mathcal{D}$ in *slices* $\{v : \lambda^h \|u\| \leqslant \|v\| < \lambda^{h+1} \|u\|\}$. Let v_1, v_2 be two distinct points in the same slice, with $h \geq 0$, and assume $\|v_1\| \leqslant \|v_2\|$; by definition, we also have $\|v_2\| \leqslant \lambda \|v_1\|$. Now we use inequality (5.6) as follows:

$$\|v_1\| \leqslant c_1 + 2 \Big\| \|v_2\| \psi(v_2) - \|v_1\| \psi(v_1) \Big\|$$

$$\leqslant c_1 + 2\Big(\|v_2\| \big\| \psi(v_2) - \psi(v_1) \big\| + \big\| (\|v_2\| - \|v_1\|) \psi(v_1) \big\| \Big)$$

$$\leqslant c_1 + 2\Big(\|v_2\| (1/12) + (\|v_2\| - \|v_1\|) \Big)$$

$$\leqslant c_1 + 2 \|v_1\| \Big((\lambda/12) + (\lambda - 1) \Big) = c_1 + 2 \|v_1\| (\lambda(13/12) - 1).$$

---

[2] This follows from (3.12) and is implicitly a consequence of Northcott Theorem; note that (3.12) however yields estimates depending on generators for Γ, not only on r. As we shall explicitly soon recall, a uniform estimate follows from Theorem 4.18, actually for the set $U' \subset \mathbb{Q}^r$ corresponding to solutions in the larger group Γ'.

Choosing for example $\lambda = 12/11$, we have $2((13\lambda/12)-1) = 4/11 < 1$, and so we see that $\|v_1\|$ is bounded by the constant $c_4 := 11c_1/7$.

Now, set $c_5 := \max(c_3, c_4)$. By what we have just shown (with the above choices $\epsilon = 1/12$, $\lambda = 12/11$), a slice of a $(1/24)$-cone $\mathcal{D}$ can contain at most one point in U with norm $> c_5$. The number of slices with $h \geq 0$ for each cone is $\leq 1 + \log 28/\log(12/11) \leq 60$, so we get a total of ≤ 60 points in U with norm $> c_5$, for each $(1/24)$-cone $\mathcal{D}$.

To estimate the number of $(1/24)$-cones necessary to cover $\mathbb{R}^r$, we use the method already seen in the proof of Theorem 4.18. We take a maximal set of disjoint $(1/48)$-cones. Every such cone contains a ball of radius $1/48$, centered at some point of the unit ball B_1. This translated ball is contained in the ball of radius $49/48$ centered at the origin. Since these translated balls are pairwise disjoint, on considering volumes we see that the number of $(1/48)$-cones will be at most 49^r. Take now any point z with $\|z\| = 1$; its $(1/48)$-cone must intersect some of the said $(1/48)$-cones, so z will have distance $\leq 1/24$ from the center of one of the $(1/48)$-cones. This means that the $(1/24)$-cones with the same centers as the mentioned $(1/48)$-cones cover $\mathbb{R}^r$; in other words, we can cover $\mathbb{R}^r$ with at most 49^r $(1/24)$-cones.

Taking into account the above estimates, we find at most $60 \cdot 49^r$ points in U with norm $> c_5$.

On the other hand, by Theorem 4.18, the number of points of norm $\leq c_5$ is $\leq c_6 c_7^r$ for suitable absolute constants c_6, c_7. Hence we obtain an estimate $c_8 c_9^r$ for the number of points in U.

We have not quite concluded, because we have counted merely the solutions $(x, y) \in \Gamma$, whereas we want to count the solutions in the larger group Γ'. However, Γ' may be expressed as the union of the groups $\Gamma_m := \{P \in \mathbb{G}_m^2(\overline{\mathbb{Q}}) : P^{m!} \in \Gamma\}$, for $m = 1, 2, \ldots$. Each of these groups is finitely generated and of the same rank as Γ; also, we have $\Gamma_m \subset \Gamma_{m+1}$. On the other hand the above estimate depends only on r, so the number of solutions in Γ_m is $\leq c_8 c_9^r$, independently of m. Therefore the same estimate holds for the number of solutions in Γ'. (Alternatively, we could formulate the whole proof directly with the solutions in Γ'.)

This concludes the proof of Theorem 5.1. □

Remark 5.4.

(1) **Effectivity.** Note that, similarly to Thue's method, the above method does not give an algorithm to find actually the solutions. Such an algorithm follows from Baker's theory or from Bombieri's method, often mentioned in the previous chapters. See [7] and [17].

(ii) **Examples with many solutions.** The above estimate is not far from being best-possible: actually, for arbitrarily large r one can construct groups $\Gamma \subset (\mathbb{Q}^*)^2$ of rank r such that the number of solutions of $x + y = 1$ with

$(x, y) \in \Gamma$ is $> \exp(c\frac{\sqrt{r}}{\log r})$, for a certain abolute constant $c > 0$. See [17], Example 5.2.4.

We sketch very briefly the idea (due to Erdös, Stewart, Tijdeman, Zagier). For a positive integer n and a large real x we let T be the set of integers up to x whose prime factors do not exceed $x^{1/n}$. By known prime-number estimates one proves $M := \#T > 2x/(\log x)^n$ for $x > 17^n$. Considering the M^2 sums $a + b$ for $a, b \in T$, one finds a sum σ occurring at least $M^2/2x > 2x/(\log x)^{2n}$ times. Defining Γ as the subgroup of $(\mathbb{Q}^*)^2$ generated by the pairs $(p, 1)$, $(1, p)$ with a prime $p \le x^{1/n}$, together with (σ, σ), the unit equation has $> 2x/(\log x)^{2n}$ solutions in Γ. On the other hand, Γ has rank $\le 2\pi(x^{1/n}) + 1$. With the choice $n \approx \log x/2 \log\log x$ one finds the above mentioned result. See [17] for more details and examples.

5.4. An application

As we have seen in Chapter 3, some significant diophantine equations (like the Thue-Mahler's and the hyperelliptic ones) can be reduced to the S-unit Theorem. In that chapter we have shown some qualitative finiteness deductions; now that we have the sharp quantitative Theorem 5.1, we can correspondingly obtain quantitative versions of the said applications. For brevity we shall treat only the Thue-Mahler equation, referring to [17] for the hyperelliptic equation.

We let k be a number field, S be a finite set of places of k, containing the archimedean ones and we set as usual $\mathcal{O}_S := \mathcal{O}_{k,S}$. Also, we let C_1, C_2 be the absolute constants appearing in Theorem 5.1. We have:

Theorem 5.9. *Let $f(X, Y) \in \mathcal{O}_S[X, Y]$ be a homogeneous polynomial with at least three linear, pairwise non-proportional factors over k. Up to proportionality, there are at most $C_1 C_2^{2\#S-1}$ pairs $(x, y) \in \mathcal{O}_S$ such that $f(x, y) \in \mathcal{O}_S^*$.*

Proof of Theorem 5.9. Following [17], we start with a convenient (though not indispensable) coordinate change. We may assume that there is at least one point $(x_0, y_0) \in \mathcal{O}_S^2$ with $f(x_0, y_0) =: a \in \mathcal{O}_S^*$. Since f is homogeneous we may write $a = f(x_0, y_0) = Ax_0 + By_0$ for certain $A, B \in \mathcal{O}_S$. Then the substitution $X \mapsto x_0 X - BY, Y \mapsto y_0 X + AY$ is invertible over $\mathcal{O}_S$ (its determinant is $a \in \mathcal{O}_S^*$) and transforms $f(X, Y)$ into $f(x_0 X - BY, y_0 X + AY) \in \mathcal{O}_S[X, Y]$. This new polynomial has also at least three linear simple factors over k, so in the proof we may replace f by it, and assume at once that $f(1, 0) = a \in \mathcal{O}_S^*$.

This implies that the roots of $f(X, 1) = 0$ are integral over $\mathcal{O}_S$; since three distinct of them lie in k by assumption, they in fact lie in $\mathcal{O}_S$, so we may write

$$f(X, Y) = a(X - \xi_1 Y)(X - \xi_2 Y)(X - \xi_3 Y)g(X, Y),$$

where $\xi_1, \xi_2, \xi_3 \in \mathcal{O}_S$ are pairwise distinct and where $g(X, Y) \in \mathcal{O}_S[X, Y]$.

Let now $(x, y) \in \mathcal{O}_S^2$ be such that $f(x, y) \in \mathcal{O}_S^*$. The displayed shape for $f(X, Y)$ shows that all the factors on the right, evaluated at (x, y), lie in $\mathcal{O}_S$, whence in fact they must lie in $\mathcal{O}_S^*$, so in particular $\mu_i = \mu_i(x, y) :=$ $x - \xi_i y \in \mathcal{O}_S^*$, for $i = 1, 2, 3$. The three equations defining μ_1, μ_2, μ_3 also show that the determinant of the matrix with rows $(\mu_i, 1, \xi_i)$, $i = 1, 2, 3$, vanishes, which yields

$$\frac{\mu_1(\xi_3 - \xi_2)}{\mu_2(\xi_3 - \xi_1)} + \frac{\mu_3(\xi_2 - \xi_1)}{\mu_2(\xi_3 - \xi_1)} = 1. \tag{5.8}$$

Let now Γ be the subgroup of $\mathbb{G}_m^2(k)$ generated by $(\mathcal{O}_S^*)^2$ and by the point $\left(\frac{(\xi_3 - \xi_2)}{(\xi_3 - \xi_1)}, \frac{(\xi_2 - \xi_1)}{(\xi_3 - \xi_1)}\right)$. By (one half of) Dirichlet's often mentioned theorem, $\mathcal{O}_S^*$ has rank $\leq \#S - 1$ (see also Exercise 3.55 for a proof), hence Γ has rank $\leq 2\#S - 1$. Equation (5.8) delivers a solution of $u + v = 1$ with $(u, v) \in \Gamma$. By Theorem 5.1 this equation has at most $N := C_1 C_2^{2\#S-1}$ solutions in Γ, so we deduce that the pair $\left(\frac{\mu_1}{\mu_2}, \frac{\mu_3}{\mu_2}\right)$ has also at most N possibilities. Suppose now that (x, y), (x', y') lead to the same pair, i.e. that

$$\frac{x - \xi_1 y}{x - \xi_2 y} = \frac{x' - \xi_1 y'}{x' - \xi_2 y'}, \qquad \frac{x - \xi_3 y}{x - \xi_2 y} = \frac{x' - \xi_3 y'}{x' - \xi_2 y'}.$$

Since the map $(X : Y) \mapsto (X - \xi_1 Y : X - \xi_2 Y)$ is an automorphism of $\mathbb{P}_1$, the first of the displayed equations suffices to yield $(x : y) = (x' : y')$, which means that the solutions (x, y) and (x', y') are proportional. Therefore the sought number of solutions, up to proportionality, is at most N, which proves the theorem. □

Corollary 5.10. *Let $f(X, Y) \in \mathcal{O}_S[X, Y]$ be a homogeneous polynomial of degree d, with at least three distinct roots in $\mathbb{P}_1(\overline{\mathbb{Q}})$. Up to proportionality, there are at most $C_1 C_2^{2d(d-1)(d-2)\#S-1}$ pairs $(x, y) \in \mathcal{O}_S$ such that $f(x, y) \in \mathcal{O}_S^*$.*

Proof. By a linear transformation, we may assume that $f(X, 1)$ has at least three distinct roots ξ_1, ξ_2, ξ_3 in $\overline{\mathbb{Q}}$. Let $K := k(\xi_1, \xi_2, \xi_3)$, so $[K : k] \leq d(d-1)(d-2)$. We apply Theorem 5.9 to this situation, with K in place of k and S' in place of S, where S' is the set of places of K above S. We just note that $\#S' \leq [K : k]\#S$, and the sought estimate follows. □

Corollary 5.11. *Let $f \in \mathbb{Z}[X, Y]$ be homogeneous of degree d, with at least three distinct roots in $\mathbb{P}_1(\overline{\mathbb{Q}})$, and let c be a nonzero integer having at most v distinct prime factors. Then the equation $f(x, y) = c$ has at most $2C_1 C_2^{2(v+1)d(d-1)(d-2)-1}$ solutions in integer pairs $(x, y) \in \mathbb{Z}^2$.*

Proof. Let us apply the previous corollary, with $k = \mathbb{Q}$ and S consisting of the infinite place together with the places corresponding to primes dividing c. Then $\#S \leq v + 1$, so we obtain that the number of non-proportional pairs $(x, y) \in \mathbb{Z}_S$ with $f(x, y) \in \mathbb{Z}_S^*$ is at most $C_1 C_2^{2(v+1)d(d-1)(d-2)-1}$. If $f(u, v) = c$ for integers u, v, then (u, v) is one of the pairs in question, because $c \in \mathbb{Z}_S^*$. If $f(u', v') = c$ for integers u', v' and if (u, v) is proportional to (u', v') then $u = lu', v = lv'$ for some nonzero $l \in \mathbb{Q}$, so $c = f(u, v) = l^d f(u', v') = l^d c$, whence $l^d = 1$ and $l = \pm 1$. Hence distinct integer solutions to $f(X, Y) = c$ give rise to non-proportional pairs, with the exception of solutions $(u, v), (-u, -v)$. In conclusion, the number of integer solutions is at most twice the number of non-proportional pairs, which proves the stated conclusion.   □

Remark 5.5. This last corollary shows in particular that, for fixed c, d, remarkably, the number of solutions is bounded independently of the coefficients of f. (This was first proved by Bombieri and Schmidt.) On the other hand, it is not known how far this estimate is from the truth. In this direction, see Exercise 5.5 below and the discussions in [17].

Exercise 5.4. Let $R(X) \in \mathbb{Q}(X)$ be a rational function with at least three distinct poles in $\mathbb{P}_1(\overline{\mathbb{Q}})$. Prove a bound for the number of rationals $\rho \in \mathbb{Q}$ such that $R(\rho) \in \mathbb{Z}$, depending on $\deg(R)$ and the number of primes of 'bad reduction' (*i.e.* the primes p such that the reduction of R modulo p has lower degree). (Hint: Express $R(m/n) = A(m, n)/B(m, n)$ where A, B are primitive coprime polynomials over $\mathbb{Z}$. Then note that if $R(m/n) \in \mathbb{Z}$ the prime factors of $B(m, n)$ are of bad reduction. Now apply Corollary 5.11.)

Exercise 5.5. (i) For any integer $d \geq 1$, produce a Thue Equation over $\mathbb{Z}$ of degree d and $\geq d$ distinct integer solutions.

(ii) Prove that for any $K > 0$ there exist integers $a, b, c \neq 0$ such that the Thue Equation $aX^3 + bY^3 = c$ has at least K distinct integer solutions. (Hint: Find first integers $a, b, t, t \neq 0$ such that the elliptic curve $aX^3 + bY^3 = t$ has positive rank, hence infinitely many rational solutions. Let then $N > 0$ be a common denominator for K of these solutions. Finally, put $c := N^3 t$. This appears in Silverman's book [83].)

Exercise 5.6. Let ξ be an algebraic number of degree $d \geq 3$ and let $K > 0$. Prove that there is a number $c_1(\xi)$ and an absolute constant c_2 such that the number of rationals p/q (p, q coprime integers) with $|(p/q) - \xi| < Kq^{-d}$ is at most $c_1(\xi)K^{c_2 d^3}$. (Hint: note that any approximation of the stated type gives rise to a suitable Thue Equation.)

Notes to Chapter 5

In the whole approach, we have followed [17], with a few variations.

Previously to Beukers and Schlickewei, weaker forms of Theorem 5.1 (however still with estimates depending only on r) were given by Schlickewei and Schmidt. Beukers and Schlickewei give the values $C_1 = C_2 =$

256. They had not the full Zhang's Theorem at disposal, and for the counting of small solutions used *ad hoc* arguments similar to Zagier's one for the height of algebraic solutions of $x + y = 1$.

Generalizing the result in Exercise 5.1, uniform bounds for the number of solutions of $r_n = c$, for a recurrence r_n of order d, were given by Schmidt; solving a longstanding conjecture, he proved in [72] an estimate depending only on d, with natural necessary assumptions on r_n.

The Padé approximations to special functions had been used already at least by Thue and Siegel, and more recently by Baker [6], who produced some effective results.

An upper bound for the integer solutions of a Thue Equation $f(X,Y) = c$, dependent only on the degree of f and on the number of prime factors of c was given by Bombieri and Schmidt, prior to the results on the S-unit equation. See for instance [70] for an exposition.

References

[1] F. AMOROSO and S. DAVID, Le problème de Lehmer en dimension supérieure. [The Lehmer problem in higher dimension]. *J. Reine Angew. Math.*, 513:145–179, 1999.

[2] F. AMOROSO and R. DVORNICICH, A lower bound for the height in abelian extensions. *J. Number Theory*, 80(2):260–272, 2000.

[3] F. AMOROSO and U. ZANNIER, A relative Dobrowolski lower bound over abelian extensions. *Ann. Scuola Norm. Sup. Pisa Cl. Sci. (4)*, 29(3):711–727, 2000.

[4] E. ARTIN and G. WHAPLES, Axiomatic characterization of fields by the product formula for valuations. *Bull. Amer. Math. Soc.*, 51:469–492, 1945.

[5] R. M. AVANZI and U. ZANNIER, Genus one curves defined by separated variable polynomials and a polynomial Pell equation. *Acta Arith.*, 99(3):227–256, 2001.

[6] A. BAKER, Rational approximations to $\sqrt[3]{2}$ and other algebraic numbers. *Quart. J. Math. Oxford Ser. (2)*, 15:375–383, 1964.

[7] A. BAKER, *Transcendental Number Theory*. Cambridge Mathematical Library. Cambridge University Press, second edition, 1990.

[8] F. BEUKERS and C. J. SMYTH, Cyclotomic points on curves. In *Number theory for the millennium, I*, pages 67–85. A K Peters, Ltd., 2002. Urbana, 2000.

[9] F. BEUKERS and D. ZAGIER, Lower bounds of heights on hypersurfaces. *Acta Arith.*, pages 103–111, 1997.

[10] Y. F. BILU, Limit distribution of small points on algebraic tori. *Duke Math. J.*, 89: 465–476, 1997.

[11] Y. F. BILU and D. W. MASSER, A quick proof of Sprindzhuk's decomposition theorem. In *More sets, graphs and numbers*, volume 15 of *Bolyai Soc. Math. Stud.*, pages 25–32. Springer, 2006.

[12] E. BOMBIERI, On the Thue-Siegel-Dyson theorem. *Acta Math.*, 148:255–296, 1982.

[13] E. BOMBIERI, On Weil's "théorème de décomposition". *Amer. J. Math.*, 105(2):295–308, 1983.

[14] E. BOMBIERI, The Mordell conjecture revisited. *Ann. Scuola Norm. Sup. Pisa Cl. Sci. (4)*, 18(3):615–640, 1990.

[15] E. BOMBIERI, Effective Diophantine approximation on $_m$. *Ann. Scuola Norm. Sup. Pisa Cl. Sci. (4)*, 20(1):61–89, 1993.

[16] E. BOMBIERI and P. B. COHEN, An elementary approach to effective Diophantine approximation on $_m$. In *Number theory and algebraic geometry*, number 303 in London Math. Soc. Lecture Note Ser., pages 41–62. Cambridge University Press, 2003.

[17] E. BOMBIERI and W. GUBLER, *Heights in Diophantine geometry*, volume 4 of *New Mathematical Monographs*. Cambridge University Press, 2006.

[18] E. BOMBIERI, D. W. MASSER and U. ZANNIER, Intersecting a curve with algebraic subgroups of multiplicative groups. *Internat. Math. Res. Notices*, 20:1119–1140, 1999.

[19] E. BOMBIERI, D. W. MASSER and U. ZANNIER, Finiteness results for multiplicatively dependent points on complex curves. *Michigan Math. J.*, 51(3):451–466, 2003.

[20] E. BOMBIERI, J. MUELLER and U. ZANNIER, Equations in one variable over function fields. *Acta Arith.*, 99(1):27–39, 2001.

[21] E. BOMBIERI and U. ZANNIER, Algebraic points on subvarieties of ^n_m. *Internat. Math. Res. Notices*, 7:333–347, 1995.

[22] E. BOMBIERI and U. ZANNIER, A note on heights in certain infinite extensions of . *Atti Accad. Naz. Lincei Cl. Sci. Fis. Mat. Natur. Rend. Lincei (9) Mat. Appl.*, 12:5–14, 2001.

[23] Z. I. BOREVITCH and I. R. CHAFAREVITCH, *Théorie des Nombres*. Gauthier-Villars, 1993. Reprint of the 1967 French translation.

[24] W. D. BROWNAWELL and D. W. MASSER, Vanishing sums in function fields. *Math. Proc. Cambridge Philos. Soc.*, 100(3):427–434, 1986.

[25] J. W. S. CASSELS, *An introduction to Diophantine approximation*, volume 45 of *Cambridge Tracts in Mathematics and Mathematical Physics*. Cambridge University Press, 1957.

[26] J. W. S. CASSELS, *Rational Quadratic Forms*. Academic Press, 1978.

[27] J. COATES, Construction of rational functions on a curve. *Proc. Cambridge Philos. Soc.*, 68:105–123, 1970.

[28] P. CORVAJA and U. ZANNIER, Arithmetic on infinite extensions of function fields. *Boll. Un. Mat. Ital. B (7)*, 11(4):1021–1038, 1997.

[29] P. CORVAJA and U. ZANNIER, Diophantine equations with power sums and universal Hilbert sets. *Indag. Math. (N.S.)*, 9(3):317–332, 1998.

[30] P. CORVAJA and U. ZANNIER, Some new applications of the subspace theorem. *Compositio Math.*, 131(3):319–340, 2002.

[31] P. CORVAJA and U. ZANNIER, On the length of the continued fraction for values of quotients of power sums. *J. Théor. Nombres Bordeaux*, 17(3):737–748, 2005.

[32] P. CORVAJA and U. ZANNIER, S-unit points on analytic hypersurfaces. *Ann. Sci. École Norm. Sup. (4)*, 38(1):76–92, 2005.

[33] P. CORVAJA and U. ZANNIER, *On the maximal order of a torsion point on a curve in* $\mathbb{G}_m^n$, Rend. Lincei Mat. Appl. 19 (2008), 73–78.

[34] R. CRANDALL and C. POMERANCE, *Prime numbers. A computational perspective*. Springer-Verlag, second edition, 2005.

[35] H. DAVENPORT, *Multiplicative Number Theory*, volume 74 of *Graduate Texts in Mathematics*. Springer, third edition, 2000.

[36] H. DERKSEN, A Skolem-Mahler-Lech theorem in positive characteristic and finite automata. *Invent. Math.*, 168(1):175–224, 2007.

[37] L. E. DICKSON, *History of the theory of numbers*. Chelsea Publishing Co., 1966.

[38] L. E. DICKSON, *Introduction to the Theory of Numbers*, The University of Chicago Press, 1929, 183 pp.

[39] R. DVORNICICH and U. ZANNIER, On sums of roots of unity. *Monatsh. Math.*, 129(2):97–108, 2000.

[40] B. DWORK, G. GEROTTO and F. J. SULLIVAN, *An introduction to G-functions*, volume 133 of *Ann. of Math. Studies*. Princeton University Press, 1994.

[41] B. EDIXOVEN and J.-H. EVERTSE (eds.), *Diophantine Approximation and Abelian Varieties*, volume 1566 of *Lecture Notes in Math*. Springer, 1993. Proc. conf. Soesterberg, Netherlands, 1992.

[42] H. M. EDWARDS *Fermat's last Theorem. A genetic introduction to algebraic number theory*, volume 50 of *Graduate Texts in Mathematics*. Springer-Verlag, 1996. Corrected reprint of the 1977 original.

[43] H. ESNAULT and E. VIEHWEG, Dyson's lemma for polynomials in several variables (and the theorem of Roth). *Invent. Math.*, 78(3):445–490, 1984.

[44] G. EVEREST, A. VAN DER POORTEN, I. SHPARLINSKI and T. WARD, *Recurrence sequences*. Number 104 in Mathematical Surveys and Monographs. American Mathematical Society, 2003.

[45] J.-H. EVERTSE, The number of solutions of linear equations in roots of unity. *Acta Arith.*, 89(1):45–51, 1999.

[46] G. FALTINGS, Diophantine approximation on abelian varieties. *Ann. of Math. (2)*, 133(3):549–576, 1991.

[47] A. O. GELFOND, *Transcendental and algebraic numbers*. Dover Publications Inc., 1960.

[48] P. HABEGGER, *Heights and multiplicative relations on algebraic varieties*. PhD thesis, Basel, 2007.

[49] G. H. HARDY and E. M. WRIGHT, *An introduction to the theory of numbers*. Oxford University Press, 1979.

[50] M. HINDRY and J. H. SILVERMAN, *Diophantine geometry. An introduction*. Number 201 in Graduate Texts in Mathematics. Springer-Verlag, 2000.

[51] S. KAWAGUCHI and J. H. SILVERMAN, Dynamics of projective morphisms having identical canonical height. *Proc. London Math. Soc.*, 95(3):519–544, 2007.

[52] S. LANG, Integral points on curves. *Inst. Hautes Ètudes Sci. Publ. Math.*, 6:27–43, 1960.

[53] S. LANG, *Fundamentals of Diophantine geometry*. Springer-Verlag, 1983.

[54] S. LANG, *Algebraic number theory*, volume 110 of *Graduate Texts in Mathematics*. Springer-Verlag, second edition, 1994.

[55] T. LOHER and D. W. MASSER, Uniformly counting points of bounded height. *Acta Arith.*, 111(3):277–297, 2004.

[56] R. C. MASON, *Diophantine equations over function fields*, volume 96 of *London Mathematical Society Lecture Note Series*. Cambridge University Press, 1984.

[57] D. W. MASSER, Lecture notes. (Manuscript).

[58] D. W. MASSER, Heights, transcendence, and linear independence on commutative group varieties. In F. Amoroso and U. Zannier, editors, *Diophantine approximation*, volume 1819 of *Lecture Notes in Math.*, pages 1–51. Springer, 2003. (Cetraro, 2000).

[59] L. J. MORDELL, *Diophantine Equations*, volume 30 of *Pure and Applied Mathematics*. Academic Press, 1969.

[60] T. NAGELL, L'analyse indéterminée de degré supérieur. In P. Ribenboim, editor, *Collected papers of Trygve Nagell*, volume 121 of *Queen's Papers in Pure and Applied Mathematics*. Queen's University, Kingston, 2002. Originally published in *Mémorial des sciences mathématiques*, Gauthier-Villars, 1929.

[61] W. NARKIEWICZ, *Elementary and Analytic Theory of Algebraic Numbers*. Springer Monographs in Mathematics. Springer-Verlag, third edition, 2004.

[62] C. PETSCHE, A quantitative version of Bilu's equidistribution theorem. *Int. J. Number Theory*, 1(2):281–291, 2005.

[63] M. RAYNAUD, Courbes sur une variété abélienne et points de torsion. [Curves on an abelian variety and torsion points]. *Invent. Math.*, 71(1):207–233, 1983.

[64] M. ROSEN, *Number theory in function fields*, volume 210 of *Graduate Texts in Mathematics*. Springer-Verlag, 2002.

[65] A. SCHINZEL, An improvement of Runge's theorem on Diophantine equations. *Comment. Pontificia Acad. Sci.*, II(20):1–9, 1969.

[66] A. SCHINZEL, On the product of the conjugates outside the unit circle of an algebraic number. *Acta Arith.*, 24:385–399, 1973.

[67] A. SCHINZEL, *Polynomials with special regard to reducibility*, volume 77 of *Encyclopedia of Mathematics and its Applications*. Cambridge University Press, 2000. With an appendix by Zannier, U.

[68] H. P. SCHLICKEWEI, Equations in roots of unity. *Acta Arith.*, 76(2):99–108, 1996.

[69] H. P. SCHLICKEWEI and E. WIRSING, Linear bounds for the heights of solutions of linear equations. *Invent. Math.*, 129:1–10, 1997.

[70] W. M. SCHMIDT, *Diophantine approximation*, volume 785 of *Lecture Notes in Math.* Springer, 1980.

[71] W. M. SCHMIDT, Heights of points on subvarieties of $\mathbb{G}_m^n$. In *Number theory*, volume 235 of *London Math. Soc. Lecture Note Ser.*, pages 157–187. Cambridge Univ. Press, 1996.

[72] W. M. SCHMIDT, The zero multiplicity of linear recurrence sequences. *Acta Math.*, 182(2):243–282, 1999.

[73] L. SCHNEPS, Dessins d'enfants on the Riemann sphere. In *The Grothendieck theory of dessins d'enfants*, volume 200 of *London Mathematical Society Lecture Note Series*, pages 47–77. Cambridge University. Press, 1994. Luminy, 1993.

[74] J.-P. SERRE, *Cours d'arithmétique*. Presses Universitaires de France, 1970.

[75] J.-P. SERRE, *Lie Algebras and Lie Groups*. Springer-Verlag LNM 1500, 1992.

[76] J.-P. SERRE, *Topics in Galois theory*, volume 1 of *Research Notes in Mathematics*. Jones and Bartlett Publishers, 1992.

[77] J.-P. SERRE, *Lectures on the Mordell-Weil theorem*. Aspects of Mathematics. Friedr. Vieweg & Sohn, third edition, 1997.

[78] C. L. SIEGEL, *Transcendental Numbers*. Number 16 in Ann. of Math. Studies. Princeton University Press, 1949.

[79] C. L. SIEGEL, *Gesammelte Abhandlungen. Bände I, II, III.* Springer-Verlag, 1966.

[80] C. L. SIEGEL, *Lectures on Quadratic forms*, volume 7. Tata institute of Fundamental Research, 1967.

[81] C. L. SIEGEL, Einige Erläuterungen zu Thues Untersuchungen über Annäherungswerte algebraischer Zahlen und diophantische Gleichungen. *Nachr. Akad. Wiss. Göttingen Math.-Phys. Kl. II*, pages 169–195, 1970.

[82] C. L. SIEGEL, *Avdanced Analytic Number Theory*, volume 9. Tata Institute of Fundamental Research, second edition, 1980.

[83] J. H. SILVERMAN, *The Arithmetic of Elliptic Curves*, volume 106 of *Graduate Texts in Mathematics*. Springer-Verlag, 1992. Corrected reprint of the 1986 original.

[84] J. H. SILVERMAN, On the distibution of integer points on curves of genus zero. *Theoret. Comput. Sci.*, 235(1):163–170, 2000.

[85] C. J. SMYTH, Mahler measure of one-variable polynomials: a survey. In J. McKee and C. J. Smyth, editors, *Number theory and polynomials*, LMS Lecture notes, (preprint). Conference proceedings, University of Bristol, 3-7 April 2006.

[86] W. W. STOTHERS, Polynomial identities and Hauptmoduln. *The Quarterly Journal of Mathematics. Oxford. Second Series*, 32(127):349–370, 1981.

[87] G. TROI and U. ZANNIER, Note on the density constant in the distribution of self-numbers. II. *Boll. Unione Mat. Ital. Sez. B Artic. Ric. Mat. (8)*, 2(2):397–399, 1999.

[88] E. ULLMO, Positivité et discrétion des points algébriques des courbes. *Ann. of Math. (2)*, 147(1):167–179, 1998.

[89] C. VIOLA, On Dyson's lemma. *Ann. Scuola Norm. Sup. Pisa Cl. Sci. (4)*, 12(1):105–135, 1985.

[90] P. VOJTA, *Diophantine approximations and value distribution theory*, volume 1239 of *Lecture Notes in Math*. Springer-Verlag, 1987.

[91] H. VÖLKLEIN, *Groups as Galois groups*, volume 53 of *Cambridge Studies in Advanced Mathematics*. Cambridge University Press, 1996.

[92] J. T.-Y. WANG, An effective Roth's theorem for function fields. *Rocky Mountain J. Math.*, 26(3):1225–1234, 1996. Symposium on Diophantine Problems (Boulder, CO, 1994).

[93] A. WEIL, *Number theory. An approach through history. From Hammurapi to Legendre*. Birkhäuser Boston, Inc., 1984.

[94] K. YU, p-adic logarithmic forms and group varieties. II. *Acta Arith.*, 89(4):337–378, 1999.

[95] D. ZAGIER, Algebraic numbers close to both 0 and 1. *Math. Comp.*, 61(203):485–491, 1993.

[96] U. ZANNIER, Some remarks on the s-unit equation in function fields. *Acta Arith.*, 64(1):87–98, 1993.

[97] U. ZANNIER, An effective solution of a certain Diophantine problem. *Rend. Sem. Mat. Univ. Padova*, 93:177–183, 1995.

[98] U. ZANNIER, On Davenport's bound for the degree of $f^3 - g^2$ and Riemann's existence theorem. *Acta Arith.*, 71(2):107–137, 1995.

[99] U. ZANNIER, Fields containing values of algebraic functions and related questions. In *Number theory (Paris, 1993–1994)*, volume 235 of *London Mathematical Society Lecture Note Series*, pages 199–213. Cambridge University Press, 1996.

[100] U. ZANNIER, Polynomials modulo p whose values are squares (elementary improvements on some consequences of Weil's bounds). *Enseign. Math. (2)*, 44(1-2):95–102, 1998.

[101] U. ZANNIER, *Some applications of diophantine approximation to diophantine equations*. Forum, Udine, 2003.

[102] U. ZANNIER, On the integer solutions of exponential equations in function fields. *Ann. Inst. Fourier (Grenoble)*, 54(4):849–874, 2004.

[103] U. ZANNIER, Diophantine equations with linear recurrences. an overview of some recent progress. *J. Théor. Nombres Bordeaux*, 17(1):423–435, 2005.

[104] U. ZANNIER (ed.), *Diophantine Geometry*, volume 4 of *Publications of the Scuola Normale Superiore*. Birkäuser, 2007. CRM Series.

[105] U. ZANNIER, Roth theorem, integral points and certain ramified covers of $\mathbb{P}_1$. In *Analytic Number Theory - Essays in Honour of Klaus Roth*, Cambridge University Press, 2008.

[106] S. ZHANG, Positive line bundles on arithmetic surfaces. *Ann. of Math. (2)*, 136(3):569–587, 1992.

[107] S. ZHANG, Positive line bundles on arithmetic varieties. *J. Amer. Math. Soc.*, 8(1):187–221, 1995.

Index

Appendix A
Lower bounds for the height
(by Francesco Amoroso)

A.1. Introduction

The former Manin-Mumford conjecture predicted that the set of torsion points of a curve of genus ≥ 2 embedded in its Jacobian is finite. More generally, let $\mathbb{G}$ be a semi-Abelian variety and let V be an irreducible[1] algebraic subvariety of $\mathbb{G}$, defined over some algebraically closed field K. We say that V is a *torsion variety* if V is a translate of a proper subtorus by a torsion point of $\mathbb{G}$. We also denote by V_{tors} the set of torsion points of $\mathbb{G}$ lying on V. Then we have the following generalization of the Manin-Mumford conjecture.

Theorem A.1.
  i) *If V is not a torsion variety, then the set V_{tors} of torsion points of $\mathbb{G}$ lying on V is not Zariski dense.*
  ii) *The Zariski closure of V_{tors} is a finite union of torsion varieties.*

The two assertions are clearly equivalent. Theorem A.1 was proved by Raynaud [31] when $\mathbb{G}$ is an Abelian variety, by Laurent [26] if $\mathbb{G} = \mathbb{G}_m^n$, and finally by Hindry [24] in the general situation.

 We assume from now on that all varieties are algebraic and defined over $\overline{\mathbb{Q}}$. Bogomolov [13] gave the following generalization of the former Manin-Mumford conjecture. Let C be a curve of genus ≥ 2 embedded in its Jacobian. Then $C(\overline{\mathbb{Q}})$ is discrete for the metric induced by the Néron-Tate height. In other words, Bogomolov conjecture that the set of points of "sufficiently small" height on C is finite, while the former Manin-Mumford conjecture makes a similar assertion on the set of torsion points (which are precisely the points of height zero).
More generally, let $\mathbb{G}$ be a semi-Abelian variety and let $\hat{h}$ be a normalized height on $\mathbb{G}(\overline{\mathbb{Q}})$. Hence, $\hat{h}$ is the Neron-Tate height if $\mathbb{G}$ is Abelian, and

---

[1] By irreducible we mean geometrically irreducible.

it is the Weil height if $\mathbb{G} = \mathbb{G}_m^n \hookrightarrow \mathbb{P}^n$. In particular, $\hat{h}$ is a non-negative function on $\mathbb{G}$, and $\hat{h}(P) = 0$ if and only if P is a torsion point. Given an algebraic subvariety of $\mathbb{G}$, we denote by V^* the complement in V of the Zariski closure of the set of torsion points of V. Therefore, by theorem A.1, $V \setminus V^* = \overline{V_{\text{tors}}}$ is a finite union of torsion varieties.

Theorem A.2. *Let V be an irreducible subvariety of a semi-Abelian variety $\mathbb{G}$. Then:*

i) *If V is not a torsion variety, then there exists $\theta > 0$ such that the set $V(\theta) = \{P \in V \text{ s.t. } \hat{h}(P) \leq \theta\}$ is not Zariski dense in V.*
ii) *V^* is discrete for the metric induced by $\hat{h}$, i.e.*

$$\inf\{\hat{h}(P) \text{ s.t. } P \in V^*\} > 0.$$

It is easy to see that the two assertions are equivalent. In this formulation, Theorem A.2 was proved for $\mathbb{G} = \mathbb{G}_m^n$ by Zhang (see [37]). In the Abelian case, Ullmo (see [35]) proved Bogomolov's original formulation for curves ($\dim(V) = 1$); immediately after Zhang (see [38]) proved Theorem A.2. The semi-Abelian case was solved by David and Philippon (see [21]).

In this appendix we shall describe some quantitative versions of Theorem A.2 for a torus $\mathbb{G} = \mathbb{G}_m^n$, and we sketch proofs of theorems which prove these conjectures "up to an ε".

A.2. Algebraic numbers

In this section we first recall some facts from Sections 2.1, 2.2 and 2.3, of Chapter 3.

Let $\alpha \in \overline{\mathbb{Q}}$ and let K be *any* number field containing α. We denote by $\mathcal{M}_K$ the set of places of K. For $v \in K$, let K_v be the completion of K at v and let $| \cdot |_v$ be the (normalized) absolute value of the place v. Hence

$$|\alpha|_v = |\sigma\alpha|,$$

if v is an archimedean place associated to the embedding $\sigma : K \hookrightarrow \overline{\mathbb{Q}}$. If v is a non archimedean place associated with the prime ideal $\wp$ over the rational prime p, we have

$$|\alpha|_v = p^{-\lambda/e},$$

where e is the ramification index of $\wp$ and λ is the exponent of $\wp$ in the factorization of the ideal (α) in the ring of integers of K. Thus $\|\alpha\|_v =$

$|\alpha|_v^{[K_v:Q_v]}$, in the notation of Chapter 3, Section 2.1. Our normalization agrees with the product formula

$$\prod_{v \in \mathcal{M}_K} |\alpha|_v^{[K_v:Q_v]} = 1$$

which holds for any $\alpha \in K^*$.

For further reference, we recall that for any rational place w (thus $w = \infty$ or $w = $ a prime number),

$$\sum_{v|w} [K_v : Q_v] = [K : Q].$$

We define the Weil height of α by

$$h(\alpha) = \frac{1}{[K : Q]} \sum_{v \in \mathcal{M}_K} [K_v : Q_v] \log \max\{|\alpha|_v, 1\}.$$

It is easy to see that this definition does not depend on the field K containing α; it thus defines a function $h \colon \overline{Q} \to \mathbb{R}^+$.

The Weil height of an algebraic number is related to the Mahler measure of a polynomial. Let $P \in \mathbb{C}[x]$ be non-zero; then its Mahler measure is

$$M(P) = \exp \int_0^1 \log |P\left(e^{2\pi i t}\right)| dt.$$

We also agree that $M(0) = 0$. The Mahler measure has some nice properties. It is a multiplicative function, and it is invariant by the morphism $P(x) \to P(x^l)$ ($l \in \mathbb{N}$). Let $\alpha_1, \dots, \alpha_d$ be the roots of P and let P_d be its leading coefficient. By Proposition 2.5, Chapter 3

$$M(P) = |P_d| \prod_{j=1}^d \max\{|\alpha_j|, 1\}. \tag{A.1}$$

Let K be a number field, and let $f \in K[x]$. We define:

$$\hat{h}(f) = \frac{1}{[K : Q]} \sum_{v \in \mathcal{M}_K} [K_v : Q_v] \log M_v(f),$$

where $M_v(f)$ is the maximum of the v-adic absolute values of the coefficients of f if v is non archimedean, and $M_v(f)$ is the Mahler measure of σf if v is an archimedean place associated with the embedding $\sigma \colon K \hookrightarrow \overline{Q}$. As for the Weil height, this definition does not depend on the field K containing the coefficients of f. Moreover, by the product formula, $\hat{h}(\lambda f) = \hat{h}(f)$ for any $\lambda \in K^*$. We also remark that $\hat{h}$ is an additive function. Indeed, $M_v(*)$ is a multiplicative function at least for $v \mid \infty$. By a simple exercise this property still holds for $v \nmid \infty$. By the above properties and by (A.1), $\hat{h}(f)$ is the sum of the Weil height of its roots. As a special case

$$h(\alpha) = \frac{\log M(f)}{[\mathbb{Q}(\alpha) : \mathbb{Q}]}. \tag{A.2}$$

where $f \in \mathbb{Z}[x]$ is the minimal polynomial of α over $\mathbb{Z}$ (*i.e.* f is irreducible in $\mathbb{Z}[x]$, $f(\alpha) = 0$ and its leading coefficient is positive).

Let $\|P\|_1$ be the sum of the absolute values of the coefficients of $P \in \mathbb{C}[x]$ (the "length" of P). Since the maximum of $|P|$ on the unit disk is bounded by $\|P\|_1$, we have $M(P) \le \|P\|_1$. Moreover,

$$\|P\|_1 \le 2^{\deg(P)} M(P). \tag{A.3}$$

This follows from (A.1) and from the usual formulas for the coefficients of a polynomial as symmetric functions of its roots. Inequality (A.3) implies a theorem of Northcott: the set of algebraic numbers of bounded height and degree is finite. If $h(\alpha) \le B$, by the above inequality the coefficients of the minimal polynomial of α are bounded by $2^{[\mathbb{Q}(\alpha):\mathbb{Q}]} B$. Thus the minimal polynomials of the algebraic numbers of bounded height and degree belong to a finite set.

We now state some other important properties of the height. Let α, $\beta \in \overline{\mathbb{Q}}^*$. Then $h(\alpha\beta) \le h(\alpha) + h(\beta)$. This follows from the inequality $\max\{xy, 1\} \le \max\{x, 1\} \max\{y, 1\}$ (for $x, y > 0$) applied at each place. Moreover, if β is a root of unity, $h(\alpha\beta) = h(\alpha)$. Indeed roots of unity have absolute value 1 at each place. Let $\alpha \in \overline{\mathbb{Q}}$ and $n \in \mathbb{Z}$. Then $h(\alpha^n) = |n| h(\alpha)$. If $n \ge 0$, this is obvious from the definition, while, if $n < 0$, this follows from the fact that $h(\alpha^{-1}) = h(\alpha)$, by the product formula.

This last property implies that $h(\alpha) = 0$ if and only if α is a root of unity. This is a theorem of Kronecker, and it is precisely the simplest case of Zhang's theorem on the Bogomolov toric conjecture. The problem of finding sharp lower bounds for the height of a non-zero algebraic number α which is not a root of unity is a famous problem of Lehmer. Let $f \in \mathbb{Z}[x]$ be a nonconstant irreducible polynomial. Assume that $f \ne \pm x$ and that $\pm f$ is not a cyclotomic polynomial. Lehmer (see [27]) asks whether there exists an absolute constant $C > 1$ such that $M(f) \ge C$. An equivalent formulation in terms of the height is the following. Let α be a non-zero algebraic number of degree d which is not a root of unity. Then *Lehmer's conjecture* may be stated as follows: there exists an absolute constant $c > 0$ such that

$$h(\alpha) \geq \frac{c}{d}.$$

This should be the best possible lower bound for the height (without any further assumption on α), since $h(2^{1/d}) = (\log 2)/d$. The best known result in the direction of Lehmer's conjecture is the following theorem.

Theorem A.3 (Dobrowolski, 1979). *For any algebraic number $\alpha \in \overline{\mathbb{Q}}^*$ of degree $d \geq 2$ which is not a root of unity we have*

$$h(\alpha) \geq \frac{c}{d} \left(\frac{\log d}{\log \log d} \right)^{-3}$$

for some absolute constant $c > 0$.

In the original statement [22] $c = 1/1200$; later Voutier [36] shows that one can take $c = 1/4$.

A.2.1. Sketch of the proof of Theorem A.3

We may assume that α is an algebraic integer, otherwise $h(\alpha) \geq (\log 2)/d$. Let f be its minimal polynomial over $\mathbb{Z}$ and let p be a prime number. Then, by Fermat's little theorem,

$$f(x)^p \equiv f(x^p) \bmod p\mathbb{Z}[x].$$

Thus

$$|f(\alpha^p)|_v \leq p^{-1}$$

for any $v \mid p$. Let $F \in \mathbb{Z}[x]$ be a polynomial of degree L vanishing on α with multiplicity $\geq T$ for some *parameters* L and T with $L \geq dT$. Then

$$|F(\alpha^p)|_v \leq p^{-T}$$

for any $v \mid p$. Moreover $|F(\alpha^p)|_v \leq 1$ for $v \nmid \infty$ and

$$|F(\alpha^p)|_v \leq \|F\|_1 \max(1, |\alpha|_v)^{pL}$$

if $v \mid \infty$. Assume that

$$F(\alpha^p) \neq 0. \tag{A.4}$$

Then, by the product formula,

$$0 = \sum_v \frac{[K_v : \mathbb{Q}_v]}{[K : \mathbb{Q}]} \log |F(\alpha^p)|_v$$

$$\leq \sum_{v|p} \frac{[K_v : \mathbb{Q}_v]}{[K : \mathbb{Q}]} \log |F(\alpha^p)|_v + \sum_{v|\infty} \frac{[K_v : \mathbb{Q}_v]}{[K : \mathbb{Q}]} \log |F(\alpha^p)|_v$$

$$\leq -\sum_{v|p} \frac{[K_v : \mathbb{Q}_v]}{[K : \mathbb{Q}]} T \log p + \sum_{v|\infty} \frac{[K_v : \mathbb{Q}_v]}{[K : \mathbb{Q}]} (\log \|F\|_1 + pL \log^+ |\alpha|_v)$$

$$\leq -T \log p + \log \|F\|_1 + pLh(\alpha).$$

This yields

$$h(\alpha) \geq \frac{T \log p - \log \|F\|_1}{pL}. \tag{A.5}$$

We choose $L = d$, $T = 1$ and $F = f$. The non vanishing condition (A.4) is satisfied. Indeed, if α is not a root of unity, then α^p is not a conjugate of α, otherwise $ph(\alpha) = h(\alpha^p) = h(\alpha)$ and α would be a root of unity. Thus we obtain

$$h(\alpha) \geq \frac{\log p - \log \|f\|_1}{pd}.$$

Unfortunately, $\log \|f\|_1$ can be as large as a power of d, even if the height of α is very small (see [1]). Thus, to get a positive lower bound, we must choose p to be exponential in d^c, and the argument terminates with a poor lower bound of the shape $h(\alpha) \geq e^{-d^c}$.

The use of Siegel's Lemma [15], a classical tool in diophantine approximation, improves enormously the quality of this bound. Using this lemma, we find a non-zero polynomial $F \in \mathbb{Z}[x]$ ("auxiliary function") of degree $\leq L$ vanishing on α with multiplicity $\geq T$ as required and such that

$$\log \|F\|_\infty \leq \frac{dT}{L + 1 - dT}(T \log(L + 1) + Lh(\alpha)). \tag{A.6}$$

Here $\|F\|_\infty$ denotes the maximum of the absolute values of the coefficients of F.

The proof now follows the scheme of a classical transcendence proof: choice of the auxiliary function, extrapolation and zero's lemma. During the proof we assume that the height of α is pathologically small and we argue for a contradiction.

Let A and B real functions of d. We write $A \approx B$ if and only if $cB < A < CB$ with $c, C > 0$. Similarly, $A \ll B$ (or $B \gg A$) if and only if $A \leq cB$ whith $c > 0$. We shall also denote by $c_1, \ldots, c_4$ positive constants.

• *Choice of the auxiliary function*

Since $\log \| F \|_1 \leq (L + 1) \log \| F \|_\infty$, by (A.6) we have

$$\log \| F \|_1 \leq \log(L + 1) + \frac{dT}{L + 1 - dT}(T \log(L + 1) + Lh(\alpha)) \,.$$

This inequality cannot give anything better than $\log \| F \|_1 \ll \log(L + 1)$. Therefore, it is reasonable to choose L and T in such a way that

$$\frac{dT^2}{L + 1 - dT} \approx 1,$$

say $L = dT^2$, and to assume that $Lh(\alpha) \ll T \log(L + 1)$. Assume further that $\log \log T \ll \log d$. This implies $\log(L + 1) \approx \log d$. Thus if $h(\alpha) \ll (\log d)/dT$, the lenght of the auxiliary polynomial satisfies $\log \| F \|_1 \ll \log d$.

• *Extrapolation*

We fix a third parameter N. We assume that our primes p satisfy $N/2 \leq p \leq N$. As for T, we suppose that $\log N \ll \log \log d$. We want to show that F vanishes on α^p for all p as before. Assume that for some p we have $F(\alpha^p) \neq 0$. Then, by (A.5),

$$h(\alpha) \gg \frac{T \log \log d - c_1 \log d}{NT^2 d};$$

We choose

$$T = \left\lceil \frac{2c_1 \log d}{\log \log d} \right\rceil$$

Thus

$$h(\alpha) \geq \frac{c_2(\log \log d)^2}{Nd \log d} \tag{A.7}$$

Assume by contradiction that this inequality does not hold[2]. This forces F to vanish on α^p for all $N/2 \leq p \leq N$.

---

[2] In particular this assumption implies $Lh(\alpha) \ll T \log(L + 1)$ as required before.

• *Zero's lemma and conclusion*

Since α is not a root of unity, α^{p_1} and α^{p_2} are not conjugate for primes $p_1 \neq p_2$, since otherwise $p_1 h(\alpha) = p_2 h(\alpha)$ and α would be a root of unity. Assume

$$[\mathbb{Q}(\alpha^n) : \mathbb{Q}] = d \qquad (A.8)$$

for all integer n. Let

$$\Sigma = \{\sigma(\alpha^p), \sigma \in \mathrm{Gal}(\overline{\mathbb{Q}}/\mathbb{Q}), \ p \text{ prime}, \ N/2 \leq p \leq N\} .$$

Then, by the Prime Number Theorem,

$$\#\Sigma = \sum_{N/2 \leq p \leq N} d \geq \frac{c_3 d N}{\log N}.$$

We choose N in such a way that $\frac{c_3 d N}{\log N} > L$, say

$$N = \frac{c_4 (\log d)^2}{\log \log d}.$$

Then

$$L < \#\Sigma \leq \deg F \leq L.$$

This contradiction shows that, at least if α satisfies the additional hypothesis (A.8), the inequality (A.7) holds. Thus

$$h(\alpha) \geq \frac{c_2 (\log \log d)^2}{N d \log d} \gg \frac{1}{d} \left(\frac{\log d}{\log \log d} \right)^{-3}$$

as required.

Dobrowolski's theorem is proved under the additional assumption (A.8). In the general case we proceed by induction on d. Let α be an algebraic number of degree $d \geq 1$ and assume

$$d' h(\beta) \geq \varepsilon(d') = c \left(\frac{\log 5d'}{\log \log 3d'} \right)^{-3}$$

for all algebraic numbers $\beta \in \overline{\mathbb{Q}}^*$ different from a root of unity and with $d' = [\mathbb{Q}(\beta) : \mathbb{Q}] < d$. From the first part of the proof, we can assume that for some $n > 1$ (A.8) does not hold. We follow an argument of [32]. We have $k = [\mathbb{Q}(\alpha) : \mathbb{Q}(\alpha^n)] > 1$. Let β be the norm of α from $\mathbb{Q}(\alpha)$ to $\mathbb{Q}(\alpha^n)$. Then $\beta = \zeta \alpha^k$ for some root of unity ζ and $h(\beta) = h(\alpha^k) = k h(\alpha)$. Since $d' = [\mathbb{Q}(\beta) : \mathbb{Q}] < d$ and since $t \mapsto \varepsilon(t)$ decreases,

$$dh(\alpha) = [\mathbb{Q}(\alpha^n) : \mathbb{Q}] h(\beta) \geq d' h(\beta) \geq c\varepsilon(d') \geq c\varepsilon(d). \qquad \square$$

A.2.2. Height in Abelian extensions

In some special cases, not only Lehmer's conjecture is true, but it can also be sharpened. Assume for instance that L is a totally real number field or a CM field (a totally complex quadratic extension of a totally real number field). Then, as a special case of a more general result, Schinzel proved that

$$h(\alpha) \geq \frac{1}{2} \log \frac{1+\sqrt{5}}{2} = 0.2406...$$

if $\alpha \in L^*$ and $|\alpha| \neq 1$. In particular, by Kronecker's theorem, this inequality holds if α is an algebraic integer different from zero and from a root of unity. It may happen that algebraic numbers of absolute value 1 in CM fields have arbitrary small Weil height. Let for instance $\alpha = (\sqrt{2}-i)/(\sqrt{2}+i)$. Then all the algebraic conjugates of α have absolute value 1. Thus, the same property holds for the algebraic conjugates of $\alpha^{1/d}$, where d is an arbitrary positive integer. In turns, this implies that $\mathbb{Q}(\alpha^{1/d})$ is a CM field. Nevertheless, we have $0 < h(\alpha^{1/d}) = h(\alpha)/d \to 0$ as $d \mapsto \infty$.

When the extension $L/\mathbb{Q}$ is an imaginary Galois extension, L is CM if and only if the complex conjugation lies in the center of the Galois group. Assume further that $L/\mathbb{Q}$ is Abelian. After a question of Zannier, in [8] we prove:

Theorem A.4 (A. – Dvornicich, 2000). *Let $L/\mathbb{Q}$ be an Abelian extension, and let $\alpha \in L^*$, α not a root of unity. Then*

$$h(\alpha) \geq \frac{\log 5}{12} = 0.1341...$$

The above lower bound is not far from the best possible one. Let L be the 21-th cyclotomic field. We recall that L is one of the 29 cyclotomic fields with class number one. The prime 7 splits as $(P\overline{P})^6$ in the ring of integer of L and P is a prime ideal of norm 7. Let γ be a generator of P and define $\alpha = \gamma/\overline{\gamma}$. Then

$$|\alpha|_v^{[L_v:\mathbb{Q}_v]} = \begin{cases} 7^{-1}, & \text{if } v \text{ is over } P; \\ 7, & \text{if } v \text{ is over } \overline{P}; \\ 1, & \text{otherwise.} \end{cases}$$

Thus

$$h(\alpha) = \frac{\log 7}{12}.$$

This example shows that numbers of small height in an Abelian extension are closely related to the class number problem. We can reverse the above construction and use lower bounds for the height to obtain informations on the size of the ideal class group of some fields. For instance, let L_m be the m-th cyclotomic field, and define e_m to be the exponent of its class group, *i.e.* the smallest positive integer e such that I^e is a principal ideal for all integral ideals I of L_m. By Linnik's theorem, there exists an absolute constant $c > 0$ and a prime $p \leq m^c$ which splits completely in L_m. Let P be a prime ideal of L_m over p; by definition $P^{e_m} = (\gamma)$ for some integer $\gamma \in L_m$. Define $\alpha = \gamma/\overline{\gamma}$. The above argument shows that

$$h(\alpha) = \frac{e_m \log p}{[L_m : \mathbb{Q}]} \leq \frac{e_m c \log m}{\varphi(m)} ,$$

where $\varphi(\cdot)$ is the Euler function. Since $L_m/\mathbb{Q}$ is Abelian,

$$\frac{\log 5}{12} \leq h(\alpha) \leq \frac{e_m c \log m}{\varphi(m)} .$$

We obtain:

$$e_m \geq \frac{\log 5}{12c} \times \frac{\varphi(m)}{\log m} .$$

Let K be a CM field of discriminant Δ and degree d. We assume the Generalized Riemann Hypothesis for the Dedekind zeta function of K. More sophisticated argument show (see [9]) that for any $\varepsilon > 0$ the exponent e_K of the class group of K satisfies:

$$e_K \geq \max \left\{ \frac{C \log |\Delta|}{d \log\log |\Delta|}, C(\varepsilon)d^{1-\varepsilon} \right\} ,$$

where C and $C(\varepsilon)$ are positive constants. Thus the exponent of the class group of a CM field goes to infinity with its discriminant.

We can "mix" the lower bound in Abelian extensions (Theorem A.4) with Dobrowolski's result, Theorem A.3. Let K be a fixed number field, and let L/K be an Abelian extension. In [11], we prove that for $\alpha \in L^*$ not a root of unity,

$$h(\alpha) \geq \frac{c(K)}{D} \left(\frac{\log 2D}{\log\log 5D} \right)^{-13} , \qquad (A.9)$$

where $D = [L(\alpha) : L]$ and where $c(K) > 0$. In the proof of [11], $c(K)$ depended on both the degree *and* the discriminant of K.

We come back to the lower bounds for the height on an Abelian extension L of a number field K. As a very special case of (A.9), the height in L^*, outside the set of roots of unity, is bounded from below by a positive function depending only on K. The following question arises: is it true that we can choose a function depending only on the degree $[K : \mathbb{Q}]$? In [12] we gave a positive answer to this problem. Let L/K be as before. Then for any $\alpha \in L^*$ which is not a root of unity, we have

$$h(\alpha) > 3^{-d^2-2d-6}$$

where $d = [K : \mathbb{Q}]$. This result has some amusing consequences. For instance, let L be a dihedral extension of the rational field of degree $2n$, say. Then L is an Abelian extension of its quadratic subfield K fixed by the normal cyclic group of order n. Thus for any $\alpha \in L^*$ which is not a root of unity we have

$$h(\alpha) \geq 3^{-14}.$$

A.2.3. Sketch of proof of Theorem A.4

For a natural number $m \geq 3$ we denote by ζ_m a primitive m th-root of unity, and we let $L_m = \mathbb{Q}(\zeta_m)$ be the m-th cyclotomic field. We need two lemmas. Let $p \geq 3$ be a prime number, and let $\alpha \in L_m^*$, α not a root of unity. We show that

$$h(\alpha) \geq \frac{\log(p/2)}{2p}.$$

Choosing $p = 5$, this gives, *via* Kronecker-Weber's theorem, the lower bound

$$h(\alpha) \geq \frac{\log(5/2)}{10}$$

for the height of a non-zero algebraic number α (α not a root of unity) lying in an Abelian extension. A refinement of the proof gives the more precise result of Theorem A.4.

The following simple lemma is the key argument in the proof.

Lemma A.5. *Let p be a rational prime. Then there exists $\sigma = \sigma_p \in$ Gal$(L_m/\mathbb{Q})$ with the following two properties.*

i) *If $p \nmid m$, then*

$$p \mid (\gamma^p - \sigma\gamma)$$

for any integer $\gamma \in L_m$.

ii) *If $p \mid m$, then*

$$p \mid (\gamma^p - \sigma\gamma^p)$$

for any integer $\gamma \in L_m$. Moreover, if $\sigma\gamma^p = \gamma^p$ for some $\gamma \in L_m$, then there exists a root of unity $\zeta \in L_m$ such that $\zeta\gamma$ is contained in a proper cyclotomic subextension of L_m.

Proof. Assume first that $p \nmid m$. Let $\sigma \in \mathrm{Gal}(L_m/\mathbb{Q})$ be the Frobenius automorphism defined by $\sigma \zeta_m = \zeta_m^p$. For any integer $\gamma \in L_m$, we have $\gamma = f(\zeta_m)$ for some $f \in \mathbb{Z}[x]$. Hence

$$\gamma^p \equiv f(\zeta_m^p) \equiv f(\sigma \zeta_m) \equiv \sigma \gamma \pmod{p}.$$

Assume now that $p|m$. The Galois group $\mathrm{Gal}(L_m/K_{m/p})$ is cyclic of order $k = p$ or $k = p - 1$ depending on whether $p^2|m$ or not. Let σ be one of its generators; hence $\sigma \zeta_m = \zeta_p \zeta_m$ for some primitive p-th root of unity ζ_p. For any integer $\gamma = f(\zeta_m) \in \mathbb{Z}[\zeta_m]$, we have

$$\gamma^p \equiv f(\zeta_m^p) \equiv f(\sigma \zeta_m^p) \equiv \sigma \gamma^p \pmod{p}.$$

Suppose finally that $\sigma \gamma^p = \gamma^p$: then $\sigma \gamma = \zeta_p^u \gamma$ for some integer u. It follows that $\sigma(\gamma/\zeta_m^u) = \gamma/\zeta_m^u$, hence γ/ζ_m^u belongs to the fixed field $K_{m/p}$, as desired.   □

Let $L = L_m$, and let $\sigma = \sigma_p$ be the homomorphism given by Lemma A.5. Assume first that $p \nmid m$. Let v be a place of L dividing p (thus $|p|_v = 1/p$). By the "strong approximation theorem" (see for instance [17, Chapter 2, Section 15, page 67]), we see easily that there exists an algebraic integer $\beta = \beta_v \in L$ such that $\alpha\beta$ is integer and

$$|\beta|_v = \max\{1, |\alpha|_v\}^{-1}.$$

Then

$$|(\alpha\beta)^p - \sigma(\alpha\beta)|_v \le p^{-1} \quad \text{and} \quad |\beta^p - \sigma\beta|_v \le p^{-1}.$$

Using the ultrametric inequality, we deduce that

$$
\begin{aligned}
|\alpha^p - \sigma\alpha|_v &= |\beta|_v^{-p}|(\alpha\beta)^p - \sigma(\alpha\beta) + (\sigma\beta - \beta^p)\sigma\alpha|_v \\
&\le |\beta|_v^{-p} \max\left(|(\alpha\beta)^p - \sigma(\alpha\beta)|_v, |\beta^p - \sigma\beta|_v|\sigma\alpha|_v\right) \\
&\le p^{-1} \max(1, |\alpha|_v)^p \max(1, |\sigma\alpha|_v).
\end{aligned}
$$

Suppose now that v is a finite place not dividing p. Then we have

$$|\alpha^p - \sigma(\alpha)|_v \le \max(1, |\alpha|_v)^p \max(1, |\sigma(\alpha)|_v).$$

Finally, if $v|\infty$,

$$|\alpha^p - \sigma(\alpha)|_v \le 2\max(1, |\alpha|_v)^p \max(1, |\sigma(\alpha)|_v).$$

Moreover $\alpha^p \neq \sigma\alpha$, since α is not a root of unity. We now apply the product formula to $\gamma = \alpha^p - \sigma\alpha$, using

$$\sum_{v|p} [L_v : \mathbb{Q}_v] = \sum_{v|\infty} [L_v : \mathbb{Q}_v] = [L : \mathbb{Q}].$$

We get

$$0 = \sum_{\substack{v \nmid \infty \\ v \nmid p}} \frac{[L_v : \mathbb{Q}_v]}{[L : \mathbb{Q}]} \log |\gamma|_v + \sum_{v|p} \frac{[L_v : \mathbb{Q}_v]}{[L : \mathbb{Q}]} \log |\gamma|_v$$

$$+ \sum_{v|\infty} \frac{[L_v : \mathbb{Q}_v]}{[L : \mathbb{Q}]} \log |\gamma|_v$$

$$\leq \sum_v \frac{[L_v : \mathbb{Q}_v]}{[L : \mathbb{Q}]} (p \log^+ |\alpha|_v + \log^+ |\sigma\alpha|_v)$$

$$- \sum_{v|p} \frac{[L_v : \mathbb{Q}_v]}{[L : \mathbb{Q}]} \log p + \sum_{v|\infty} \frac{[L_v : \mathbb{Q}_v]}{[L : \mathbb{Q}]} \log 2$$

$$= ph(\alpha) + h(\sigma\alpha) - \log p + \log 2$$
$$= (p + 1)h(\alpha) - \log(p/2).$$

Therefore,

$$h(\alpha) \geq \frac{\log(p/2)}{p + 1} \geq \frac{\log(p/2)}{2p}.$$

Assume now that $p \mid m$. Let v be a place of L dividing p and let $\beta = \beta_v \in L$ as in the first part of the proof. Then

$$|(\alpha\beta)^p - \sigma(\alpha\beta)^p|_v \leq p^{-1} \quad \text{and} \quad |\beta^p - \sigma\beta^p|_v \leq p^{-1}.$$

Using the ultrametric inequality, we find

$$|\alpha^p - \sigma\alpha^p|_v = |\beta|_v^{-p} |(\alpha\beta)^p - \sigma(\alpha\beta)^p + (\sigma\beta^p - \beta^p)\sigma\alpha^p|_v$$
$$\leq p^{-1} \max(1, |\alpha|_v)^p \max(1, |\sigma\alpha|_v)^p.$$

Moreover, we can assume $\alpha^p \neq \sigma\alpha^p$. Otherwise, by lemma A.5, there would exist a root of unity $\zeta \in L$ such that $\zeta\alpha$ is contained in a proper cyclotomic subextension of L; hence $h(\alpha) = h(\zeta\alpha)$ and, by induction, $h(\zeta\alpha) \geq \frac{\log(p/2)}{2p}$. Applying the product formula to $\gamma = \alpha^p - \sigma\alpha^p$ as in the first part of the proof, we get

$$0 \leq ph(\alpha) + ph(\sigma\alpha) - \log p + \log 2 = 2ph(\alpha) - \log(p/2).$$

Again

$$h(\alpha) \geq \frac{\log(p/2)}{2p}.$$

A.3. Subvarieties of $\mathbb{G}_m^n$

We consider a torus $\mathbb{G}_m^n$, and we fix the "standard embedding" $\iota\colon \mathbb{G}_m^n \hookrightarrow \mathbb{P}^n$,

$$\iota(x_1, \ldots, x_n) = (1 : x_1 : \cdots : x_n).$$

By a subvariety of $\mathbb{G}_m^n$ we mean an algebraic subvariety V defined over some number field K. The degree of V is the degree of its Zariski closure in $\mathbb{P}^n$. We shall say that V is irreducible if its Zariski closure is geometrically irreducible. Similarly, we say that V is irreducible over K if its Zariski closure is irreducible over K.

We recall some definitions from Chapter 4, Section 2.2. Given $\lambda \in \mathbb{Z}^n$ and $\mathbf{x} = (x_1, \ldots, x_n)$ we set $\mathbf{x}^\lambda = x_1^{\lambda_1} \cdots x_n^{\lambda_n}$. Given any m-tuple of vectors $\lambda_1, \ldots, \lambda_m \in \mathbb{Z}^n$ we define a regular map $\varphi\colon \mathbb{G}_m^n \to \mathbb{G}_m^m$ by $\varphi(\mathbf{x}) := (\mathbf{x}^{\lambda_1}, \ldots, \mathbf{x}^{\lambda_m})$. This map is an algebraic group homomorphism, called *monoidal*. When $m = n$, the homomorphism φ is invertible if and only if $\det(\lambda_1, \ldots, \lambda_m) = \pm 1$; in this case it is called a *monoidal automorphism* of $\mathbb{G}_m^n$. If $\det(\lambda_1, \ldots, \lambda_m) \neq 0$ the kernel of φ is finite; we shall say that φ is finite. We shall often use a special finite monoidal morphism. Let $l \in \mathbb{N}$. We denote by $[l]\colon \mathbb{G}_m^n \to \mathbb{G}_m^n$ the "multiplication" by $[l]$, *i.e.* the morphism $\mathbf{x} \mapsto \mathbf{x}^l = (x_1^l, \ldots, x_n^l)$. Thus the kernel $\mathrm{Ker}[l]$ is the set of l-torsion points. It is a subgroup isomorphic to $(\mathbb{Z}/l\mathbb{Z})^n$.

By algebraic subgroup of $\mathbb{G}_m^n$ we mean a closed algebraic subvariety stable under the group operations. An irreducible algebraic subgroup is called a torus. Any algebraic subgroup is a finite disjoint union of translates of a torus. Given an algebraic subgroup H we denote by H^0 its connected component containing the neutral element. Let $\Lambda \subseteq \mathbb{Z}^n$ be a subgroup. Then

$$H_\Lambda = \{\mathbf{x} \in \mathbb{G}_m^n, \ \forall \lambda \in \Lambda, \ \mathbf{x}^\lambda = 1\}$$

is an algebraic group. Moreover, $\Lambda \mapsto H_\Lambda$ is a bijection between subgroups of $\mathbb{Z}^n$ and algebraic subgroups of $\mathbb{G}_m^n$.

Let V be an irreducible subvariety of $\mathbb{G}_m^n$. We define its stabilizer to be

$$\mathrm{Stab}(V) = \{\boldsymbol{\alpha} \in \mathbb{G}_m^n \text{ s.t. } \boldsymbol{\alpha}V = V\}.$$

Thus

$$\mathrm{Stab}(V) = \bigcap_{\mathbf{x}\in V} \mathbf{x}^{-1} V.$$

This shows that $\mathrm{Stab}(V)$ is an algebraic subgroup of dimension $\leq \dim(V)$. We remark that equality of the dimensions holds if and only if V is a translate of a torus.

Let l be a positive integer. We are interested in relations between the degree of V and the degrees of $[l]^{-1}V = \{\boldsymbol{\alpha} \in \mathbb{G}_m^n \text{ s.t. } \boldsymbol{\alpha}^l \in V\}$ and of $[l]V = \{\boldsymbol{\alpha}^l \text{ s.t. } \boldsymbol{\alpha} \in V\}$.

Proposition A.6. *We have*

$$\deg([l]^{-1}V) = l^{\operatorname{codim}(V)} \deg(V)$$

and

$$\deg([l]V) = \frac{l^{\dim(V)} \deg(V)}{|\operatorname{Ker}[l] \cap \operatorname{Stab}(V)|}. \tag{A.10}$$

Proof. This is a special case of a general result of [24]. We give a sketch of the proof. Let us prove the first formula. For a hypersurface, this statement is clear. Indeed, let f be an equation of V. Then $f(\mathbf{x}^l)$ is an equation of $[l]^{-1}V$. We consider the general case. Let d be the dimension of V and let $W_1, \ldots, W_d$ be generic hypersurfaces of degree $D_1, \ldots, D_d$ such that $X = V \cap W_1 \cap \cdots \cap W_d$ is a finite set of $\deg(V)D_1 \cdots D_d$ points. Then $[l]^{-1}X = [l]^{-1}V \cap [l]^{-1}W_1 \cap \cdots \cap [l]^{-1}W_d$ is a set of cardinality $l^n|X|$. On the other hand, for what we have seen for hypersurfaces, this set has cardinality $\deg([l]^{-1}V)l^d D_1 \cdots D_d$. Thus $\deg([l]^{-1}V) = l^{n-d} \deg(V)$ as required.

The equality (A.10) follows from the previous one. Indeed $[l]^{-1}[l]V = \operatorname{Ker}[l]V$ and $\operatorname{Ker}[l]V$ is a union of

$$\frac{l^n}{|\operatorname{Ker}[l] \cap \operatorname{Stab}(V)|}$$

distinct components. Thus

$$l^{\operatorname{codim}(V)} \deg([l]V) = \deg([l]^{-1}[l]V) = \frac{l^n \deg([l]V)}{|\operatorname{Ker}[l] \cap \operatorname{Stab}(V)|}. \qquad \Box$$

A.3.1. Heights of subvarieties

Let $\boldsymbol{\alpha} = (\alpha_0 : \cdots : \alpha_n) \in \mathbb{P}^n(K)$ and let K be any number field containing $\alpha_0, \ldots, \alpha_n$. We define the Weil height of $\boldsymbol{\alpha}$ by:

$$h(\boldsymbol{\alpha}) = \frac{1}{[K : \mathbb{Q}]} \sum_{v \in \mathcal{M}_K} [K_v : \mathbb{Q}_v] \log \max\{|\alpha_0|_v, \ldots, |\alpha_n|_v\}.$$

As for the height of algebraic numbers, this definition does not depend on the number field K; moreover, by the product formula, it does not depend on the projective coordinates of $\boldsymbol{\alpha}$.

This provides a height function $\hat{h}(x_1, \ldots, x_n) = h(1 : x_1 : \cdots : x_n)$ on $\mathbb{G}_m^n(\overline{\mathbb{Q}})$. The following properties hold:

i) the function $\hat{h}$ is a positive function on $\mathbb{G}_m^n(\overline{\mathbb{Q}})$, vanishing only on its torsion points;

ii) $\hat{h}(\alpha\beta) \leq \hat{h}(\alpha) + \hat{h}(\beta)$. Moreover, if ζ is a torsion point, $\hat{h}(\zeta\alpha) = \hat{h}(\alpha)$. If $n \in \mathbb{N}$ then $\hat{h}(\alpha^n) = n\hat{h}(\alpha)$;

iii) a subset of $\mathbb{G}_m^n(\overline{\mathbb{Q}})$ of bounded height and bounded degree is finite (Northcott's theorem)

The proofs are similar to those in dimension 1.

On hypersurfaces we have a "natural" definition of height arising from an extension of the Mahler measure to polynomials in several variables. Let $P \in \mathbb{C}[x_1^{\pm 1}, \ldots, x_n^{\pm 1}]$; we define its Mahler measure as:

$$M(P) = \exp \int_0^1 \cdots \int_0^1 \log |P\left(e^{2\pi it_1}, \ldots, e^{2\pi it_n}\right)| \, dt_1 \ldots dt_n$$

and we make the convention $M(0) = 0$. As in dimension 1, the Mahler measure is a multiplicative function. Moreover, if $\varphi(\mathbf{x}) = (\mathbf{x}^{\lambda_1}, \ldots, \mathbf{x}^{\lambda_m})$ is a finite monoidal morphism, then $M(P(\mathbf{x})) = M(P(\mathbf{x}^\lambda))$. Let K be a number field and let $f \in K[\mathbf{x}]$ be a polynomial. We define

$$\hat{h}(f) = \frac{1}{[K : \mathbb{Q}]} \sum_{v \in \mathcal{M}_K} [K_v : \mathbb{Q}_v] \log M_v(f),$$

where $M_v(f)$ is the maximum of the v-adic absolute values of the coefficients of f if v is non archimedean, and $M_v(f)$ is the Mahler measure of σf if v is an archimedean place associated with the embedding $\sigma : K \hookrightarrow \overline{\mathbb{Q}}$. As for the Weil height, this definition does not depend on the field K containing the coefficients of f and $\hat{h}$ defines a positive and additive function on $\overline{\mathbb{Q}}[\mathbf{x}]$. Let

$$V = \{\alpha \in \mathbb{G}_m^n \text{ s.t. } f(\alpha) = 0\}$$

be a hypersurface in $\mathbb{G}_m^n$ defined by some square-free polynomial $f \in K[\mathbf{x}]$. We define the normalized height of V as

$$\hat{h}(V) = \hat{h}(f).$$

This definition does not depend on the equation we choose for V. Let $\varphi : \mathbb{G}_m^n \to \mathbb{G}_m^n$ be a finite monoidal morphism. We also remark that $\hat{h}(\varphi^{-1}(V)) = \hat{h}(V)$.

Following Schinzel, we say that an irreducible $f \in \mathbb{Z}[\mathbf{x}]$ is an *extended cyclotomic polynomial* if there exist a cyclotomic polynomial ϕ and λ, $\mu \in \mathbb{Z}^n$ such that

$$f(\mathbf{x}) = \pm \mathbf{x}^\lambda \phi(\mathbf{x}^\mu) .$$

In other words, an irreducible polynomial $f \in \mathbb{Z}[\mathbf{x}]$ is extended cyclotomic if and only if the hypersurface $\{f = 0\}$ in $\mathbb{G}_m^n$ is a union of torsion varieties. In this context, Zhang's theorem on the toric Bogomolov conjecture can be paraphrased as follows. Let $f \in \mathbb{Z}[\mathbf{x}]$ be irreducible. Then $M(f) = 1$ if and only if $f = \pm x_j$ or if f is an extended cyclotomic polynomial. This result was proved earlier in [14], [25] and [34] independently.

The normalized height of an irreducible hypersurface has a nice behaviour under the action of pull back and pull out by multiplication by $[l]$. Indeed

$$\hat{h}([l]^{-1}V) = \hat{h}(V)$$

and

$$\hat{h}([l]V) = \frac{l^n \hat{h}(V)}{|\text{Ker}[l] \cap \text{Stab}(V)|}.$$

The first equality is a special case of the invariance of $\hat{h}(V)$ under inverse image by finite monoidal morphisms. The second equality follows from the first one and from the additivity of $\hat{h}$, exactly as the corresponding formulas for the degree.

The normalized height of a hypersurface can be computed as a limit. Let $f \in \mathbb{C}[\mathbf{x}]$. From inequality (A.3) we deduce by induction on n (see [28] for details)

$$\|f\|_1 \leq 2^{d_1 + \cdots + d_n} M(f),$$

where $d_1, \ldots, d_n$ are the partial degrees of f. Let $\| \cdot \|$ be any norm on $\mathbb{C}[\mathbf{x}]$ such that

$$\log \|f\| = \log \|f\|_1 + O(\deg f) \tag{A.11}$$

We define a height on hypersurfaces of $\mathbb{G}_m^n$ by choosing the norm $\| \cdot \|$ at the archimedean places. Let as before

$$V = \{\alpha \in \mathbb{G}_m^n \text{ s.t. } f(\alpha) = 0\}$$

be a hypersurface in $\mathbb{G}_m^n$ defined by some square-free polynomial $f \in K[\mathbf{x}]$. Let us define

$$h(V) = \frac{1}{[K : \mathbb{Q}]} \sum_{v \in \mathcal{M}_K} [K_v : \mathbb{Q}_v] \log H_v(f),$$

where $H_v(f) = M_v(f)$ if v is non archimedean, and $H_v(f) = \|\sigma f\|$ if v is an archimedean place associated with the embedding $\sigma : K \hookrightarrow \overline{\mathbb{Q}}$. Then,

$$\hat{h}(V) = h(V) + O(\deg(V)). \tag{A.12}$$

Let l be a positive integer. Using the relations between degrees and heights of V and $[l]V$ we see that

$$\hat{h}(V) = \frac{\hat{h}([l]V)\deg(V)}{l\deg([l]V)}.$$

Thus, replacing in (A.12) V by $[l]V$,

$$\hat{h}(V) = \frac{h([l]V)\deg(V)}{l\deg([l]V)} + O(l^{-1}\deg(V)).$$

This shows

$$\lim_{l\mapsto\infty} \frac{h([l]V)\deg(V)}{l\deg([l]V)} = \hat{h}(V).$$

The last formula suggests a "simple" definition of normalized height on subvarieties of $\mathbb{G}_m^n$, due to Philippon [30]. We start by choosing a height on subvarieties. Let V be a d dimensional irreducible subvariety and let F be the Chow form of its Zariski closure in $\mathbb{P}^n$. The Chow form is an irreducible multihomogeneous polynomial $F(u_0^1, ..., u_n^1, ..., u_0^{d+1}, ..., u_n^{d+1})$ vanishing precisely if the intersection of V with the linear space

$$\{\mathbf{x} \in \mathbb{P}^n \text{ s.t. } u_0^1 x_0 + \cdots + u_n^1 x_n = \cdots = u_0^{d+1} x_0 + \cdots + u_n^{d+1} x_n = 0\}$$

is non empty. We define a height $h(V)$ as the height of the hypersurface in $\mathbb{G}_m^{(d-1)n}$ defined by $\{F = 0\}$, where one choose any reasonable norm at the archimedean places (*i.e.* a norm satisfying (A.11)). David and Philippon (see [20]) prove that the limit

$$\hat{h}(V) = \lim_{l\to+\infty} \frac{h([l]V)\deg(V)}{l\deg([l]V)}$$

exists. We can see (compute the Chow form) that this definition of normalized height specializes to the previous ones if V is a point or if V is a hypersurface (see [20]). Moreover:

i) the function $\hat{h}(\cdot)$ is non-negative;
ii) for every $l \in \mathbb{N}$ we have

$$\hat{h}([l]^{-1}V) = l^{\operatorname{codim}(V)-1}\hat{h}(V)$$

and

$$\hat{h}([l]V) = \frac{l^{\dim(V)+1}\hat{h}(V)}{|\operatorname{Ker}[l]\cap\operatorname{Stab}(V)|}.$$

iii) for every torsion point ζ we have $\hat{h}(\zeta V) = \hat{h}(V)$.

For further details on the construction of the normalized height on tori and Abelian varieties, see [30].

A.3.2. Small height problems

Using properties ii) and iii) of the normalized height, we see that a torsion variety $V = \zeta H$ has height zero. Indeed, if ζ is a torsion point and H is a subtorus, then $\hat{h}(\zeta H) = \hat{h}(H)$ and $\hat{h}(H) = \hat{h}([l]H) = l\hat{h}(H)$ for any $l \in \mathbb{N}$ (since $H = [l]H$ and $|\mathrm{Ker}[l] \cap H| = l^{\dim(H)}$).

Are torsion varieties the only varieties of zero height? The answer is positive; more precisely, this question is equivalent to the multiplicative analogue of the former Bogomolov's conjecture. Let us define, for $\theta > 0$,

$$V(\theta) = \{ \alpha \in V \text{ s.t. } \hat{h}(\alpha) \le \theta \}$$

and let $\hat{\mu}^{\mathrm{ess}}(V)$ ("essential minimum") be the infimum of the set of $\theta > 0$ such that $V(\theta)$ is Zariski dense in V. Theorem A.2 asserts that $\hat{\mu}^{\mathrm{ess}}(V) = 0$ if and only if V is torsion. By a special case of an inequality of Zhang (see [37, Theorem 5.2.]), we have, for an irreducible V,

$$\hat{\mu}^{\mathrm{ess}}(V) \le \frac{\hat{h}(V)}{\deg(V)} \le (\dim(V) + 1)\hat{\mu}^{\mathrm{ess}}(V). \qquad (A.13)$$

This inequality shows that $\hat{h}(V) = 0$ if and only if $\hat{\mu}^{\mathrm{ess}}(V) = 0$. The problem of finding sharp lower bounds for $\hat{\mu}^{\mathrm{ess}}(V)$ for non-torsion subvarieties of $\mathbb{G}_m^n$ is a generalization of Lehmer's problem. Lower bounds for the essential minimum of a non-torsion subvariety will depend on some geometric invariants of V, for instance its degree. Moreover, if we do not make any further geometric assumption on the variety, such a bound must also depend on its field of definition ("arithmetic case"). Indeed, let H be a proper subtorus of $\mathbb{G}_m^n$ and let α_n be a sequence of non-torsion points whose height tends to zero (for instance, $\alpha_n = (2^{1/n}, \ldots, 2^{1/n})$). Then, the varieties $V_n = H\alpha_n$ have fixed degree $\deg(H)$ and essential minimum $\hat{\mu}^{\mathrm{ess}}(V_n) \le \hat{h}(\alpha_n) \to 0$. In spite of that, if we further assume that V is not a translate of a proper subtorus (even by a point of infinite order), then Bombieri and Zannier [16] proved that the essential minimum of V can be bounded from below only in terms of the degree of V ("geometric case").

As an exercice, we remark that this result of Bombieri and Zannier gives an alternative proof of Schinzel's result stated in Section A.2.2. Let L be a CM field and let $\alpha \in L^*$ such that $|\alpha| \ne 1$. We consider the curve $C \subseteq \mathbb{G}_m^2$ defined by the equation

$$f(x, y) = (x - \alpha)y - (\overline{\alpha}x - 1).$$

Since $\overline{\alpha} \ne \alpha^{-1}$ this curve is irreducible. Moreover, it is easy to see that C is not a translate of a subgroup. By the quoted result of Bombieri and

Zannier, $\hat{h}(C) \geq c > 0$ for some c which *does not depend* on α. Let v be an archimedean place associate with the embedding σ. Then

$$\log M_v(f) = \log M(x - \sigma\alpha) + \log M\left(y - \frac{\sigma(\overline{\alpha})x - 1}{x - \sigma\alpha}\right)$$

where we have extended $M(\cdot)$ to $\mathbb{C}(x, y)$ by multiplicativity. By (A.1), $\log M(x - \sigma\alpha) = \log^+|\sigma\alpha|$ and

$$\log M\left(y - \frac{\sigma(\overline{\alpha})x - 1}{x - \sigma\alpha}\right) = \int_0^1 \log^+\left|\frac{\sigma(\overline{\alpha})e^{2\pi it} - 1}{e^{2\pi it} - \sigma\alpha}\right| dt.$$

This last quantity is zero. Indeed $\sigma(\overline{\alpha}) = \overline{\sigma(\alpha)}$ (recall that L is CM) and, for $z, \beta \in \mathbb{C}$ with $|z| = 1$,

$$\left|\frac{\overline{\beta}z - 1}{z - \beta}\right| = 1.$$

Thus $\log M_v(f) = \log^+|\alpha|$ for $v \mid \infty$. Let now $v \nmid \infty$. An easy computation shows that

$$M_v(f) = \max(|\alpha|_v, |\overline{\alpha}|_v, 1) \leq \max(|\alpha|_v, 1)\max(|\overline{\alpha}|_v, 1).$$

Putting all together we get $0 < c \leq \hat{h}(C) \leq 2h(\alpha)$.

Let V be a subvariety of $\mathbb{G}_m^n$. We define the "absolute obstruction index" $\omega(V)$ of V as the minimum of $\deg(Z)$ where Z is a hypersurface containing V. Similarly, we define the "rational obstruction index" $\omega_{\mathbb{Q}}(V)$ as the minimum of $\deg(Z)$ where Z is a hypersurface defined over $\mathbb{Q}$ containing V. For instance, let α be an algebraic number of degree d. Then $\omega_{\mathbb{Q}}(\alpha) = d$. More generally, let $\boldsymbol{\alpha} \in \mathbb{G}_m^n(\overline{\mathbb{Q}})$. Then, by standard linear algebra,

$$\omega_{\mathbb{Q}}(\boldsymbol{\alpha}) \leq n[\mathbb{Q}(\boldsymbol{\alpha}) : \mathbb{Q}]^{1/n} \tag{A.14}$$

Even more generally, let V be a subvariety of $\mathbb{G}_n^m$. Then, if V is irreducible,

$$\omega(V) \leq n \deg(V)^{1/\text{codim}(V)}.$$

Similarly, if V is defined and irreducible over the rational field, $\omega_{\mathbb{Q}}(V) \leq n \deg(V)^{1/\text{codim}(V)}$. Both inequalities are special cases of a result of Chardin [18].

It turns out that $\omega_{\mathbb{Q}}(V)$, and not the degree of V, is the right invariant to formulate the sharpest conjectures on $\hat{\mu}^{\text{ess}}(V)$ in the "arithmetic case". Similarly, $\omega(V)$ is the right invariant in the "geometric case". Although, in order to get statements depending on ω we need to assume, in the geometric case, not only that V is not a translate but also that V is not contained in any proper translate. Indeed, consider a curve $C \subseteq \mathbb{G}_m^{n-1}$. Let $C' = C \times \{1\} \subseteq \mathbb{G}_m^n$ and choose, for $l \in \mathbb{N}$, an irreducible component V_l of $[l]^{-1}C'$. Then $\hat{\mu}^{\text{ess}}(V_l) \mapsto 0$, while $\omega(V_l) = 1$ since V_l is contained in the hypersurface $x_n = 1$. We shall say that an irreducible variety V is "transverse" if it is not contained in any proper translate. Similarly, in the arithmetic case we need to assume that V is not in a torsion variety. Such a V will be called a "weak-transverse" variety. Let $\alpha \in \mathbb{G}_m^n(\overline{\mathbb{Q}})$. We remark that the 0-dimensional variety $\{\alpha\}$ is weak-transverse if and only if $\alpha_1, \ldots, \alpha_n$ are multiplicatively dependent.

Let V be a weak-transverse subvariety of $\mathbb{G}_m^n$. In [2] we conjecture that

$$\hat{\mu}^{\text{ess}}(V) \geq \frac{c(n)}{\omega_{\mathbb{Q}}(V)} \tag{A.15}$$

for some $c(n) > 0$. Observe that this conjecture generalizes Lehmer's one. In [2] (case $\dim V = 0$), [3] (case $\operatorname{codim} V = 1$) and [4] (general case) we prove the following analogue of Dobrowolski's theorem on $\mathbb{G}_m^n$.

Theorem A.7. *Let V be a weak-tranverse subviety of $\mathbb{G}_m^n$ of codimension k. Let us assume that V is not contained in any torsion variety. Then there exist two positive constants $c(n)$ and $\kappa(k) = (k+1)(k+1)!^k - k$ such that*

$$\hat{\mu}^{\text{ess}}(V) \geq \frac{c(n)}{\omega(V)} \left(\log 3\omega_{\mathbb{Q}}(V)\right)^{-\kappa(k)}.$$

Sometimes this theorem produces lower bounds for the height of algebraic numbers which are even sharper than what is expected by Lehmer's conjecture. Let $\alpha_1, \ldots, \alpha_n$ multiplicatively independent algebraic numbers of height $\leq h$, lying in a number field of degree d. Then $\hat{\mu}^{\text{ess}}(\alpha) \leq h$ and, by (A.14),

$$\omega_{\mathbb{Q}}(\alpha) \leq nd^{1/n}.$$

Thus, by Theorem A.7,

$$h \geq \frac{c(n)}{d^{1/n}} (\log 3d)^{-\kappa(n)}.$$

for some $c(n) > 0$.

Assuming that the subvariety V is tranverse, we now look for lower bounds for $\hat{\mu}^{\text{ess}}(V)$ which do not depend on the field of definition of V (geometric case). In [5] we conjecture that

$$\hat{\mu}^{\text{ess}}(V) \geq \frac{c(n)}{\omega_{\overline{\mathbb{Q}}}(V)}.$$

for some $c(n) > 0$. In the same paper we prove:

Theorem A.8. *Let V be a transverse subvariety of $\mathbb{G}_m^n$ of codimension k. Then there exist two positive constants $c(n)$ and $\lambda(k) = \left(9(3k)^{(k+1)}\right)^k$ such that*

$$\hat{\mu}^{\mathrm{ess}}(V) \geq \frac{c(n)}{\omega_{\overline{\mathbb{Q}}}(V)} \left(\log 3\omega_{\overline{\mathbb{Q}}}(V)\right)^{-\lambda(k)}.$$

The proofs of Theorems A.7 and A.8 require several technical tools. By contradiction, we assume in both proofs that the essential minimum is sufficiently small. We then start following the usual steps of a transcendence proof: interpolation (construction of an auxiliary function), extrapolation and zero estimates. Concerning the last step, in both cases these proofs become very technical. In diophantine analysis a classical zero lemma (as [29]) is normally enough to conclude the proof. On the contrary, in [2] we need a more complicated zero lemma. As a consequence, this forces to extrapolate over different sets of primes. In Dobrowolski's proof one construct, using Siegel's Lemma, an auxiliary function F which vanishes on α. Then we extrapolate by proving that F must also vanish at α^p at least for small primes p. In the proof of Theorem A.7 (in the 0 dimensional case which is the hardest one) we construct an auxiliary function vanishing on α and then we extrapolate by proving that F must also vanish at $\alpha^{p_1 \cdots p_n}$ for p_j small primes. The zero lemma we alluded before shows that for some $l = p_1 \cdots p_n$ the obstruction index $\omega_{\mathbb{Q}}(\alpha^l)$ is pathologically smaller than $\omega_{\mathbb{Q}}(\alpha)$. Unfortunately, it seems hard to find lower bound for $\omega_{\mathbb{Q}}(\alpha^l)$ in terms of $\omega_{\mathbb{Q}}(\alpha)$. Thus, we cannot easily conclude easily the proof. To avoid this problem, we start again the whole construction replacing α with α^l. To ensure that the process end at some moment, we need a cumbersome induction ("descent step").

The situation is quite similar in the original proof of Theorem A.8. We construct again an auxiliary function vanishing on V and then we extrapolate by proving that F must also vanish on $\ker[p_1 \cdots p_n]V$ for p_j small primes. We need again a variant of a zero lemma which use the fact that our set of translation (the union of $\ker[p_1 \cdots p_n]$) is actually a union of big subgroups. Using this new zero lemma we succeed to show that again for some $l = p_1 \cdots p_n$ the obstruction index $\omega_{\mathbb{Q}}([l]V)$ is pathologically small than $\omega_{\mathbb{Q}}(V)$. As in the arithmetic situation, we cannot conclude easily and we need again a cumbersome descent step.

In [10] we recently succeed to drastically simplify the proof of Theorem A.8. The new proof encodes the classical diophantine analysis in an inequality involving some parameters, the essential minimum of a subvariety of $\mathbb{G}_m^n$ and two Hilbert's functions. To decode the diophantine information we use a sharp lower bound for the Hilbert function due to Chardin and Philippon [19]. Finally, a delicate reduction process allows us to obtain the desired result. Possibly, this new method also applies in the arithmetic case.

We now consider the problem of the description of small points. Let V be a non-torsion variety of $\mathbb{G}_m^n$ and define

$$V^* = V \setminus \bigcup_{\substack{B \subseteq V \\ B \text{ torsion}}} B.$$

By the former Manin-Mumford conjecture, V^* is a Zariski open set, indeed $V \setminus V^*$ is a finite union of torsion varieties. As mentioned in the introduction, an equivalent version of Theorem A.2 says that the height on $V^*(\overline{\mathbb{Q}})$ is bounded from below by a positive quantity:

$$\hat{\mu}^*(V) = \inf_{\alpha \in V^*} \hat{h}(\alpha) > 0.$$

Remark that obviously $\hat{\mu}^*(V) \leq \hat{\mu}^{\text{ess}}(V)$. Hence one could hope, in analogy to (A.15), that

$$\hat{\mu}^*(V) \geq \frac{c(n)}{\omega_{\mathbb{Q}}(V)}$$

for some constant $c(n) > 0$. This lower bound is not always true, as the following example shows.

Let α_k be a sequence of algebraic numbers whose height is positive and tends to zero as $k \to +\infty$. Let us consider

$$V_k = \{(\alpha_k, x_2, x_3) \in \mathbb{G}_m^3 \text{ s.t. } \alpha_k^2 + \alpha_k^3 - x_2 - x_3 = 0\}.$$

The height of $\boldsymbol{\alpha}_k = (\alpha_k, \alpha_k^2, \alpha_k^3) \in V_k \setminus V_k^*$ tends to zero and $\omega_{\mathbb{Q}}(V) \leq 3$, since

$$V_k \subseteq \{x_1^2 + x_1^3 - x_2 - x_3 = 0\}.$$

In [6] we formulate the following conjecture. Let V be a non-torsion variety of $\mathbb{G}_m^n$ defined by equation of degree $\leq \delta$ with integer coefficients. Then there exists a constant $c(n) > 0$ such that

$$\hat{\mu}^*(V) \geq \frac{c(n)}{\delta}.$$

In the same article, using a variant of Theorem A.7 and an additional induction, we prove this conjecture up to a logarithmic factor:

$$\hat{\mu}^*(V) \geq \frac{c(n)}{\delta} (\log 3\delta)^{-\kappa(n)}.$$

for some $c(n) > 0$ and where $k(n)$ is as in Theorem A.7.

We make a similar analysis in the geometric case. Let V be a tranverse subvariety of $\mathbb{G}_m^n$ and define, as in [16],

$$V^0 = V \setminus \bigcup_{B \subseteq V} B.$$

where the union is now on the set of translates B of tori of dimension 1. Again $V \setminus V^0$ is an open set (see [16] and [33]); Bombieri and Zannier prove that, outside a finite set, the height on V^0 is bounded from below by a positive quantity depending on the degree of V (and not on its field of definition). More precisely, assume that V is defined by equation of degree $\leq \delta$. Schmidt [33] proves that the set of points $\boldsymbol{\alpha} \in V^0$ such that $\hat{h}(\boldsymbol{\alpha}) < q^{-1}$ is finite, of cardinality $\leq q$, where

$$q = \exp\left((4n)^{2\delta(2n)^\delta} \right).$$

David and Philippon [20] improve this result. They show that the above assertion still hold choosing:

$$q = n \left(2^{n+4d+22} D (\log(D+1))^{2/3} \right)^{7^d},$$

where D is the degree of the Zariski closure of V in $(\mathbb{P}^1)^n \subseteq \mathbb{P}^{2n-1}$ and d is the dimension of V.

Using a variant of Theorem A.8 and an additional induction, in [7] we prove that, for all but finitely many $\boldsymbol{\alpha} \in V^0(\overline{\mathbb{Q}})$,

$$\hat{h}(\boldsymbol{\alpha}) \geq c(n)\delta^{-1} (\log(3\delta))^{-\lambda(n-1)},$$

where $c(n) > 0$ and $\lambda(k) = \left(9(3k)^{(k+1)} \right)^k$.

The proof of [7] gives no information on the cardinality of the set of points of pathologically small height. The new method introduced in [10] leads us to a complete quantitative description of the small points of a variety V. As a corollary of the main result of [10] we have:

Theorem A.9. *Let* $V \subseteq \mathbb{G}_m^n$ *be a (not necessarily irreducible) variety of dimension d defined by equation of degree $\leq \delta$. Let*

$$\theta = \delta \left(200 n^5 \log(n^2 \delta) \right)^{dn(n-1)}.$$

Then,

$$|V^0(\theta^{-1})| \leq \theta^n.$$

Results of this shape have several applications. Using, among other deep ingredients, Schmidt's bound for the number of small points in V^0, Evertse, Schlickewei and Schmidt [23] prove an uniform bound for the number of arithmetic progression in the Skolem-Mahler-Lech theorem (Chapter 4, Theorem 4.72). They show that the set of zeros of a simple linear recurrence sequence in $\overline{\mathbb{Q}}$ of order $n \geq 1$ is the union of at most $\exp\left((6n)^{3n}\right)$ arithmetic progressions. For this kind of application V is a linear variety. Thus $\delta = 1$ and the important dependance is on n. Using Theorem A.9 instead of Schmidt's bound we can replace $\exp\left((6n)^{3n}\right)$ by $(8n)^{4n^5}$ in the result of Evertse, Schlickewei and Schmidt.

References

[1] F. AMOROSO, *Sur des polynômes de petites mesures de Mahler*, C. R. Acad. Sci. Paris Sér. I Math. **321**, no. 1 (1995), 11–14.

[2] F. AMOROSO and S. DAVID, *Le problème de Lehmer en dimension supérieure"*, J. Reine Angew. Math. **513** (1999), 145–179.

[3] F. AMOROSO and S. DAVID, *Minoration de la hauteur normalisée des hypersurfaces*, Acta Arith. **92**, 4 (2000), 340–366.

[4] F. AMOROSO and S. DAVID, *Densité des points à cordonnées multiplicativement indépendantes*, Ramanujan J. **5** (2001), 237–246.

[5] F. AMOROSO and S. DAVID, *Minoration de la hauteur normalisée dans un tore*, Journal de l'Institut de Mathématiques de Jussieu **2**, no. 3 (2003), 335–381.

[6] F. AMOROSO and S. DAVID, *Distribution des points de petite hauteur dans les groupes multiplicatifs*, Ann. Scuola Norm. Sup. Pisa Cl. Sci. (5) **3** (2004), 325–348.

[7] F. AMOROSO and S. DAVID, *Points de petite hauteur sur une sous-variété d'un tore*, Compos. Math. **142** (2006), 551–562.

[8] F. AMOROSO and R. DVORNICICH, *A Lower Bound for the Height in Abelian Extensions*, J. Number Theory **80**, no. 2 (2000), 260–272.

[9] F. AMOROSO and R. DVORNICICH, *Lower bounds for the height and size of the ideal class group in CM fields*, Monatsh. Math. **138**, No.2 (2003), 85–94.

[10] F. AMOROSO and E. VIADA, *Small points on subvarieties of a torus*, preprint 2008.

[11] F. AMOROSO and U. ZANNIER, *A relative Dobrowolski's lower bound over Abelian extensions*, Ann. Scuola Norm. Sup. Pisa Cl. Sci. (4) **29** (2000), 711–727.

[12] F. AMOROSO and U. ZANNIER, *Some remarks on relative Lehmer*, preprint 2008.

[13] F. A. BOGOMOLOV, *Points of finite order on an Abelian variety*, Math. USSR Izv. **17** (1981), 55–72.

[14] D. BOYD, *Kronecker's theorem and Lehmer's problem for polynomials in several variables*, J. of Number Theory **13** (1980), 116–121.

[15] E. BOMBIERI and J. VAALER, *Siegel's lemma*, Invent. Math. **73** (1983), 11–32.

[16] E. BOMBIERI and U. ZANNIER, *Algebraic points on subvarieties of* $\mathbb{G}_m^n$, Internat. Math. Res. Notices **7** (1995), 333–347.

[17] J. W. S. CASSELS and A. FRÖHLICH, "Algebraic Number Theory"; Proceedings of an instructional conference organized by the London Mathematical Society, Academic Press, London-New-York, 1967.

[18] M. CHARDIN, *Une majoration de la fonction de Hilbert et ses conséquences pour l'interpolation algébrique*, Bulletin de la Société Mathématique de France, **117** (1988), 305–318.

[19] M. CHARDIN and P. PHILIPPON, *Régularité et interpolation*, J. Algebr. Geom. **8**, no. 3 (1999), 471–481; *erratum*, ibidem, **11** (2002), 599–600.

[20] S. DAVID and P. PHILIPPON, *Minorations des hauteurs normalisées des sous-variétés des tores*, Ann. Scuola Norm. Sup. Pisa Cl. Sci. (4) **28** (1999), 489–543; Errata, ibidem, **29** (2000), 729–731.

[21] S. DAVID and P. PHILIPPON, *Sous variétés de torsion des variétés semi-abéliennes*, C. R. Acad. Sci. Paris. Série I **331** (2000), 587–592.

[22] E. DOBROWOLSKI, *On a question of Lehmer and the number of irreducible factors of a polynomial*, Acta Arith. **34** (1979), 391–401.

[23] J.-H. EVERTSE, H.-P. SCHLICKEWEI and W. SCHMIDT, *Linear equations in variables which lie in a finitely generated group*, Annals of Math., **155** (2002), 807–836.

[24] M. HINDRY, *Autour d'une conjecture de S. Lang*, Invent. Math. **94** (1988), 575–603.

[25] W. LAWTON, *A generalization of a theorem of Kronecker*, Journal of the Science Faculty of the Chiangmai University (Thaïlande) **4** (1977), 15–23.

[26] M. LAURENT, *Equations diophantiennes exponentielles*, Invent. Math. **78** (1984), 299–327.

[27] D. H. LEHMER, *Factorization of certain cyclotomic functions*, Ann. of Math. **34** (1933), 461–479.

[28] M. MIGNOTTE, "Mathematics for Computer Algebra", Springer-Verlag, New York, 1992, xiv+346.

[29] P. PHILIPPON, *Lemmes de zéros dans les groupes algébriques commutatifs*, Bull. Soc. Math. France **114** (1986), 353–383.

[30] P. PHILIPPON, *Sur des hauteurs alternatives*, I, II and III, Math. Ann. **289** (1991), 255–283; Ann. Inst. Fourier **44** no. 4 (1994), 1043–1065; J. Math. Pures Appl. **74** no. 4 (1995), 345–365.

[31] M. RAYNAUD, *Courbes sur une variété abélienne et points de torsion*, Invent. Math. **71** (1983), 207–233.

[32] U. RAUSCH, *On a theorem of Dobrowolski about the product of conjugate numbers*, Colloq. Math. **50** (1985), 137–142.

[33] W. M. SCHMIDT, *Heights of points on subvarieties of* $\mathbb{G}_m^n$, In: "Number Theory 93-94", S. David (ed.), London Math. Soc. Ser., Vol. 235, Cambridge University Press, 1996.

[34] C. J. SMYTH, *A Kronecker-type theorem for complex polynomials in several variables*, Canad. Math. Bull. **24** (1981), 447–452. Errata, *ibidem*, **25**, 504 (1982).

[35] E. ULLMO, *Positivité et discrétion des points algébriques des courbes*, Ann. of Math. **147** no. 1 (1998), 81–95.

[36] P. VOUTIER, *An effective lower bound for the height of algebraic numbers*, Acta Arith. **74** (1996), 81–95.

[37] S. ZHANG, *Positive line bundles on arithmetic varieties*, J. Amer. Math. Soc. **8**, no. 1 (1995), 187–221.

[38] S. ZHANG, *Equidistribution of small points on Abelian varieties*, Ann. of Math. **147** no. 1 (1998), 159–165.

LECTURE NOTES

This series publishes polished notes dealing with topics of current research and originating from lectures and seminars held at the Scuola Normale Superiore in Pisa.

Published volumes

1. M. TOSI, P. VIGNOLO, *Statistical Mechanics and the Physics of Fluids*, 2005 (second edition). ISBN 978-88-7642-144-0

2. M. GIAQUINTA, L. MARTINAZZI, *An Introduction to the Regularity Theory for Elliptic Systems, Harmonic Maps and Minimal Graphs*, 2005. ISBN 978-88-7642-168-8

3. G. DELLA SALA, A. SARACCO, A. SIMIONIUC, G. TOMASSINI, *Lectures on Complex Analysis and Analytic Geometry*, 2006. ISBN 978-88-7642-199-8

4. M. POLINI, M. TOSI, *Many-Body Physics in Condensed Matter Systems*, 2006. ISBN 978-88-7642-192-0

   P. AZZURRI, *Problemi di Meccanica*, 2007. ISBN 978-88-7642-223-2

5. R. BARBIERI, *Lectures on the ElectroWeak Interactions*, 2007. ISBN 978-88-7642-311-6

6. G. DA PRATO, *Introduction to Stochastic Analysis and Malliavin Calculus*, 2007. ISBN 978-88-7642-313-0

   P. AZZURRI, *Problemi di meccanica*, 2008 (second edition). ISBN 978-88-7642-317-8

   A. C. G. MENNUCCI, S. K. MITTER, *Probabilità e informazione*, 2008 (second edition). ISBN 978-88-7642-324-6

7. G. DA PRATO, *Introduction to Stochastic Analysis and Malliavin Calculus*, 2008 (second edition). ISBN 978-88-7642-337-6

8. U. ZANNIER, *Lecture Notes on Diophantine Analysis*, 2014 (revised edition). ISBN 978-88-7642-341-3, e-ISBN: 978-88-7642-517-2

9. A. LUNARDI, *Interpolation Theory*, 2009 (second edition). ISBN 978-88-7642-342-0

10. L. AMBROSIO, G. DA PRATO, A. MENNUCCI, *Introduction to Measure Theory and Integration*, 2012.
ISBN 978-88-7642-385-7, e-ISBN: 978-88-7642-386-4

11. M. GIAQUINTA, L. MARTINAZZI, *An Introduction to the Regularity Theory for Elliptic Systems, Harmonic Maps and Minimal Graphs*, 2012 (second edition). ISBN 978-88-7642-442-7, e-ISBN: 978-88-7642-443-4
G. PRADISI, *Lezioni di metodi matematici della fisica*, 2012.
ISBN: 978-88-7642-441-0

12. G. BELLETTINI, *Lecture Notes on Mean Curvature Flow, Barriers and Singular Perturbations*, 2013.
ISBN 978-88-7642-428-1, e-ISBN: 978-88-7642-429-8

13. G. DA PRATO, *Introduction to Stochastic Analysis and Malliavin Calculus*, 2014. ISBN 978-88-7642-497-7, e-ISBN: 978-88-7642-499-1

14. R. SCOGNAMILLO, U. ZANNIER, *Introductory Notes on Valuation Rings and Function Fields in One Variable*, 2014. ISBN 978-88-7642-500-4, e-ISBN: 978-88-7642-501-1

Volumes published earlier

G. DA PRATO, *Introduction to Differential Stochastic Equations*, 1995 (second edition 1998). ISBN 978-88-7642-259-1

L. AMBROSIO, *Corso introduttivo alla Teoria Geometrica della Misura ed alle Superfici Minime*, 1996 (reprint 2000).

E. VESENTINI, *Introduction to Continuous Semigroups*, 1996 (second edition 2002). ISBN 978-88-7642-258-4

C. PETRONIO, *A Theorem of Eliashberg and Thurston on Foliations and Contact Structures*, 1997. ISBN 978-88-7642-286-7

Quantum cohomology at the Mittag-Leffler Institute, a cura di Paolo Aluffi, 1998. ISBN 978-88-7642-257-7

G. BINI, C. DE CONCINI, M. POLITO, C. PROCESI, *On the Work of Givental Relative to Mirror Symmetry*, 1998. ISBN 978-88-7642-240-9

H. PHAM, *Imperfections de Marchés et Méthodes d'Evaluation et Couverture d'Options*, 1998. ISBN 978-88-7642-291-1

H. CLEMENS, *Introduction to Hodge Theory*, 1998. ISBN 978-88-7642-268-3

Seminari di Geometria Algebrica 1998-1999, 1999.

A. LUNARDI, *Interpolation Theory*, 1999. ISBN 978-88-7642-296-6

R. SCOGNAMILLO, *Rappresentazioni dei gruppi finiti e loro caratteri*, 1999.

S. RODRIGUEZ, *Symmetry in Physics*, 1999. ISBN 978-88-7642-254-6

F. STROCCHI, *Symmetry Breaking in Classical Systems*, 1999 (2000).
ISBN 978-88-7642-262-1

L. AMBROSIO, P. TILLI, *Selected Topics on "Analysis in Metric Spaces"*, 2000. ISBN 978-88-7642-265-2

A. C. G. MENNUCCI, S. K. MITTER, *Probabilità ed Informazione*, 2000.

S. V. BULANOV, *Lectures on Nonlinear Physics*, 2000 (2001).
ISBN 978-88-7642-267-6

Lectures on Analysis in Metric Spaces, a cura di Luigi Ambrosio e Francesco Serra Cassano, 2000 (2001). ISBN 978-88-7642-255-3

L. CIOTTI, *Lectures Notes on Stellar Dynamics*, 2000 (2001).
ISBN 978-88-7642-266-9

S. RODRIGUEZ, *The Scattering of Light by Matter*, 2001.
ISBN 978-88-7642-298-0

G. DA PRATO, *An Introduction to Infinite Dimensional Analysis*, 2001.
ISBN 978-88-7642-309-3

S. SUCCI, *An Introduction to Computational Physics: – Part I: Grid Methods*, 2002. ISBN 978-88-7642-263-8

D. BUCUR, G. BUTTAZZO, *Variational Methods in Some Shape Optimization Problems*, 2002. ISBN 978-88-7642-297-3

A. MINGUZZI, M. TOSI, *Introduction to the Theory of Many-Body Systems*, 2002.

S. SUCCI, *An Introduction to Computational Physics: – Part II: Particle Methods*, 2003. ISBN 978-88-7642-264-5

A. MINGUZZI, S. SUCCI, F. TOSCHI, M. TOSI, P. VIGNOLO, *Numerical Methods for Atomic Quantum Gases*, 2004. ISBN 978-88-7642-130-0

Fotocomposizione "CompoMat" Loc. Braccone, 02040 Configni (RI) Italia
Finito di stampare nel mese di agosto 2014

Printed in the United States
By Bookmasters